Progress in Inflammation Research

Series Editor

Prof. Dr. Michael J. Parnham
Senior Scientific Advisor
PLIVA dd
Prilaz baruna Filipovica 25
10000 Zagreb
Croatia

Advisory Board

Forthcoming titles:
Cancer and Inflammation, D.W. Morgan, M. Nakada, U. Forssmann (Editors), 2004
Antirheumatic Therapy: Actions and Outcomes,
R.O. Day, D.E. Furst, P.L. van Riel (Editors), 2004
Antibiotics as Anti-inflammatory and Immunomodulatory Agents,
B. Rubin, J. Tamaoki (Editors), 2004

(Already published titles see last page.)

Cytokines and Joint Injury

Wim B. van den Berg
Pierre Miossec

Editors

Springer Basel AG

Editors

Prof. Dr. Wim B. van den Berg
Rheumatology Research Lab
University Medical Center Nijmegen, 189
Geert Grooteplein 26–28
6500 HB Nijmegen
The Netherlands

Prof. Dr. Pierre Miossec
Université Claude Bernard – Lyon 1
Hôpital Edouard Herriot
Service de Rhumatologie
Unité d'Immunologie Clinique
69437 Lyon
France

A CIP catalogue record for this book is available from the Library of Congress, Washington D.C., USA

Bibliographic information published by Die Deutsche Bibliothek
Die Deutsche Bibliothek lists this publication in the Deutsche Nationalbibliografie;
detailed bibliographic data is available in the internet at http://dnb.ddb.de

ISBN 978-3-0348-9609-2 ISBN 978-3-0348-7883-8 (eBook)
DOI 10.1007/978-3-0348-7883-8

Originally published by Birkhäuser Verlag, Basel, Switzerland in 2004
Printed on acid-free paper produced from chlorine-free pulp. TCF ∞
Cover design: Markus Etterich, Basel
Cover illustration: Expression of receptor activator of nuclear factor kappaB (RANK) at sites of focal bone erosion (see p. 274). With the friendly permission of Erik Lubberts and Wim B. van den Berg.

ISBN 978-3-0348-9609-2

9 8 7 6 5 4 3 2 1 www.birkhasuer-science.com

Contents

List of contributors

Brad Bolon, GEMpath Inc, 2540 N 400 W, Cedar City, UT 84720-8400, USA; e-mail: bradgempath@aol.com

Danielle Burger, Clinical Immunology Unit (Hans Wilsdorf Laboratory), Division of Immunology and Allergy, University Hospital, 24 rue Micheli-du-Crest, 1211 Geneva 14, Switzerland; e-mail: danielle.burger@hcuge.ch

Tim E. Cawston, Rheumatology: School of Clinical and Medical Sciences, The Medical School, University of Newcastle, Framlington Place, Newcastle upon Tyne, NE2 4HH, UK; e-mail: T.E.Cawston@ncl.ac.uk

Carlo Chizzolini, Clinical Immunology Unit (Hans Wilsdorf Laboratory), Division of Immunology and Allergy, University Hospital, 24 rue Micheli-du-Crest, 1211 Geneva 14, Switzerland; e-mail: chizzolini@medecine.unige.ch

Ernest Choy, Academic Department of Rheumatology, 2nd Floor, Weston Education Centre, King's College London, Cutcombe Road, London SE5 9PJ, UK; e-mail: ernest.choy@kcl.ac.uk

Jean-Michel Dayer, Clinical Immunology Unit (Hans Wilsdorf Laboratory), Division of Immunology and Allergy, University Hospital, 24 rue Micheli-du-Crest, 1211 Geneva 14, Switzerland; e-mail: jean-michel.dayer@hcuge.ch

Charles A. Dinarello, Department of Medicine, Division of Infectious Diseases, University of Colorado Health Sciences Center, 4200 East Ninth Ave., B168, Denver, CO 80262, USA

Ulrich Feige, Department of Inflammation Research, Amgen Inc., One Amgen Center Drive, M/S 29-M-B, Thousand Oaks, CA 91320-1799, USA; e-mail: ufeige@amgen.com

Cem Gabay, Division of Rheumatology, University Hospital of Geneva & Department of Pathology-Immunology, University of Geneva School of Medicine, 26 avenue Beau-Séjour, 1211 Geneva 14, Switzerland; e-mail: cem.gabay@hcuge.ch

J. Alastair Gracie, Centre for Rheumatic Diseases, Division of Immunology, Infection and Inflammation, University of Glasgow, Glasgow Royal Infirmary, 10 Alexandra Parade, Glasgow G31 2ER, UK; e-mail: j.a.gracie@clinmed.gla.ac.uk

Sarah L. Jongbloed, Centre for Rheumatic Diseases, Division of Immunology, Infection and Inflammation, University of Glasgow, Glasgow Royal Infirmary, 10 Alexandra Parade, Glasgow G31 2ER, UK; e-mail: sj32w@clinmed.gla.ac.uk

Tadamitsu Kishimoto, Laboratory of Immune Regulation, Graduate School of Frontier Biosciences, Osaka University, 1-3 Yamada-oka, Suita-City, Osaka, 565-0871, Japan; e-mail: kishimot@imed3.med.osaka-u.ac.jp

Paul J. Kostenuik, Department of Metabolic Disorders, Amgen Inc., One Amgen Center Drive, M/S 15-1-A, Thousand Oaks, CA 91320-1799, USA;
e-mail: paulk@amgen.com

Erik Lubberts, University Medical Center Nijmegen, Rheumatology Research and Advanced Therapeutics, 189, Geert Grooteplein 26-28, 6500 HB Nijmegen, The Netherlands; e-mail: E.Lubberts@reuma.umcn.nl

Sir Ravinder N. Maini, The Kennedy Institute of Rheumatology, Faculty of Medicine, Imperial College, 1 Aspenlea Road, London W6 8LH, UK;
e-mail: r.maini@imperial.ac.uk

Iain B. McInnes, Centre for Rheumatic Diseases, Division of Immunology, Infection and Inflammation, University of Glasgow, Glasgow Royal Infirmary, 10 Alexandra Parade, Glasgow G31 2ER, UK; e-mail: i.b.mcinnes@clinmed.gla.ac.uk

Pierre Miossec, Department of Immunology and Rheumatology, Hôpital Edouard Herriot, 69437 Lyon Cedex 03, France; e-mail: miossec@univ-lyon1.fr

Norihiro Nishimoto, Laboratory of Immune Regulation, Graduate School of Frontier Biosciences, Osaka University, 1-3 Yamada-oka, Suita-City, Osaka, 565-0871, Japan; e-mail: norihiro@fbs.osaka-u.ac.jp

Pascale Roux-Lombard, Clinical Immunology Unit (Hans Wilsdorf Laboratory), Division of Immunology and Allergy, University Hospital, 24 rue Micheli-du-Crest, 1211 Geneva 14, Switzerland; e-mail: pascale.roux-lombard@hcuge.ch

Andrew D. Rowan, Rheumatology: School of Clinical and Medical Sciences, The Medical School, University of Newcastle, Framlington Place, Newcastle upon Tyne, NE2 4HH, UK

Victoria Shalhoub, Department of Metabolic Disorders, Amgen Inc., One Amgen Center Drive, M/S 15-1-A, Thousand Oaks, CA 91320-1799, USA;
e-mail: shalhoub@amgen.com

Wim B. van den Berg, University Medical Center Nijmegen, Rheumatology Research Lab, 189, Geert Grooteplein 26-28, 6500 HB Nijmegen, The Netherlands;
e-mail: W.vandenBerg@reuma.umcn.nl

Hilary Wilson, Centre for Rheumatic Diseases, Division of Immunology, Infection and Inflammation, University of Glasgow, Glasgow Royal Infirmary, 10 Alexandra Parade, Glasgow G31 2ER, UK

Debra Zack, Department of Inflammation Research, Amgen Inc., One Amgen Center Drive, M/S 29-M-B, Thousand Oaks, CA 91320-1799, USA;
e-mail: dzack@amgen.com

Preface

The purpose of this volume in the *Progress in Inflammation Research* series is to provide the biomedical and clinical researcher with a state-of-the-art insight in the role of cytokines in joint inflammation and joint destruction. This is of relevance for better understanding of key processes in diseases such as rheumatoid arthritis (RA) and osteoarthritis (OA). Apart from the impact of old and novel cytokines on joint tissues, the various chapters address the issue of targeted therapy with biological response modifiers and future interventions with carefully designed inhibitors. Special attention is given to elements of synovial cell activation, cell–cell interaction, cytokine interplay as well as mechanisms of cartilage destruction and bone erosion.

In addition to an outline of the role of established cytokines, such as TNF, IL-1 and IL-6, new information is given on the novel cytokines IL-15, IL-17 and IL-18 and their positioning in the complex cytokine interplay. Cytokine regulation of destructive enzymes, RANKL, the endogenous inhibitor OPG and their crucial roles as central players in joint erosion are highlighted. Together, the chapters provide a complete and balanced view on pivotal cytokines and joint pathology.

January 2004

Wim B. van den Berg
Pierre Miossec

The role of tumour necrosis factor in rheumatoid arthritis

Ravinder N. Maini

The Kennedy Institute of Rheumatology Division, Faculty of Medicine, Imperial College,
1 Aspenlea Road, London W6 8LH, UK

Introduction

More than a century has passed since 'Coley's toxins' obtained from filtered media of bacterial cultures were utilized to treat sarcomas. The rationale was based on the apparent regression of a sarcoma in a patient who developed erysipelas [1–3]. The results of the intervention, though apparently occasionally successful and undoubtedly toxic, inspired later work which led to the description in 1975 of tumour necrosis factor (TNF). TNF referred to a soluble serum factor induced by hyperimmunization of mice with Bacillus Calmetre-Guerin and bacterial endotoxin. This factor induced necrosis of methylcholanthrene-induced sarcomas in mice [4]. Initial interest therefore focussed on the tumourocidal properties of TNF, but its role in immunity and inflammation only became apparent after isolation and characterisation of the molecule. In this review the focus will be on the biology of TNFα and a historical approach to the definition of TNFα as a therapeutic target that has heralded three blockbuster drugs and consolidated the future of 'biodrugs' as an important sector of future pharmaceutical products.

TNFα, its receptors and ligand-receptor interactions

In a series of papers published in the 1980s the cDNA of TNFα and its receptor were cloned, expressed and the protein products sequenced and tested in biological systems by a number of laboratories [5–10]. From these studies followed the realisation that TNFα was identical to cachectin and related to lymphotoxin (also know as TNFβ), and that both TNFα and lymphotoxin could bind to two TNF receptors, referred to as TNF RI (or p55-TNF) and TNF RII (or p75-TNF R) [11–17] (Fig. 1). In the following pages and following recent convention, the acronym TNF is used without the α to describe TNFα and not the growing family of TNF related molecules.

Cytokines and Joint Injury, edited by Wim B. van den Berg and Pierre Miossec

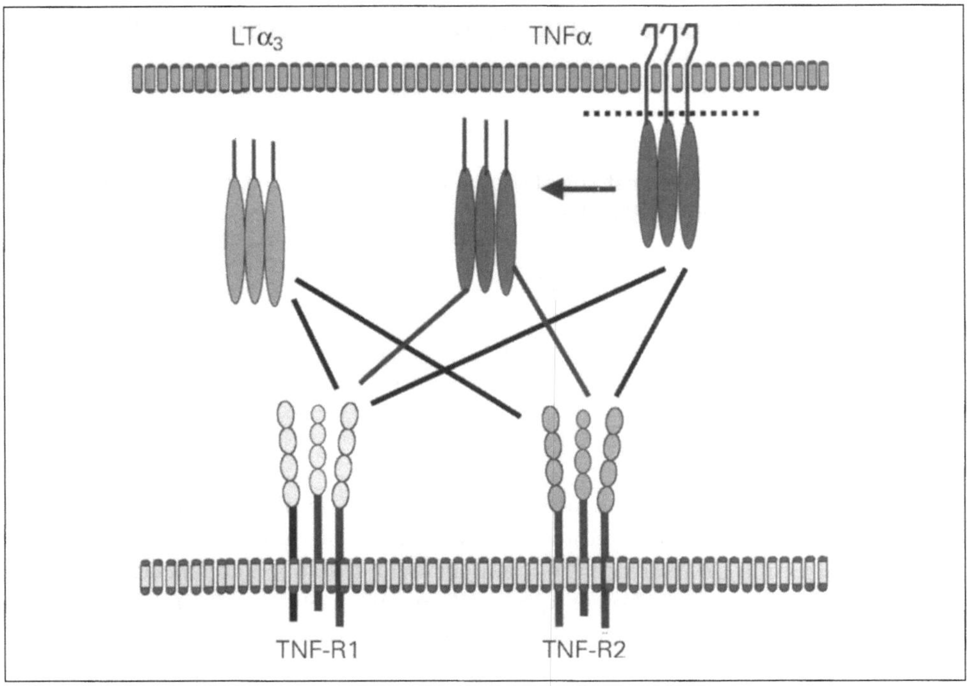

Figure 1
TNF and lymphotoxin bind to TNF receptor 1 and TNF receptor 2.

Subsequent structural analysis revealed that secreted TNF and lymphotoxin, in their naturally occurring soluble form, consisted of a pear-shaped trimer formed by non-covalent linkage of newly synthesised monomers. However, TNF exists as a biologically active entity as a membrane-bound precursor as well as a soluble molecule and is produced by many cell types. Lymphotoxin exists only in a soluble form and, unlike TNF, is produced exclusively by T cells. Soluble TNF is released by enzymatic cleavage mediated by a membrane metalloproteinase, known as TNF-converting enzyme (TACE), later termed ADAM-17, a member of the disintegrin and metalloproteinases (adamolysin) family.

The newly-synthesised TNF monomer of 26 KDa is cleaved to a 17 KDa mature form. The TNF trimer interacts with its receptor *via* binding sites located in the three grooves between the subunits [18]. Since the TNF receptors are being constantly shed in a soluble form, the binding of soluble TNF receptors to TNF leads to neutralisation of TNF bioactivity and acts as a homeostatic negative feedback mechanism (Fig. 2). Binding of TNF to TNF receptors causes aggregation of the monomers in the plasma membrane, and recruitment of proteins such as TNF-R-

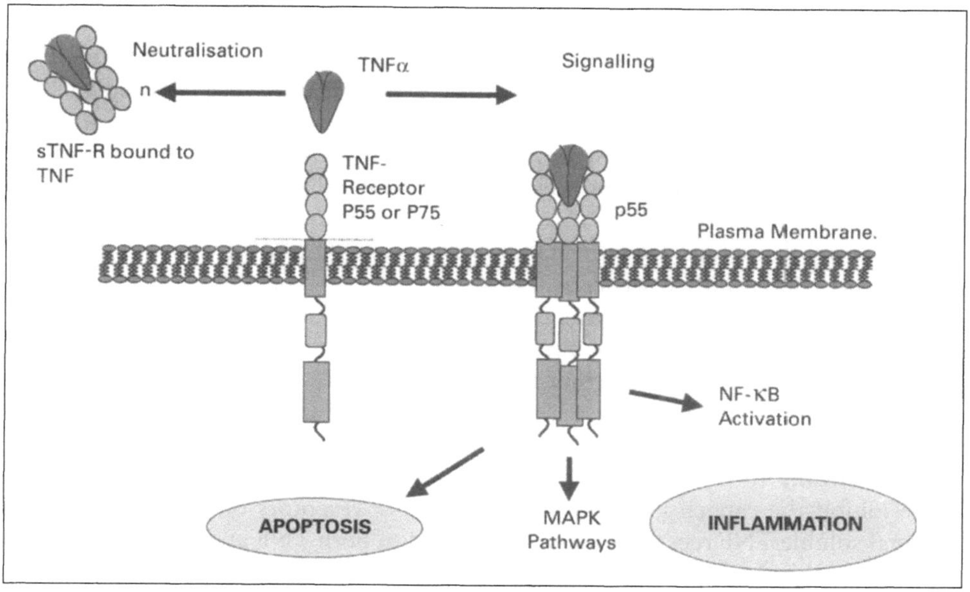

Figure 2
TNF trimer when bound to soluble TNF receptor neutralizes its bioactivity
Binding of TNF trimer to cell surface receptor leads to aggregation and signalling. Activation of MAP kinase pathways and NF-κB results in inflammatory responses. Activation of caspases leads to apoptosis.

associated factor 2 and Fas-associated death domain that initiate intracellular signalling. Intracellular signalling activates several pathways including three MAP-kinases and NF-κB [19, 20], which in turn result in an inflammatory response *via* gene transcription of proinflammatory molecules (Fig. 2). Recent interest has focussed on developing small molecular weight compounds that inhibit signalling pathways employed by TNF [21].

Alternatively, signalling may eventuate in apoptosis *via* activation of caspases (Fig. 2). There appears to be a reciprocal relationship between inflammatory and apoptotic signalling so that only one of them predominates.

TNF production is regulated at the transcriptional, post-transcriptional and translational levels. At the level of mRNA, production of TNF is regulated by proteins that bind to AU-rich elements (AREs) in the 3' untranslated region. These protein-mRNA interactions curtail or prolong the half-life of message. An example of one such regulatory protein is tristetraprolin (TTP), a member of the zinc-finger protein family, which by binding to AREs increases the stability of TNF mRNA. TTP knockout mice overproduce TNF and have a phenotype characterised by an inflammatory polyarthritis, dermatitis and autoimmunity [22]. The importance of AREs in stabilizing mRNA and biosynthesis of TNF was more directly demonstrated by Kollias and colleagues [23]. In their experiment, deletion of AREs produced a mouse phenotype which developed polyarthritis and inflammatory bowel disease. Interestingly, backcrossing this mouse with immunodeficient RAG-1-mice (thus creating a double deficiency) abolishes the inflammatory bowel disease but perpetuates polyarthritis, thus emphasising the role of T and B cells in full expression of inflammatory bowel disease, and sufficiency of overproduction of TNF alone in the development of rheumatoid-like arthritis.

The critical interaction between TNF receptors and its ligand has been investigated further. It appears that the three terminal domains of the receptor are necessary for binding to TNF, and a truncated molecule of the receptor lacking the proximal C'-domain close to the plasma membrane retains its binding affinity for TNF [24]. The anti-TNF specific monoclonal antibodies, infliximab and adalimumab, bind specifically to the TNF trimer to epitopes that also bind to the receptor and hence inhibit the capacity of TNF to bind to TNF receptors on cells (Fig. 3). The terminal soluble TNF receptor domains can bind to both TNF and lymphotoxin and similarly inhibit the ligand binding to its receptor. The inhibitory property has been exploited to construct the soluble p75-TNF receptor-immunoglobulin fusion protein etanercept (the receptor acts as an inhibitor and the immunoglobulin portion improves its pharmacokinetics). In rheumatoid arthritis (RA) TNF is overproduced and little, if any, lymphotoxin is expressed at the protein level in diseased tissues [25] and hence therapy has focussed on targeting TNF.

A TNF molecule containing single or double mutations has been recently constructed that inactivates naturally occurring TNF by sequestration [26]. The mutant TNF forms heterotrimer TNF molecules by displacing one or more naturally occurring monomers, and prevents binding of TNF to the receptor and its signalling function. It has been suggested that one of the mutant homotrimer constructs can also occupy the cell-bound TNF receptor without initiating signalling, and thus inhibit the binding of TNF to its receptors in a fashion similar to the antagonistic activity of IL-1ra [27]. An engineered TNF variant used as a PEG-therapeutic molecule reduced joint swelling when administered to the rat collagen induced arthritis model after onset of symptoms [26].

Biosynthesis of TNF in RA and experimental arthritis

Biosynthesis of TNF in the joints in RA appears to be predominantly located in macrophages in the lining layer and intima of the synovial membrane. In addition, TNF is produced by cells in pannus invading cartilage [28, 29]. In the collagen-induced model of inflammatory and destructive arthritis in DBA/1 mice, a study of cellular kinetics shows sequential production of cytokines with TNF secreting cells

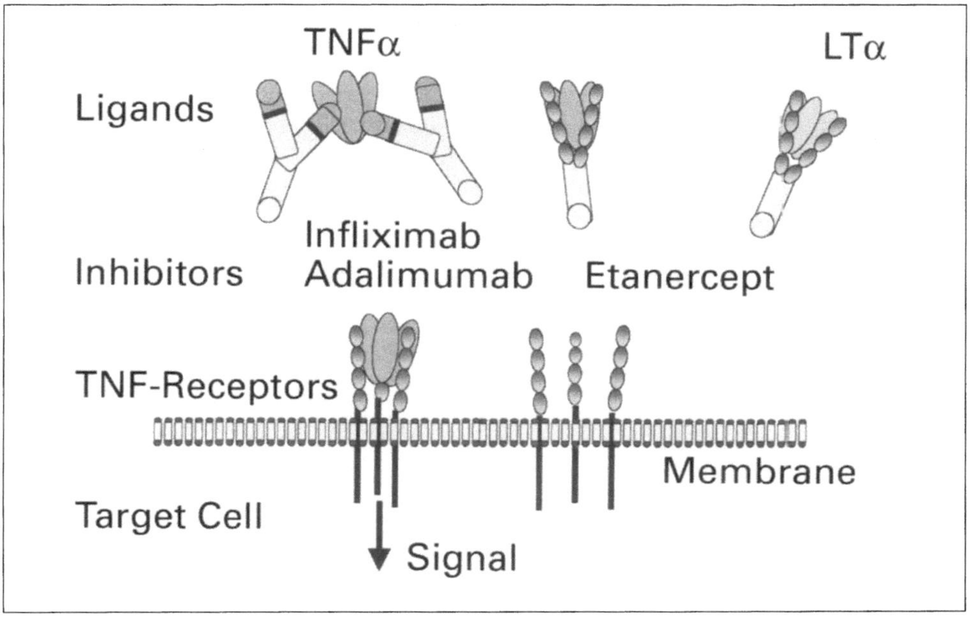

Figure 3
Inhibitors of TNF and lymphotoxin
Infliximab and adalimumab are examples of monoclonal antibodies that specifically bind and inhibit TNF only. Etanercept is an example of a soluble TNF receptor fusion protein, which binds to TNF and lymphotoxin and inhibits both. Binding to ligands prevents access to TNF receptors and signalling.

appearing before cells producing IL-1β and IL-6 in the early phase of onset of clinical arthritis [30]. These cells (mostly macrophages) are located in the lining layer, intima and pannus tissue contiguous with cartilage and bone (Fig. 4) supporting the hypothesis that local paracrine activity of TNF amplifies the cytokine network in this model of RA. A study of murine models has also demonstrated that mesenchymal precursor cells migrate into the joints from the bone marrow *via* enlarged bony canals before leukocyte infiltration is established in joints of mice immunized with adjuvant alone or with collagen II. This adjuvant-induced cellular infiltration, presumably signalling *via* TOLL-receptors, is abrogated by anti-TNF treatment and, as predicted, is absent in TNF p55/75 knock-out mice [31]. Thus it appears that in CIA, TNF is implicated in recruitment of mesenchymal cells in the pre-arthritic innate-immune conditioning phase and recruitment of T and B cells in the immune-mediated inflammatory and destructive chronic phase of arthritis.

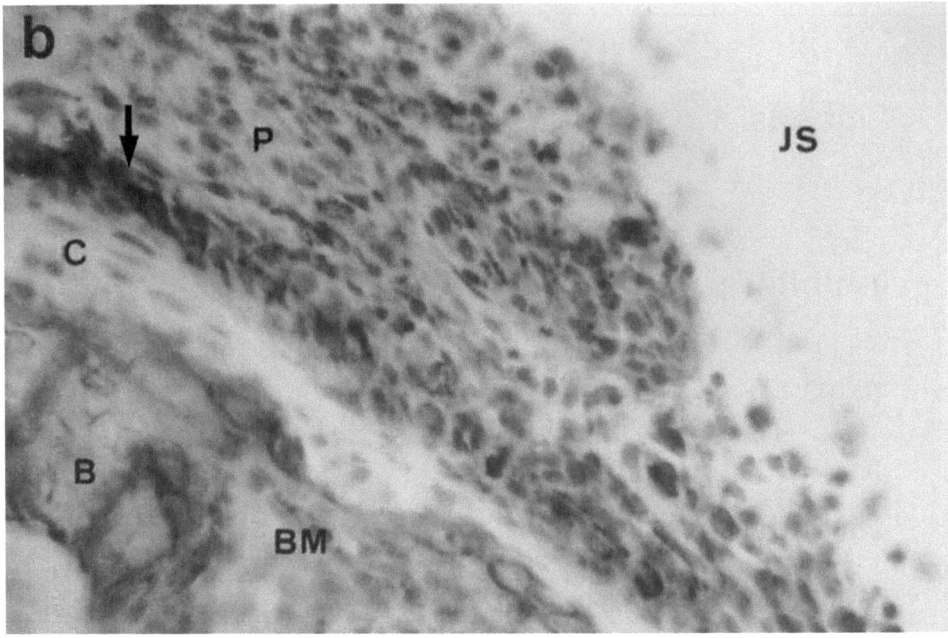

Figure 4
TNF producing cells in pannus invading cartilage in collagen induced arthritis
The photomicrograph shows an immunostained section of a mouse joint with TNF producing cells (↓) in close proximity and overlying cartilage (C). Pannus Tissue (P) is shown as well as bone (B) and bone marrow (BM).

TNF as a therapeutic target: preclinical rationale

Access to molecular probes to an increasing list of cytokines from the mid-1980s made it possible to describe their expression profile in RA. Essentially the list includes many proinflammatory cytokines, chemokines and growth factors as well as inhibitors and anti-inflammatory cytokines [32]. The known biological activities of each of these molecules appears relevant to our understanding of the disease process and the complex inter-connectivity of immunological, inflammatory, tissue destructive and remodelling responses that are characteristic of RA. However, two basic principles have emerged that help us to conceptualise a coherent biological model of understanding the role of these molecules in disease.

First, that chronic inflammation is perpetuated by the action of a range of these molecules despite attempts at homeostasis by reversing or inhibiting their action by other biologically active molecules produced simultaneously. This loss of homeosta-

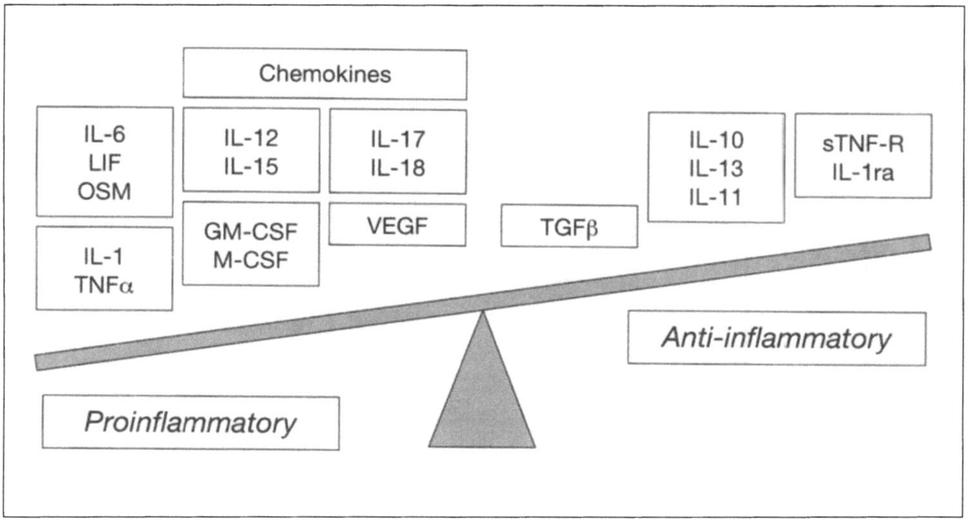

Figure 5
Loss of homeostatic balance between the expression of cytokines, chemokines and growth factors and their inhibitors in rheumatoid joints.

sis gives rise to an imbalance between a proinflammatory and a counter-inflammatory drive (Fig. 5).

Second, that despite the apparent overlapping activities, and hence the inbuilt redundancy of the cytokines expressed in RA, there appears to be a dominant role for TNF in regulating the inflammatory response. Other cytokines may also play a more restricted role in disease pathogenesis in man that may be predicted from *in vitro* experiments and study of animal models. Despite these reservations a key role played by TNF in RA did emerge from a number of observations made on tissues *in vitro* and in animal models. Most significant were the following:

- Demonstration of overproduction of TNF and TNF receptors in adjacent cells in areas of inflammation and tissue destruction in the synovium and pannus [28, 33].
- Demonstration that TNF regulates the production of IL-1 *in vitro*. This was shown by obtaining disaggregated cells of the synovial membrane from RA patients and by measuring specific suppression of IL-1 production by anti-TNF antibody [25]. Later this observation was extended to the production of GM-CSF, IL-6 and IL-8 [34]. These results underpinned the original hypothesis that TNF was of pivotal importance as a controller of inflammation (Fig. 6).

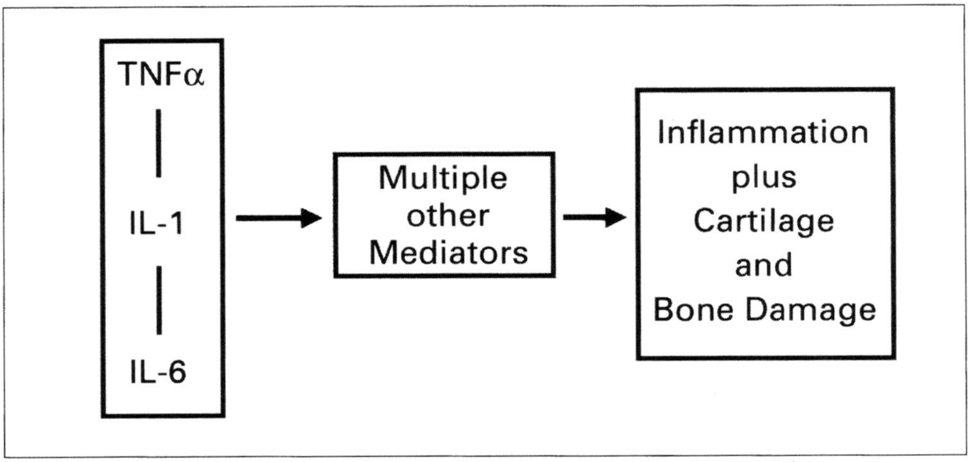

Figure 6
The key role of TNF in the cytokine network and recruitment of multiple mediators in inducing inflammation plus cartilage and bone damage
This concept has been validated in RA by clinical trials of agents that inhibit TNF, IL-1 and IL-6.

- The significance of the role of TNF in a physiological system was investigated in the collagen-induced arthritis model of RA. In this model, treatment with a monoclonal anti-TNF antibody initiated immediately after the onset of clinical arthritis reduced inflammation and prevented serious cartilage and bone damage [35, 36] (Fig. 7). Around the same time Kollias and colleagues demonstrated the arthritogenic properties of human TNF in a mouse model over-expressing the TNF gene modified at the 3' end to stabilize mRNA expression [37]. Moreover, the arthritis was prevented by treatment with anti-TNF antibody commenced soon after birth.
- Taken together these preclinical data provided the rationale for targeting TNF in the early 1990s.

Anti-TNF therapy provides clues to role of TNF in RA

Therapeutic intervention clinical trials have provided an opportunity to gain insights into the role of TNF in RA by investigating the mechanism of action of anti-TNF agents (for reviews see [38] and [39]). Already from the very first 'open-label' clinical trial of infliximab (Remicade®) in 1992/93 (then known as cA2), used in the treatment of RA, a profound reduction in the acute phase C-reactive protein, asso-

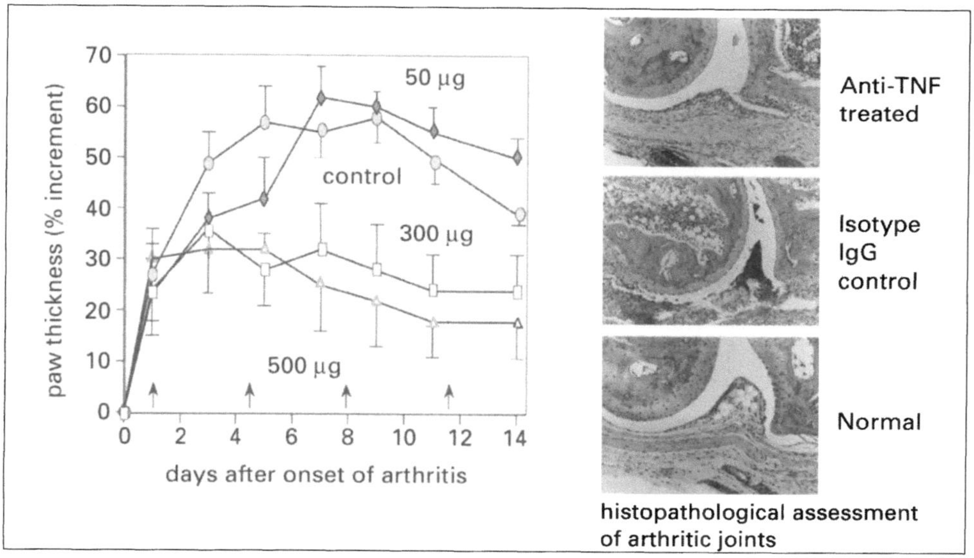

Figure 7

Monoclonal anti-TNF antibody reduces the severity of collagen-induced arthritis in DBA-1 mice

The antibody was administered after signs of inflammation first appeared. A dose related inhibition of inflammation (paw thickness), and destruction of cartilage and bone is shown (from Williams, Feldmann, Maini (1992). Proc Natl Acad Sci USA *89: 9784).*

ciated with a reduction in its induced interleukin-6 (IL-6) and reduction in rheumatoid factor, was observed [40].

A randomised placebo-controlled trial of infliximab [41] further substantiated the role TNF plays in controlling IL-6 production and serum concentrations (Fig. 8). Stability of IL-6 levels in placebo-treated patients and a dose-related reduction in patients receiving infliximab clearly demonstrated that neutralisation of TNF (presumably in rheumatoid joints) had a rapid effect on IL-6 production and a consequential reduction of CRP, serum amyloid A protein and haptoglobin production by the liver [42]. In the same trial, a reduction of blood levels of IL-1β following infliximab was noted in a subgroup of patients treated in Erlangen, Germany [43]. More recently it was shown that TNF blockade with infliximab downregulates the production of IL-18 [44] and etanercept downregulates IL-13 [45]. Thus the TNF mediated regulation of the cytokine network postulated from *in vitro* experiments [25] appeared to have a clinical relevant counterpart *in vivo*.

A striking feature of anti-TNF therapy is the rapid reduction in swelling and tenderness of joints. Serial tissue biopsies and imaging technologies have provided evi-

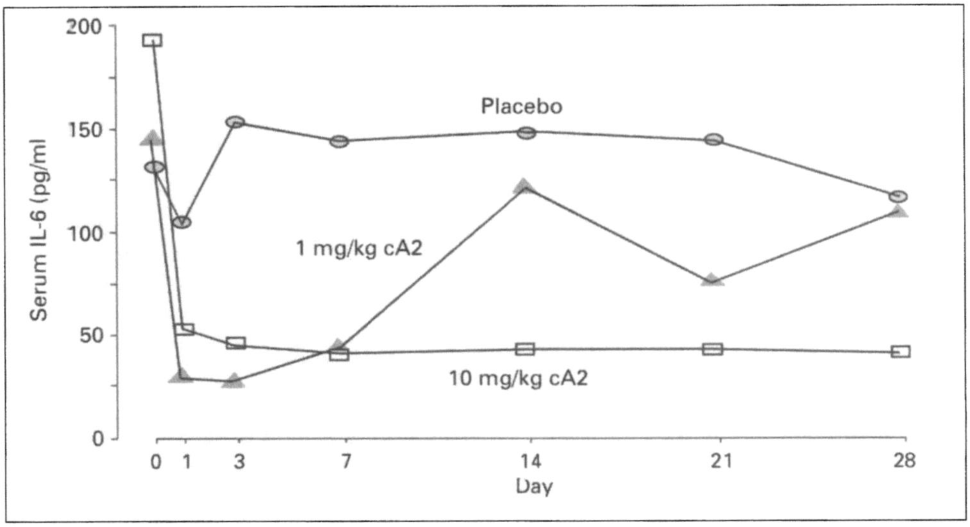

Figure 8
Reduction of serum IL-6 following administration of infliximab (cA2), 1 mg/kg or 10 mg/kg compared with no change following placebo (Based on: Charles et al (1999) J Immunol *163: 1521–1528, including unpublished data.)*

dence that following infliximab this is due to a reduction in the mass of inflammatory tissue [46]. Since infliximab is capable of mediating cell death of TNF expressing cells *via* complement mediated or antibody-dependant cell cytoxicity *in vitro* [47, 48], the possibility has to be considered that such a mechanism could operate *in vivo*. However, no increase in apoptotic cell death is evident in serial biopsies from rheumatoid patients following infliximab [49]. Moreover, etanercept is not lytic *in vitro* but has a clinical action on inflammation similar to that observed with infliximab suggesting other mechanisms are involved. On the other hand, in infliximab treated Crohn's disease patients, apoptosis of T cells in the lamina propria have been documented [50].

In contrast to the uncertainty about cell death, there is ample evidence that infliximab does reduce cell recruitment into inflamed joints. Thus a single infusion of infliximab at 1 or 10 mg/kg leads to decreased serum concentrations of soluble ICAM-1 and soluble E-selectin at the same time as a transient and significant dose-related increase in lymphocyte counts [51]. This indirect evidence, suggesting decreased adhesiveness of blood vessels and retention of lymphocytes in the circulation, is supported by reduction of immunoreactive adhesion molecules E-selectin and VCAM-1 in serial synovial biopsies following infliximab [52]. The possibility

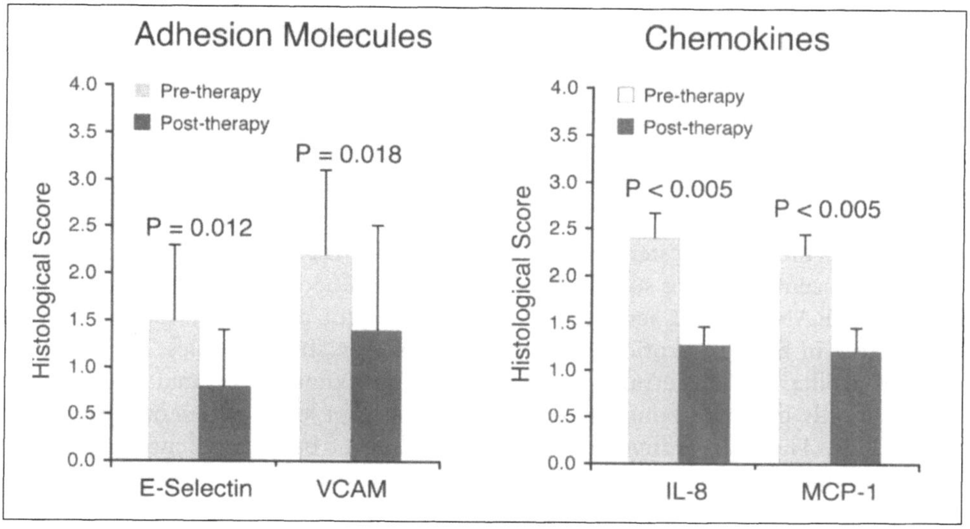

Figure 9
Infliximab therapy reduces expression of adhesion molecules E-selectin and VCAM-1 and chemokines IL-8 and MCP1 in synovial biopsies from RA patients before and after a single treatment with infliximab (Based on: Tak PP et al (1996) Arthritis Rheum *39: 1077–1081, and Taylor PC et al (2000)* Arthritis Rheum *43: 38–47.)*

that cell recruitment and trafficking are affected seemed likely since serial biopsies also demonstrated a reduction in chemokines IL-8 and MCP-1 [53] (Fig. 9).

Direct evidence supporting this hypothesis is forthcoming from experiments demonstrating a reduced uptake following infliximab of autologous indium-111 labelled polymorphonuclear cells in the knees, wrists and hands visualised by a gamma camera [53].

Angiogenesis is a characteristic early and ongoing feature of rheumatoid synovitis and plays a part in increased delivery of inflammatory cells and mediators of inflammation and tissue destruction. Vascular endothelium growth factor (VEGF) concentrations are increased in rheumatoid joints and serum of RA patients and the latter correlates with joint damage assessed by radiographs at one year [54]. Reduction of VEGF following infliximab thereby has been observed [55] and likely contributes to reduced vascularity and reduced $\alpha v\beta 3$ immunostaining of new blood vessel formation in tissue biopsies [38]. It may also explain the reduction in vascular signals detected by colour Doppler sonograpy [56] and reduced vascular leakage of gadolinium by magnetic resonance imaging (MRI) [57]. A reduction in angiogenesis, as well as decreased production of matrix metalloproteinases [58] following anti-TNF therapy, may explain the inhibition of joint space narrowing (cartilage loss) and bone erosions that have been documented in clinical trials [59, 60].

Since bone damage in RA is mediated by osteoclasts, anti-TNF therapy may additionally influence osteoclast formation and activity by reducing recruitment of monocyte precursors. Anti-TNF therapy could also block differentiation signals to osteoclasts directly delivered by TNF and indirectly by inactivation of TNF-induced expression of RANK-ligand on T cells and synoviocytes. Indirect evidence of an effect on this biological system in RA has been demonstrated by the observation that elevated concentrations of soluble RANK-ligand in blood are normalised by infliximab [61]. RANK-ligand activates the RANK receptor on osteoclasts and plays a critical role in bone resorption [62]. In the transgenic strain of mice Tg197 developed by Kollias [37], neutralisation of TNF by infliximab instituted after disease onset not only inhibited bone and cartilage damage but led to regeneration of these tissues [63]. The regenerative effect of remodelling of bone is accentuated by co-administration of infliximab and osteoprotegerin in this mouse model, thus blocking TNF and RANK–RANK-ligand interactions [64, 65].

Why does anti-TNF therapy require continuing treatment? The answer to this is partially answered by the finding that TNF blockade is due to trapping of bioactive TNF in a complex of TNF and inhibitor that is detectable as increased immunoreactive TNF in the circulation following therapy, and not due to abolition of biosynthesis. It is also apparent that while TNF-driven proinflammatory cytokine production is reduced, it is rarely completely suppressed. At the same time, TNF inhibition also reduces soluble TNF receptor and IL-1ra levels [42] and hence the imbalance between pro- and anti-inflammatory effects of TNF and IL-1 is not restored to normal. A recent study of cytokines in blood following adalimumab has similarly shown lack of restoration of homeostasis since reduction of IL-6 was observed but not soluble TNF receptors or IL-1Ra concentrations. Moreover, no reduction in TNF was apparent in synovial tissue biopsies [66].

The lack of universal response to TNF blockade is not understood and does not correlate with baseline features of disease activity or serological status. However, there is a relationship between high TNF production in synovial biopsies and a good clinical response [67]. There is some evidence of differences in the genotype of responders and non-responders [68, 69] but insufficient with current knowledge to be of predictive value in individual patients.

TNF modulates immune function in RA

Suppression of delayed hypersensitivity skin reactions and lymphocyte activation to recall antigens is a recognised feature of RA. Altered T cell responses (endogenous disease-related immunosuppression) may result from chronic exposure to TNF. *In vitro* experiments have shown this to be a consequence of downregulation of the ζ

chain of the T cell receptor complex (reviewed by Cope [70]). In RA patients treated with infliximab, depressed lymphocyte proliferative responses *in vitro* to a recall antigen (tuberculin PPD) and phytohaemaglutin were restored to normality [71], suggesting recovery of memory function. However, in patients treated with etanercept, mitogen induced responses *in vitro* and skin delayed hypersensitivity reactions were not significantly different from patients treated with placebo [72]. It is possible to envisage that if in patients with expanded autoreactive T cell clones (for example to antigens in the central nervous system or to nucleosomal antigens) treatment with anti-TNF agents leads to restoration of suppressed function of the T cell receptor, it could unmask autoimmune reactions and disease (such as appearance of antinuclear antibodies, SLE and multiple sclerosis) as a rare undesirable side effect.

It has been proposed that IL-15 expressed by macrophages in rheumatoid joints provides the signal for TNF overproduction [73]. Chronic stimulation of T cells with a mixture of cytokines, including TNF IL-15, IL-2 and IL-6, induces a phenotype resembling activated T cells in RA joints [74]. Through cell to cell contact mediated interactions with macrophages in an experimental approach described by Dayer [75, 76] these cytokine-conditioned T cells activate macrophages to produce proinflammatory cytokines (IL-1, TNF and metalloproteinases (MMP-1 and MMP-3) *via* NF-κβ pathway [77]. This proinflammatory cytokine-dependent mechanism of T cell activation thus acts as a positive feedback amplification loop and shifts the balance from adaptive immune (T cell receptor dependent) responses to the innate immune (cytokine dependent) pathway in chronic RA. Recovery of regulatory suppressor T cell phenotype ($CD4^+$ $CD25^+$) and function *in vitro* following anti-TNF therapy has recently been reported [78] adding to the complexity of effects of TNF on immune function in RA.

It is noteworthy that in a Phase III trial, infliximab treatment, but not placebo plus methotrexate, leads to a reduction in rheumatoid factor concentrations [79]. The reduction is evident even in the patients in whom a rise in IgM class of anti-ds-DNA antibodies is documented in a controlled study [80]. The apparent paradox may be explained by the reduction by infliximab, on the one hand, of the cytokines that stimulate B cell activity (e.g., TNF and IL-6) involved in rheumatoid factor production. On the other hand, infliximab reduces CRP (and other acute phase proteins) very rapidly thus impairing their role in binding to and clearing nucleosomal antigens released by tissue destruction in active RA, and thereby promoting antigen-induced antinuclear antibody formation [80].

Anti-TNF therapy in the clinic

At the end of 2003 infliximab and etanercept are already well established in the clinic, with an estimated 0.5 million patients exposed to them, the majority for RA.

Adalimumab became available in 2003 and sales figures suggest it is already penetrating an expanding market. The absence of head-to-head trials does not make it possible to compare their efficacy with confidence. Nevertheless, a recently published analysis suggests similar efficacy of the three drugs [81]. Comparison of safety beyond that observed in clinical trials is also problematical since safety profiles post-marketing may reflect the volume of sales and the differences in populations exposed rather than real differences in action, e.g., more patients worldwide (especially Europe) have been exposed to infliximab than to etanercept which has thus far been more extensively used in North America. Salient features of the three drugs are discussed here.

Lessons learned from clinical trials of infliximab

By the end of the 1980s sufficient evidence had accumulated to hypothesize that TNF may be a good therapeutic target. The history of the antecedents and early experience of collaboration in an investigator-led study, between the Kennedy Institute and industry, which led to a trial of TNF blockade in RA patients for the first time, has been recently documented [82]. In the first open-label trial, 20 patients with RA with active disease, despite prior treatment with multiple drugs over the previous decade, were treated with divided doses over two weeks (total/wk 20 mg/kg), of a chimeric monoclonal antibody cA2, now known as infliximab or Remicade® [40].

This trial demonstrated all the features recognised as characteristic of anti-TNF therapy, e.g., a rapid amelioration of signs and symptoms, improvement in physical function and a trend towards normalisation of abnormal laboratory measurements in the majority of patients. However it was soon apparent that the remarkable suppression of inflammation only lasted a few weeks and repeated long-term therapy would be required for this therapeutic. Seven patients were therefore retreated following recurrence of disease activity up to four times over a period of observation extending over a year [41]. Each repeat infusion was accompanied by a rapid therapeutic effect of the same magnitude as the initial treatment, thus encouraging the belief that long-term treatment with a biological might be feasible. However, with successive exposures, the duration of control diminished in a proportion of patients raising the possibility that tachyphylaxis might limit long-term therapy in some patients. An approach to circumventing the problem was raised by the finding of synergy of action of a combination of anti-TNF and anti-CD4 monoclonal antibodies in the murine collagen-induced arthritis model [35]. In this model the benefit was partially attributed to improved pharmacokinetics of the xenogenic monoclonal antibody to suppression of an anti-globulin response.

In the 1990s methotrexate treatment of RA was gaining credibility as the 'gold standard' disease modifying anti-rheumatoid drug. Since methotrexate had anti-inflammatory and immunosuppressive properties, it was reasoned that a combination of methotrexate and infliximab might reproduce the added benefit observed in a model of RA. A parallel arm randomised controlled study was preformed on methotrexate treated patients with an incomplete control of RA [83]. Three groups were each treated with infliximab monotherapy or combination of infliximab and methotrexate *versus* placebo infusions and methotrexate, the dose of methotrexate being fixed at 7.5 mg/week. The results of this trial demonstrated that the chimeric infliximab as monotherapy lost efficacy with repeated therapy at a low dose (1 mg/kg) and was immunogenic in ~50% of patients treated at this low dose at four-weekly intervals. Efficacy was retained and immunogenicity of infliximab reduced at 3 mg/kg and 10 mg/kg doses. In contrast, combination therapy was efficacious and comparatively non-immunogenic at all three doses of infliximab. Closer analysis showed that a greater suppression of disease activity, lasting for a longer duration in the follow-up period after cessation, was obtained in patients on combination therapy. The improved efficacy was related to improved pharmacokinetics of combination of infliximab 1 mg/kg plus methotrexate, but at higher doses of infliximab was probably associated with improved pharmacodynamics, since pharmacokinetics were similar [83]. These findings established the strategy for long-term trials and combination of infliximab and methotrexate is the licensed indication of this anti-TNF drug.

The greater efficacy of a combination of anti-TNF therapy with methotrexate has now been demonstrated for etanercept (presented at EULAR 2003, Lisbon) and adalimumab [84]. Since failure to respond to methotrexate is used as a prime indication for anti-TNF drugs, the addition of anti-TNF drugs to methotrexate as a management strategy is now widely used in clinical practice.

A Phase III randomised placebo controlled trial tested the efficacy of a combination of infliximab and methotrexate at a median dose of 15 mg/wk (the ATTRACT trial) in patients with moderate-to-severe RA with longstanding disease and significant disability despite prior treatment with multiple DMARDS. The three primary endpoints have demonstrated similar efficacy of infliximab at all four doses used (3 mg/kg and 10 mg/kg every four or eight weeks) *versus* placebo plus methotrexate in control of signs and symptoms at 24 weeks [79], radiographic scores of joint damage at 54 weeks [59] and physical function at 102 weeks [85].

The details of the results are published and are the subjects of other reviews, but a number of conclusions of the efficacy of infliximab are summarised here:

- The control of disease activity is significant and durable over two years in ~50% of patients although a decline occurs between 28 and 54 weeks in the proportion of responders at the lowest dose (3 mg/kg every eight weeks).
- There is marked inhibition of radiographic progression of joint damage at one year (both joint space narrowing of cartilage loss and bone erosions) while pro-

gression is seen in patients on placebo and continuing methotrexate therapy in a population of patients with significant joint destruction at baseline.
- In almost half the patients, radiographs show a reduction in joint erosions suggesting the possibility that healing may occur.
- There is significant improvement in physical function, vitality and social functioning sustained over two years despite a significant impairment and disability at baseline.
- A significant reduction in CRP and rheumatoid factor is apparent by 28 weeks and maintained thereafter.
- There appears to be inhibition of joint damage in infliximab-treated patients despite continuing disease activity suggesting dissociation in the mechanisms causing joint damage and inflammation.
- A recently completed randomised placebo controlled trial of a combination of infliximab and methotrexate in early RA patients, not previously exposed to methotrexate, has demonstrated superiority over methotrexate monotherapy (up to 20 mg/week in clinical and radiographic measurements [86]).

Etanercept is efficacious as monotherapy and in combination with methotrexate long-term

Phase II and III clinical trials of etanercept for the treatment of RA began after initial infliximab trials and progressed rapidly to demonstrate efficacy of monotherapy in DMARD unresponsive randomised placebo-controlled trials [87, 88]. An optimum dose of 25 mg twice a week by subcutaneous injection was established. This p75-TNF receptor Fcγ1 fusion protein is not entirely non-immunogenic but the induced immune response is claimed not to limit efficacy for long-term therapy.

In a randomised controlled trial in patients with persistent disease activity despite methotrexate therapy, the addition of etanercept was more efficacious than continuing methotrexate monotherapy [89]. A further trial in early RA demonstrated monotherapy with etanercept to be superior to methotrexate monotherapy (up to 20 mg/week) in controlling signs and symptoms at six and 12 months and in retarding radiographic scores of joint damage and improving physical function at the end of two years [60, 90]. In addition, patients with longer standing disease, treated in a parallel arm randomised study with a combination of etanercept and methotrexate, proved to be more efficacious than monotherapy with either etanercept or methotrexate (the results of TEMPO trial presented at EULAR 2003).

Etanercept therapy has proved efficacious in a subset of clinical trial patients continuously treated for a median of 44 months with good tolerability [91]. In a proportion of these patients it proved possible to reduce concomitant therapy with corticosteroids and methotrexate.

Etanercept is the first anti-TNF drug licensed for the treatment of RA, is also efficacious in the treatment of juvenile chronic arthritis [92] and the only anti-TNF drug licensed for this disorder to date.

In a recent study some non-responders to etanercept achieved satisfactory responses to infliximab and *vice versa* [93]. The practical implication is that switching from one anti-TNF drug to an alternative may be justified in patients to whom other effective options may not be available.

Adalimumab

Adalimumab, a fully human IgG1 monoclonal anti-TNF antibody derived by phage technology, is the latest to gain approval for the treatment of RA. In randomised clinical trials involving over 2000 patients to date, efficacy in improving symptoms, signs, physical function and inhibiting structural damage for monotherapy and in combination with methotrexate [84, 94–96] has been observed. Combination therapy shows greater efficacy than monotherapy.

Adalimumab is administered by subcutaneous injection at a dose of 20–40 mg every two weeks and the time interval between injections of this drug increases the option of regimens of anti-TNF drugs available to patients.

The clinical trial data also shows the superiority of anti-TNF therapy over methotrexate monotherapy in early RA, especially in preventing joint damage, in studies extending one to two years. Intuitively speaking, it argues for the early use of these drugs and studies are being planned for testing intervention within the first 12 weeks of onset of symptoms. Since the prognosis of individual patients cannot be accurately predicted, and anti-TNF therapy is expensive and does carry an element of risk, the pharmacoeconomic and risk-benefit issues will need to be addressed in ongoing and future studies.

Safety issues

These have been addressed in reviews and consensus statements by expert groups [97] and will not be detailed here. Risk benefit analyses are being carried out and in general, despite the potential for toxicity, anti-TNF drugs provide an important therapeutic opportunity for patients [98]. In this section comments are restricted to some pertinent issues.

TNF and lymphotoxin play an important role in host defence and an increased susceptibility to infection in anti-TNF treated patients was anticipated. The reactivation of latent tuberculosis is the best recognised problem with most case reported with infliximab in the post-marketing period [99]. Whether this is due, for example, to biological properties of infliximab, its dosing schedule and pharmacokinetics or a higher background of past infections in patients exposed to infliximab in greater numbers such as in Spain, is not clear. The Spanish Society of Rheumatology database has shown a higher incidence in rheumatoid patients than the normal population prior to, and an increase following, the introduction of infliximab [100, 101]. Encouragingly, there has been a striking reduction in the final year of the registry since the institution of guidelines for screening latent tuberculosis by clinical history, tuberculin testing and chest radiographs and pre-treatment with anti-tuberculosis drugs before anti-TNF therapy.

Bacterial pneumonias, sepsis, listirosis, coccidiomycosis, histoplasmosis and pneumocystosis have all been reported following anti-TNF therapy. Consensus guidelines on prevention and early detection and treatment with appropriate screening and monitoring have been proposed [97] and are the key to reducing risk of infection.

There is clear evidence of serological induction of antinuclear and anti-ds-DNA antibodies emerging from clinical trials, but drug-induced lupus syndrome is rare in comparison. For example, in one study the induced anti-ds-DNA antibodies in 14% of patients were of IgM class and appeared within 4–10 weeks of commencement of infliximab [80]. In contrast in the ATTRACT trial, only 1 of 340 patients treated developed symptoms of lupus [79]. Pre-existing antinuclear antibody by the immunofluorescent test is not predictive of drug-induced lupus. Other autoimmune complications have been described, notably exacerbation and new onset of multiple sclerosis (MS) [102]. Rare cases of neutropenia and aplastic anaemia with etanercept have been described but it is not known whether these have an autoimmune pathogenesis.

TNF is over-expressed in patients with congestive cardiac failure and is implicated in impaired cardiac function and cachexia. Clinical trials of infliximab and etanercept did not show significant benefit and, surprisingly, infliximab was associated with increased hospitalisation and mortality [103]. New onset cardiac failure has been reported in etanercept and infliximab treated patients, apparently previously free of heart disease [104]. Based on this evidence, treatment of patients with symptomatic heart failure is not recommended.

The possibility that all anti-TNF drugs increase the incidence of lymphomas has been raised by an increase in the standardisation incidence ratio derived by a comparison with occurrence of lymphomas recorded in cancer registries. In post-marketing studies reports of lymphoma have been reported with infliximab and etanercept [105]. The data was reviewed by the Food and Drug Administration (FDA) of the USA in 2003 with the knowledge that the increase in B cell lymphomas in RA treated with anti-TNF drugs fell within the confidence intervals reported in rheumatoid patients never exposed to these drugs. The increased incidence of lymphoma in RA may in part be explained by prior exposure to cytotoxic-immunosuppressive drugs such as azathioprine and methotrexate, and in part to severe rheumatoid itself, which is associated with an increase in occurrence of lymphomas. The FDA and European regulatory agencies await prospective data from registries in USA and Europe before further definitive analysis.

How effective is anti-TNF therapy and what are the future prospects?

The clinical trial evidence strongly supports the contention that anti-TNF therapy, particularly in combination with methotrexate, provides a relatively safe and effective treatment option for patients with moderate and severe disease unresponsive to conventional DMARDs. Such patients are spiralling further into progressive joint damage, loss of quality of life and reduced life expectancy, and anti-TNF therapy appears to make a substantial impact on at least the former two outcomes in about 50% of such patients. Studies on long-term retention rates suggest that the majority of patients remain on therapy, and the best responders in one study continued to benefit for up to 47 months [91, 106, 107].

In the future, cheaper manufacturing costs should reduce the cost of anti-TNF therapies and make them more affordable and hence accessible to patients. Since RA has a complex multi-step pathogenesis, it is likely that simultaneous targeting of more than one molecule will be required to cure the disease. However, for the present we can anticipate anti-TNF therapies being used for many other immune-inflammatory diseases [108]. Already infliximab is proven to be effective and licensed for the treatment of Crohn's disease and ankylosing spondylitis and etanercept for psoriatic arthritis and ankylosing spondylitis. Both are efficacious for the treatment of psoriasis and encouraging data is emerging for uveitis and vasculitis.

Acknowledgements
I would like to thank my colleagues Marc Feldmann, Peter Taylor, Richard Williams, Andrew Cope, Fionula Brennan for their collaboration with our work on the definition of TNF as a therapeutic target and understanding the biology of the importance of cytokines in inflammatory diseases. The Kennedy Institute is supported by the Arthritis Research Campaign and The Kennedy Institute of Rheumatology Trust.

References

1 Coley WB (1891) Contribution to the knowledge of sarcoma. *Ann Surg* 14: 199–220
2 Coley WB (1893) The treatment of malignant tumors by repeated inoculations of erysipelas: with a report of ten original cases. *Am J Med Sci* 105: 487–511
3 Coley Nauts H, Fowler GA, Bogatko FH (1953) A review of the influence of bacterial

infection and of bacterial products (Coley's toxins) on malignant tumors in man. *Acta Medica Scand* (Suppl) 274: 29–97
4 Carswell EA, Old LJ, Kassel RJ, Green S, Fiore N, Williamson B (1975) An endotoxin-induced serum factor that causes necrosis of tumors. *Proc Natl Acad Sci USA* 72: 3666–3670
5 Pennica D, Nedwin GE, Hayflick JS, Seeburg PH, Derynck R, Palladino MA, Kohr WJ, Aggarwal BB, Goeddel DV (1984) Human tumor necrosis factor: precursor structure expression and homology to lymphotoxin. *Nature* 312: 724–729
6 Pennica D, Hayflick JS, Bringman TS, Palladino MA, Goeddel DV (1985) Cloning and expression in *Escherichia coli* of the cDNA for murine tumor necrosis factor. *Proc Natl Acad Sci USA* 82: 6060–6064
7 Beutler B, Greenwald D, Hulmes JD, Chang M, Pan YC, Mathison J, Ulevitch R, Cerami A (1985) Identity of tumour necrosis factor and the macrophage-secreted factor cachectin. *Nature* 316: 552–554
8 Beutler B, Milsark IW, Cerami A (1985). Tumour necrosis factor: production, distribution and metabolic fate *in vivo*. *J Immunol* 135: 3972–3977
9 Nedwin GE, Naylor SL, Sakaguchi AY , Smith D, Jarrett-Nedwin J, Pennica D, Goeddel DV, Gray PW (1985) Human lymphotoxin and tumor necrosis factor genes: structure, homology and chromosomal localization. *Nucleic Acids Res* 13: 6361–6373
10 Shirai T, Yamaguchi H, Ito H, Todd CW, Wallace RB (1985) Cloning and expression in *Escherichia coli* of the gene for human tumour necrosis factor. *Nature* 313: 803–806
11 Seckinger P, Isaaz S, Dayer JM (1989) Purification and biologic characterisation of specific tumor necrosis alpha inhibitor. *J Biol Chem* 264: 11966–11973
12 Engelmann H, Novick D, Wallach D (1990) Two tumor necrosis factor-binding proteins purified from human urine. Evidence for immunological cross-reactivity with cell surface tumor necrosis factor receptors. *J Biol Chem* 265: 1531–1536
13 Gray PW, Barrett K, Chantry D, Turner M, Feldmann M (1990) Cloning of human tumor necrosis factor (TNF) receptor cDNA and expression of recombinant soluble TNF-binding protein. *Proc Natl Acad Sci USA* 87: 7380–7384
14 Loetscher H, Schlaeger EJ, Lahm HW, Pan Y-CE, Lesslauer W, Brockhaus M (1990) Purification and partial amino acid sequence analysis of two distinct tumor necrosis factor receptors from HL60 cells. *J Biol Chem* 265: 20131–20138
15 Smith CA, Davis T, Anderson D, Solam L, Beckmann MP, Jerzy R, Dower SK, Cosman D, Goodwin RG (1990) A receptor for tumor necrosis factor defines an unusual family of cellular and viral proteins. *Science* 248: 1019–1023
16 Aderka D, Engelmann H, Maor Y, Brakebusch C, Wallach D (1992) Stabilization of the bioactivity of tumour necrosis factor by its soluble receptors. *J Exp Med* 175: 323–329
17 Aggarwal BB, Vilcek J (eds) (1992) *Tumor Necrosis Factors: Structure, Function and Mechanism of Action*. Dekker, New York
18 Zhang M, Tracey KJ (1998) Tumor Necrosis Factor. In: Thomson A (ed): *The Cytokine Handbook*. 3rd ed. Academic Press, London, San Diego, 517–548

19 Chen G, Goeddel DV(2002) TNFR1 signalling: a beautiful pathway. *Science* 296: 1634–1635

20 Band V, Karin M (2001) Signal transduction by tumor necrosis factor and its relatives. *Trends Cell Biol* 11: 372–377

21 Foxwell B, Andreakos E, Brennan F, Feldmann M, Smith C, Conron M (2003) Prospects for the development of small molecular weight compounds to replace anti-tumour necrosis factor biological agents. *Ann Rheum Dis* 62 (Suppl II): ii90–ii93

22 Taylor GA, Carballo E, Lee DM, Lai WS, Thompson MJ, Patel DD, Schenkman DI, Gilkeson GS, Broxmeyer HE, Haynes BF, Blackhear PJ (1996) A pathogenetic role of TNFα in the syndrome of cachexia, arthritis, and autoimmunity resulting from tristetraprolin (TTP) deficiency. *Immunity* 4: 445–454

23 Kontoyiannis D, Pasparakis M, Pizarro TT, Cominelli F, Kollias G (1999) Impaired on/off regulation of TNF biosynthesis in mice lacking TNF AU-rich elements: implications for joint and gut-associated immunopathologies. *Immunity* 10: 387–398

24 Corcoran A, Barrett K, Turner M, Brown A, Kissonerghis AM, Gadnell M, Gray PW, Chernajovsky Y, Feldmann M (1994) Characterization of ligand binding by the human p55 tumour-necrosis factor receptor: involvement of individual cysteine-rich repeats. *Eur J Biochem* 223: 831–840

25 Brennan FM, Chantry D, Jackson A, Maini R, Feldmann M (1989) Inhibitory effect of TNFα antibodies on synovial cell interleukin-1 production in rheumatoid arthritis. *Lancet* 2: 244–247

26 Steed PM, Tansey MG, Zalevsky J, Zhukovsky EA, Desjarlais JR, Szymkoswki DE, Abbott C, Carmichael D, Chan C, Cherrry L et al (2003) Inactivation of TNF signaling by rationally designed dominant-negative TNF variants. *Science* 301: 1895–1898

27 Abraham E (2003) Computational design of variant TNF molecules: A novel methodology of inhibition of proinflammatory cascades. Sci STKE pe51

28 Chu CQ, Field M, Feldmann M, Maini RN (1991) Localization of tumor necrosis factor a in synovial tissues and at the cartilage-pannus junction in patients with rheumatoid arthritis. *Arthritis Rheum* 34: 1125–1132

29 Chu CQ, Field M, Allard S, Abney E, Feldmann M, Maini RN (1992) Detection of cytokines at the cartilage/pannus junction in patients with rheumatoid arthritis: implications for the role of cytokines in cartilage destruction and repair. *Br J Rheumatol* 31: 653–661

30 Marinova-Mutafchieva L, Williams RO, Mason LJ, Mauri C, Feldmann M, Maini RN (1997) Dynamics of proinflammatory cytokine expression in the joints of mice with collagen-induced arthritis (CIA). *Clin Exp Immunol* 107: 507–512

31 Marinova-Mutafchieva L, William RO, Funa K, Maini RN, Zvaifler N (2002) Inflammation is preceded by tumour necrosis factor-dependent infiltration of mesenchymal cells in experimental arthritis. *Arthritis Rheum* 46: 507–513

32 Feldmann M, Brennan FM, Maini RN (1996) Role of cytokines in rheumatoid arthritis. *Ann Rev Immunol* 14: 397–440

33 Cope A, Aderka D, Doherty M, Engelmann H, Gibbons D, Jones AC, Brennan FM,

Maini RN, Wallach D, Feldmann M (1992) Increased levels of soluble tumour necrosis factor (TNF) receptors in the sera and synovial fluids of patients with rheumatic diseases. *Arthritis Rheum* 35: 1160–1169
34 Butler DM, Maini RN, Feldmann M, Brennan FM (1995) Modulation of proinflammatory cytokine release in rheumatoid synovial membrane cell cultures. Comparison of monoclonal anti-TNFα antibody with the IL-1 receptor antagonist. *Eur Cytokine Network* 6: 225–230
35 Williams RO, Feldmann M, Maini RN (1992) Anti-tumor necrosis factor ameliorates joint disease in murine collagen-induced arthritis. *Proc Natl Acad Sci USA* 89: 9784–9788
36 Thorbecke GJ, Shah R, Leu CH, Kuruvilla AP, Hardison AM, Palladino MA (1992) Involvement of endogenous tumour necrosis factor α and transforming growth factor β during induction of collagen type II arthritis in mice. *Proc Natl Acad Sci USA* 89: 7375–7379
37 Keffer J, Probert L, Cazlaris H, Georgopoulos S, Kaslaris E, Lioussis D, Kollias G (1991) Transgenic mice expressing human tumour necrosis factor: a predictive genetic model of arthritis. *EMBO J* 10: 4025–4031
38 Feldmann M, Maini RN (2001) Anti-TNFα therapy of rheumatoid arthritis: What have we learned? *Annu Rev Immunol* 19: 163–196
39 Maini RN, Feldmann M (2002) How does infliximab work in rheumatoid arthritis? *Arthritis Res* 4 (Suppl 2): S22–S28
40 Elliott MJ, Maini RN, Feldmann M, Long-Fox A, Charles P, Katsikis P, Brennan FM, Walker J, Bijl H, Ghrayeb J (1993) Treatment of rheumatoid arthritis with chimeric monoclonal antibodies to tumour necrosis factor α. *Arthritis Rheum* 36: 1681–1690
41 Elliott MJ, Maini RN, Feldmann M, Long-Fox A, Charles P, Bijl H, Woody JN (1994) Repeated therapy with monoclonal antibody to tumour necrosis factor α (cA2) in patients with rheumatoid arthritis. *Lancet* 344: 1125–1127
42 Charles P, Elliott MJ, Davis D, Potter A, Kalden JR, Antoni C, Breedveld FC, Smolen JS, Eberl G, De Woody K et al (1999) Regulation of cytokines, cytokine inhibitors and acute phase proteins following anti-TNFα therapy in rheumatoid arthritis. *J Immunol* 163: 1521–1528
43 Lorenz HM, Antoni C, Valerius T, Repp R, Grunke M, Schwerdtner N, Nusslein H, Woody J, Kalden JR, Manger B (1996) *In vivo* blockade of TNFα by intravenous infusion of a chimeric monoclonal TNFα antibody in patients with rheumatoid arthritis – Short term cellular and molecular effects. *J Immunol* 156: 1646–1653
44 Pittoni V, Bombardieri M, Spinelli FR (2002) Anti-tumor necrosis factor (TNF) alpha treatment of rheumatoid arthritis (infliximab) selectively down regulates the production of interleukin (IL) 18 but not of IL-12 and IL-13. *Ann Rheum Dis* 61: 723–725
45 Tokayer A, Carsons SE, Chokshi B (2002) High levels of interleukin 13 in rheumatoid arthritis sera are modulated by tumor necrosis factor antagonist therapy: associated with dendritic cell growth activity. *J Rheumatol* 29: 454–461
46 Maini RN, Elliott MJ, Brennan FM, Williams RO, Chu CQ, Paleolog E, Charles PJ,

Taylor PC, Feldmann M (1995) Monoclonal anti-TNFα antibody as a probe of pathogenesis and therapy of rheumatoid disease. *Immunol Rev* 144: 195–223
47 Scallon BJ, Moore MA, Trinh H, Knight DM, Ghrayeb J (1995) Monoclonal antibody cA2 binds recombinant transmembrane TNFα and activates immune effector functions. *Cytokine* 7: 251–259
48 Scallon B, Cai A, Solowski N, Rosenberg A, Song XY, Shealy D, Wagner C (2002) Binding and functional comparisons of two types of tumor necrosis factor antagonists. *J Pharmacol Exp Ther* 301: 418–426
49 Smeets TJ, Kraan MC, van Loon ME, Tak PP (2003) Tumor necrosis factor alpha blockade reduces the synovial cell infiltrate early after initiation of treatment, but apparently not by induction of apoptosis in synovial tissue. *Arthritis Rheum* 48: 2155–2162
50 Van Deventer SJH (2001) Transmembrane TNF inductuion of apoptosis and the efficacy of TNF-targeting therapies in Crohn's disease. *Gastroenterology* 121: 1242–1245
51 Paleolog EM, Hunt M, Elliott MJ, Feldmann M, Maini RN, Woody JN (1996) Deactivation of vascular endothelium by monoclonal anti-tumor necrosis factor α antibody in rheumatoid arthritis. *Arthritis Rheum* 39: 1082–1091
52 Tak PP, Taylor PC, Breedveld FC, Smeets TJ, Daha MR, Kluin PM, Meinders AE, Maini RN (1996) Decrease in cellularity and expression of adhesion molecules by anti-tumor necrosis factor α monoclonal antibody treatment in patients with rheumatoid arthritis. *Arthritis Rheum* 39: 1077–1081
53 Taylor PC, Peters AM, Paleolog E, Chapman PT, Elliott MJ, McCloskey R, Feldmann M, Maini RN (2000) Reduction of chemokine levels and leukocyte traffic to joints by tumor necrosis factor α blockade in patients with rheumatoid arthritis. *Arthritis Rheum* 43: 38–47
54 Ballara S, Taylor PC, Reusch P, Marme D, Feldmann M, Maini RN, Paleolog EM (2001) Raised serum vascular endothelial growth factor levels are associated with destructive change inflammatory arthritis. *Arthritis Rheum* 44: 2055–2064
55 Paleolog EM, Young S, Stark AC, McCloskey RV, Feldmann M, Maini RN (1998) Modulation of angiogenic vascular endothelial growth factor (VEGF) by TNFα and IL-1 in rheumatoid arthritis. *Arthritis Rheum* 41: 1258–1265
56 Taylor PC, Steuer A, Gruber J, et al (2004) Comparison of ultrasonographic assessment of synovitis and joint vascularity with radiological evaluation in a randomized, placebo-controlled study of infliximab therapy in early rheumatoid arthritis. *Arthritis Rheum; in press*
57 Kalden-Nemeth D, Grebmeier J, Antoni C, Manager B, Wolf F, Kalden JR (1997) NMR monitoring of rheumatoid arthritis patients receiving anti-TNF monoclonal antibody therapy. *Rheumatol Int* 16: 249–222
58 Brennan FM, Browne KA, Green PA, Jasper J-M, Maini RN, Feldmann M (1997) Reduction of serum matrix metalloproteinase 1 and matrix-metalloproteinase 3 in RA patients following anti-TNFα (cA2) therapy. *Br J Rheumatol* 36: 643–650
59 Lipsky PE, van der Heijde D, St Clair EW, Furst DE, Breedveld FC, Kalden JR, Smolen

JS, Weisman M, Emery P, Feldmann M et al (2000) Infliximab and methotrexate in the treatment of rheumatoid arthritis. *New Engl J Med* 343(22): 1594–1602

60 Bathon JM, Martin RW, Fleischmann RM, Tesser JR, Schiff MH, Keystone EC, Genovese MC, Wasko MC, Moreland LW, Weaver AL et al (2000) A comparison of etanercept and methotrexate in patients with early rheumatoid arthritis. *New Engl J Med* 343: 1586–1593

61 Ziolkowska M, Kurowska M, Radzikowska A, Lulszczykiewicz G, Wiland P, Dziewczopolski W, Filipowicz-Sosnowska A, Pazdur J, Szechinski J, Kowalczewski J et al (2002) High levels of osteoprotegerin and soluble receptor activator of nuclear factor kappa B ligand in serum of rheumatoid arthritis patients and their normalization after anti-tumor necrosis factor alpha treatment. *Arthritis Rheum* 46: 1744–1753

62 Teitelbaum SL (2000) Bone resorption by osteoclasts. *Science* 289: 1504–1508

63 Shealy DJ, Wooley PH, Emmell E, Volk A, Rosenberg A, Treacy G, Wagner CL, Mayton L, Griswold DE, Song XY (2002) Anti-TNFα antibody allows healing of joint damage in polyarthritic transgenic mice. *Arthritis Res* 4: R7

64 Redlich K, Gortz B, Hayer S, Zwerina J, Doerr N, Kostenuik P, Bergmeister H, Kollias G, Steiner G, Smolen JS, Schett G (2004) Repair of local bone erosions and reversal of systemic bone loss upon therapy with anti-tumour necrosis factor in combination with osteoprotegerin or parathyroid hormone in tumor necrosis factor-mediated arthritis. *Am J Pathol* 164(2): 543–555

65 Zwerina J, Hayer S, Tohidast-Akrad M, Bergmeister H, Redlich K, Feige U, Dunstan C, Kollias G, Steiner G, Smolen J, Schett G (2004) Single and combination inhibition of tumour necrosis factor, interleukin-1, and RANKL pathways in tumor necrosis factor-induced arthritis: effects on synovial inflammation, bone erosion, and cartilage destruction. *Arthritis Rheum* 50(1): 277–290

66 Barrera P, Joosten LA, den Broeder AA, van de Putte LB, van Riel PL, van den Berg WB (2001) Effects of treatment with a fully human anti-tumour necrosis factor alpha monoclonal antibody on the local and systemic homeostasis of interleukin 1 and TNFα in patients with rheumatoid arthritis. *Ann Rheum Dis* 60: 660–669

67 Ulfgren A-K, Andersson U, Engstrom M, Klareskog L, Maini RN, Taylor PC (2000) Systemic anti-tumor necrosis factor α therapy in rheumatoid arthritis downregulates synovial tumor necrosis factor a synthesis. *Arthritis Rheum* 43: 2391–2396

68 Padyukov L, Lampa J, Heimburger M, Ernestam S, Cederholm T, Lundkvist I, Andersson P, Hermansson Y, Harju A, Klareskog L, Bratt J (2003) Genetic markers for the efficacy of tumour necrosis factor blocking therapy in rheumatoid arthritis. *Ann Rheum Dis* 62: 536–539

69 Mugnier B, Balandraud N, Darque A, Roudier C, Roudier J, Reviron D (2003) Polymorphism at position -308 of the tumor necrosis factor alpha gene influences outcome of infliximab therapy in rheumatoid arthritis. *Arthritis Rheum* 48: 1849–1852

70 Cope AP (2002) Studies of T cell activation in chronic inflammation. *Arthritis Res* 4 (Suppl 13): S197–S211

71 Cope AP, Londei M, Chu NR, Cohen SB, Elliott MJ, Brennan FM, Maini RN, Feldmann

M (1994) Chronic exposure to tumor necrosis factor (TNF) *In vitro* impairs the activation of T cells through the T cell receptor/CD3 complex; reversal *in vivo* by anti-TNF antibodies in patients with rheumatoid arthritis. *J Clin Invest* 94: 749–760
72 Moreland LW, Bucy RP, Weinblatt M, Mohler KM, Spencer-Green GT, Chatham WW (2002) Immune function in patients with rheumatoid arthritis treated with etanercept. *Clin Immunol* 103: 13–21
73 McInnes IB, Leung BP, Sturrock RD, Field M, Liew FY (1997) Interleukin-15 mediates T cell-dependent regulation of tumor necrosis factor-alpha production in rheumatoid arthritis [see comments]. *Nat Med* 3: 189–195
74 Brennan FM, Hayes AL, Ciesielski CJ, Green P, Foxwell BM, Feldmann M (2002) Evidence that rheumatoid arthritis synovial T cells are similar to cytokine-activated T cells: involvement of phosphatidylinositol 3-kinase and nuclear factor κβ pathways in tumor necrosis factor α production in rheumatoid arthritis. *Arthritis Rheum* 46: 31–41
75 Vey E, Burger D, Dayer JM (1996) Expression and cleavage of tumor necrosis factor-alpha and tumor necrosis factor receptors by human monocytic cell lines upon direct contact with stimulated T cells. *Eur J Immunol* 26: 2404–2409
76 Burger D, Dayer JM (2002) The role of human T-lymphocyte-monocyte contact in inflammationn and tissue destruction. *Arthritis Res* 4(Suppl 3): S169–S176
77 Bondeson J, Foxwell B, Brennan F, Feldmann M (1999) Defining therapeutic targets by using adenovirus: blocking NF-κβ inhibits both inflammatory and destructive mechanisms in rheumatoid synovium but spares anti-inflammatory mediators. *Proc Natl Acad Sci USA* 96: 5668–5673
78 Ehrenstein MR, Evans JG, Singh A, Moore S, Isenberg DA, Mauri C (2003) Expansion of regulatory ($CD4^+CD25hi$) T lymphocytes in the peripheral blood of patients with rheumatoid arthritis who respond to anti-TNF therapy. *Am Coll Rheumatol*: Abst 537
79 Maini RN, St Clair EW, Breedveld F, Furst D, Kalden J, Weisman M, Smolen J, Emery P, Harriman G, Feldmann M, Lipsky P (1999) Infliximab (chimeric anti-tumour necrosis factor a monoclonal antibody) *versus* placebo in rheumatoid arthritis patients receiving concomitant methotrexate: a randomised phase III trial. *Lancet* 354: 1932–1939
80 Charles PJ, Smeenk RJT, DeJong J, Feldmann M, Maini RN (2000) Assessment of antibodies to double-stranded DNA induced in rheumatoid arthritis patients following treatment with infliximab, a monoclonal antibody to tumor necrosis factor α. *Arthritis Rheum* 43: 2383–2390
81 Hochberg MC, Tracey JK, Hawkins-Holts M, Flores RH (2003) Comparison of the efficacy of the tumour necrosis factor alpha blocking agents adalimumab, etanercept, and infliximab when added to methotrexate in patients with active rheumatoid arthritis. *Ann Rheum Dis* 62 (Suppl 2): ii13–ii16
82 Feldmann M, Maini RN (2003) TNF defined as a therapeutic target for rheumatoid arthritis and other autoimmune diseases. *Nature Med* 9: 1245–1250
83 Maini RN, Breedveld FC, Kalden JR, Smolen JS, Davis D, Macfarlane JD, Antoni C,

Leeb B, Elliott MJ, Woody JN et al (1998) Therapeutic efficacy of multiple intravenous infusions of anti-tumor necrosis factor a monoclonal antibody combined with low dose weekly methotrexate in rheumatoid arthritis. *Arthritis Rheum* 41: 1552–1563
84 Weinblatt M, Keystone EC, Furst DE, Moreland LW, Weisman MH, Birbara CA, Teoh LA, Fischkoff SA, Chartash EK (2003) Adalimumab, a fully human anti-tumor necrosis factor a monoclonal antibody, for the treatment of rheumatoid arthritis in patients taking concomitant methotrexate: The ARMADA trial. *Arthritis Rheum* 48: 35–45
85 Maini RN, Breedveld FC, Kalden JR et al (2004) Sustained improvement in physical function, structural damage, and signs and symptoms through 2 years in rheumatoid arthritis patients treated with infliximab (Remicade(R)) and methotrexate. *Arthritis Rheum; in press*
86 Smolen JS, Van der Heijde DM, Emery P et al (2003) Methotrexate naive early rheumatoid arthritis (RA) patients with higher baseline ESR and CRP derive greater radiographic benefit from initiating treatment with a combination of infliximab and methotrexate. *Ann Rheum Dis* (Supp Abs)
87 Moreland LW, Baumgartner SW, Schiff MH, Tindall EA, Fleischmann RM, Weaver AL, Ettlinger RE, Cohen S, Koopman WJ, Mohler K et al (1997) Treatment of rheumatoid arthritis with a recombinant human tumor necrosis factor receptor (p75)-Fc fusion protein. *New Engl J Med* 337: 141–147
88 Moreland LE, Schiff MH, Baumgartner SW, Tindall EA, Fleischmann RM, Bulpitt KJ, Weaver AL, Keystone EC, Furst DE, Mease PJ et al (1999) Etanercept therapy in rheumatoid arthritis. *Ann Internal Med* 130: 478–486
89 Weinblatt ME, Kremer JM, Bankhurst AD, Bulpitt KJ, Fleischmann RM, Fox RI, Jackson CG, Lange M, Burge DJ (1999) A trial of etanercept, a recombinant tumor necrosis factor receptor: Fc fusion protein, in patients with rheumatoid arthritis receiving methotrexate. *New Engl J Med* 340: 253–259
90 Genovese MC, Bathon JM, Martin RW, Fleischman RM, Tesser JR, Schiff MH, Keystone EC, Wasko MC, Moreland LW, Weaver AL et al (2003) Etanercept *versus* methotrexate in patients with early rheumatoid arthritis: two-year radiographic and clinical outcomes. *Arthritis Rheum* 46: 1443–1450
91 Kremer JM, Weinblatt ME, Bankhurst AD, Bulpitt KJ, Fleischmann RM, Jackson CG, Atkins KM, Feng A, Burge DJ (2003) Etanercept added to background methotrexate therapy in patients with rheumatoid arthritis: continued observations. *Arthritis Rheum* 48: 1493–1499
92 Lovell DJ, Giannini EH, Reiff A, Cawkwell GD, Silverman ED, Nocton JJ, Stein LD, Gedalia A, Ilowite NT, Wallace CA et al (2000) Etanercept in children with polyarticular juvenile rheumatoid arthritis. Pediatric Rheumatology Collaborative Study Group. *N Engl J Med* 342: 763–769
93 Van Vollenhoven R, Harju A, Brannemark S, Klareskog L (2003) Treatment with infliximab (Remicade) when etanercept (Enbrel) has failed or *vice versa*: data from STURE

registry showing that switching tumour necrosis factor alpha blockers can make sense. *Ann Rheum Dis* 62: 1195–1198

94 Weisman MH, Moreland LW, Furst DE, Weinblatt ME, Keystone EC, Paulus HE, Teoh LS, Velagapudi RB, Noertersheuser PA, Granneman GR et al (2003) Efficacy, pharmacokinetic, and safety assessment of adalimumab, a fully human anti-tumor necrosis factor-alpha monoclonal antibody, in adults with rheumatoid arthritis receiving concomitant methotrexate: a pilot study. *Clin Ther* 25: 1700–1721

95 van de Putte LB, Rau R, Breedveld FC, Kalden JR, Malaise MG, Van Riel PL, Schattenkirchner M, Emery P, Burmester GR, Zeidler H et al (2003) Efficacy and safety of the fully human anti-tumour necrosis factor alpha monoclonal antibody adalimumab (D2E7) in DMARD refractory patients with rheumatoid arthritis: a 12 week, phase II study. *Ann Rheum Dis* 62: 1168–1177

96 Machold KP, Smolen JS (2003) Adalimumab – a new TNFα antibody for treatment of inflammatory joint disease. *Expert Opin Biol Ther* 3: 351–360

97 Furst DE, Breedveld FC, Kalden JR, Smolen JS, Burmester GR, Dougados M, Emery P, Gibofsky A, Kavenaugh AF, Keystone EC et al (2003) Updated consensus statement on biological agents for the treatment of rheumatoid arthritis and other immune mediated inflammatory diseases (May 2003). *Ann Rheum Dis* 62 (Suppl 11): ii2–ii9

98 Mikuls TR, Moreland LW (2003) Benefit-risk assessment of infliximab in the treatment of rheumatoid arthritis. *Drug Saf* 26: 23–32

99 Keane J, Gershon SK, Wise RP, Mirabile-Levens E, Kasznica J, Schwieterman WD, Siegel JN, Braun MM (2001) Tuberculosis associated with infliximab, a tumor necrosis factor alpha-neutralizing agent. *N Engl J Med* 345: 1098–1104

100 Carmona L, Hernandez-Garcia C, Vadillo C, Pato E, Balsa A, Gonzalez-Alvaro I, Belmonte MA, Tena X, Sanmarti R (2003) Increased risk of tuberculosis in patients with rheumatoid arthritis. *J Rheumatol* 30: 1397–1399

101 Gomez-Reino JJ, Carmona L, Valverde VR, Mola EM, Montero MD, BIOBADASER Group (2003) Treatment of rheumatoid arthritis with tumor necrosis factor inhibitors may predispose to significant increase in tuberculosis risk: a multicenter active-surveillance report. *Arthritis Rheum* 48: 2085–2091

102 Robinson WH, Genovese MC, Moreland LW (2001) Demyelinating and neurologic events reported in association with tumor necrosis factor a antagonism: By what mechanisms could tumor necrosis factor α antagonists improve rheumatoid arthritis but exacerbate multiple sclerosis? *Arthritis Rheum* 44: 1977–1983

103 Chung ES, Packer M, Lo KH, Fasanmade AA, Willerson JT (2003) Randomized, double-blind, placebo-controlled, pilot trial of infliximab, a chimeric monoclonal antibody to tumor necrosis factor-alpha, in patients with moderate-to-severe heart failure: results of the Anti-TNF Therapy Against Congestive Heart Failure (ATTACH) trial. *Circulation* 107: 3133–3140

104 Kwon HJ, Cote TR, Cuffe MS, Kramer JM, Braun MM (2003) Case reports of heart failure after therapy with a tumor necrosis factor antagonist. *Ann Intern Med* 138: 807–811

105 Brown SL, Greene MH, Gershon SK, Edwards ET, Braun MM (2002) Tumor necrosis factor against therapy and lymphoma development. *Arthritis Rheum* 46: 3151–3158

106 Flendrie M, Creemers MC, Welsing PM, den Broeder AA, van Riel PL (2003) Survival during treatment with tumour necrosis factor blocking agents in rheumatoid arthritis. *Ann Rheum Dis* 62 (suppl 2): ii30–ii33

107 Geborek P, Crnkic M, Petersson IF, Saxne T, Group. SSAT (2002) Etanercept, infliximab, and leflunomide in established rheumatoid arthritis: clinical experience using a structured follow up programme in southern Sweden. *Ann Rheum Dis* 61: 793–798

108 Kalden JR (2002) Emerging role of anti-tumor necrosis factor therapy in rheumatic diseases. *Arthritis Res* 4 (Suppl 2): S34–S40

The IL-1 family: The role of IL-1 and IL-18 in inflammation

Charles A. Dinarello

Department of Medicine, Division of Infectious Diseases, University of Colorado Health Sciences Center, 4200 East Ninth Ave., B168, Denver, CO 80262, USA

Introduction

IL-1 and its related family member IL-18 are primarily proinflammatory cytokines by their ability to stimulate the expression of genes associated with inflammation and autoimmune diseases. The most salient and relevant properties of IL-1 in inflammation are the initiation of cyclooxygenase type 2 (COX-2), type 2 phospholipase A and inducible nitric oxide syntha se (iNOS). This accounts for the large amount of prostaglandin-E_2 (PGE_2), platelet activating factor, leukotrienes and nitric oxide (NO) produced by cells exposed to IL-1 or in animals or humans injected with IL-1. Another important proinflammatory property of IL-1 is its ability to increase the expression of adhesion molecules such as intercellular adhesion molecule-1 (ICAM-1) on mesenchymal cells and vascular-cell adhesion molecule-1 (VCAM-1) on endothelial cells. This latter property promotes the infiltration of inflammatory and immunocompetent cells into the extravascular space. IL-1 is also an angiogenic factor and plays a role in tumor metastasis and blood vessel supply [1]. Mice lacking IL-1 receptors fail to develop proliferative lesions of vascular smooth muscle cells in mechanically injured arteries. In humans with rheumatoid arthritis (RA), the inflammation and joint destructive nature of the disease is reduced with systemic injections of the IL-1 receptor antagonist (IL-1Ra), a member of the IL-1 family that prevents IL-1 activity. However, in addition to these and other proinflammatory properties, IL-1 is also an adjuvant during antibody production and acts on bone marrow stem cells for differentiation in the myeloid series. In fact, unlike TNF, IL-1 is a bone marrow stimulant and was used to treat patients with bone marrow suppression in order to reduce the nadir of thrombocytopenia [2].

In terms of the role of IL-1 in human disease, specific blockade of the IL-1 receptor type I (the ligand binding chain of the heterodimeric IL-1 receptor signaling complex) with the naturally occurring IL-1Ra in patients with RA has resulted in reduced disease activity and reduced joint destruction [3–6]. IL-1Ra has been approved for use in the United States, Canada and Europe for the treatment of RA and over 50,000 patients receive daily treatment; the results support the essential inflammatory and tissue remodeling functions of IL-1.

Cytokines and Joint Injury, edited by Wim B. van den Berg and Pierre Miossec

Processing and secretion of the IL-1 family members: a unique evolution for cytokines

IL-1α, IL-1β, IL-1Ra and IL-18 are unique in the cytokine families. IL-1Ra, the only naturally occurring cytokine receptor antagonist, is readily secreted from macrophagic cells. Evolving from a common gene, IL-1β is found diffusely in the cells whereas IL-1Ra possesses a clear signal peptide and is secreted *via* the Golgi apparatus in a typical fashion similar to most secreted proteins. Secretion of IL-1Ra provides a mechanism to reduce IL-1 activity in health and disease. Indeed, mice deficient in IL-1Ra develop spontaneous disease [7, 8]. On the other hand, the agonist members of the IL-1 family are not readily secreted and require a series of intracellular steps before being biologically active cytokines.

IL-1α, IL-1β and IL-18 are each initially synthesized as precursor molecules without a signal peptide. In the case of IL-1α, the precursor form (31 kDa) is biologically active when inserted into the cell membrane as a surface protein [9, 10] as well as an intracellular cytokine [11, 12]. Evidence also supports the concept that the intracellular IL-1α can bind to nucleic acid and act as a transcription factor. This is due to the nuclear localization site in the N-terminus of the IL-1α precursor. IL-1α can also be cleaved from its membrane inserted form by calcium activated calpain [13]. The role of cell-associated IL-1α in disease is unclear, but mice deficient in IL-1α are protected in some models of inflammation and tumor growth [1, 14]. IL-1Ra blocks the activity of extracellular IL-1β and membrane IL-1α.

The IL-1β converting enzyme

After removal of N-terminal amino acids of IL-1β and IL-18 by a specific intracellular protease, the resulting active peptides are called "mature" or active forms of the cytokines. The precursor forms of IL-1β (31 kDa) and IL-18 (24 kDa) are biologically inactive and require cleavage by the intracellular cysteine protease, termed IL-1β converting enzyme (ICE). ICE is also termed caspase-1 [15], the first member of a large family of intracellular cysteine proteases with important roles in programmed cell death. However, there is little evidence that ICE (caspase-1) participates in programmed cell death [16]. Rather, ICE seems to be primarily used by the cell to cleave the IL-1β and IL-18 precursors. ICE cleaves both the IL-1β as well as the IL-18 precursors immediately following the aspartic acid in the P1 position. As a result of cleavage, a mature form of IL-1β of 17.5 kDa and 18 kDa for IL-18 are generated. Specific inhibitors of ICE effectively reduce the secretion of IL-1β as well as IL-18 and as a result, reduce the biological activities of either cytokine [17]. Oral inhibitors of ICE are in clinical trials in patients with RA. Although ICE is primarily responsible for cleavage of the IL-1β precursor intracellularly, other proteases such as proteinase-3 can process the IL-1β precursor extracellularly into an active cytokine [18]. An extracellular protease, which cleaves IL-18 into an active cytokine, has not been reported. However, treatment of epithelial cells with proteinase-3, a neutral serine protease found in phagocytic cells, increases the release of biologically active IL-18 [19].

The P2X-7 receptor and secretion of IL-1β and IL-18

The addition of extracellular ATP to stimulated human monocytes results in the rapid release of mature IL-1β within minutes; a receptor-mediated event has been proposed. Purinergic receptors (adenosine is a purine) have been studied since these are found on monocytes and macrophages releasing IL-1β and 18. For ATP, this receptor is designated P2X-7. The anti-RA drug, tenidap, had been shown to affect this receptor, and tenidap reduced the secretion of IL-1β [20]. When triggered by millimolar concentrations of ATP, reversible pores form in the plasma membrane and due to ion fluxes, the electrical potential of the membrane is transiently lost. In the presence of endotoxin, this activation by ATP triggers the release of mature IL-1β [21]. A monoclonal antibody to the P2X-7 receptor prevents the release of mature IL-1β from activated macrophages [22]. Importantly, the secretion of IL-1β *via* activation of the P2X-7 receptor by ATP is independent of ICE, since highly specific inhibitors of ICE prevent processing of the IL-1β and IL-18 precursors but have no effect on release of the IL-1β precursor or the release of lactic dehydrogenase in cells stimulated with ATP [23]. Triggering of the P2X-7 receptor is specific for the release of mature IL-1β and IL-18, but does not result in the release of TNFα, ATP or nigericin, which rapidly results in the loss of intracellular potassium, and also stimulates the release of IL-1 and IL-18 in endotoxin-stimulated whole blood cultures [24]. Convincing evidence for a role of the P2X-7R in the post-translational processing and secretion of IL-1β is found in mice deficient in this receptor. Similar to wild-type macrophages, P2X-7R-deficient macrophages synthesize PGE_2 and the IL-1β precursor in response to endotoxins. However, when activated by ATP, P2X-7R-deficient macrophages do not process or release IL-1β [25]. Similar results have been observed with IL-18. *In vivo*, wild-type and P2X-7R-deficient release the same amounts of IL-6 into the peritoneal cavity, but there is no release of mature IL-1β from P2X-7R-deficient mice.

IL-1 and mutations in the chronic inflammatory autoimmune syndrome-1 (CIAS-1) locus

Much basic biology is learned from mutations resulting in a particular disease-associated phenotype. In terms of inflammatory diseases, the autosomal dominantly inherited disease called Muckle-Wells syndrome has been linked to single amino acid mutations in the NALP3/CIAS-1/PYPAF-1 gene [26]. [NALP gene = NACHT leucine rich repeats, PYRIN-containing proteins; CIAS-1 = chronic inflammatory autoimmune syndrome.]

NALP-3 indicates three leucine rich repeats; NALP-4 contains four. The gene has been mapped to a location on chromosome 1q44. The earliest reports of the syndrome date to the early 1970s and Muckle wrote of the syndrome in 1979 [27]. Because this is an autosomal dominant disease, variable expression or intermittent expression may account for a spectrum of clinical manifestations and also for the periodic nature of the disease. Although Muckle-Wells syndrome is a rare disease, it has the hallmarks of systemic and local inflammation common to many inflammatory diseases, but particularly to RA and autoimmune diseases. For example, the syndrome is characterized by recurrent fevers, urticaria, painful and in some cases deforming arthropathies, periodic synovitis and amyloid nephropathy. Most early reports describe generalized amyloid depositions in all tissues and death resulting in renal failure. Congenital deafness is also associated with the syndrome [28]. Some patients have all the manifestations of the disease without renal amyloidosis [29]. Mutations in the same gene are also the linked to another inflammatory disease called familial cold autoinflammatory syndrome (FCAS) [30]. Patients with this syndrome exhibit urticarial eruptions and systemic symptoms when exposed to cold. Another systemic inflammatory disease of neonates has mutations in this same gene [31]. Termed neonatal-onset multisystem inflammatory disease (NOMID), or CINCA (chronic infantile neurologic cutaneous articular) syndrome, patients exhibit fever, chronic meningitis, uveitis, sensorineural hearing loss, urticarial skin rash and a characteristic deforming arthropathy [31]. An unusual finding of a genetic study in four of these patients was the absence of the mutations in parental DNA supporting the conclusion that in some cases the disease is from a *de novo* mutation [31]. Similar to many systemic diseases, elevated circulating levels of IL-1β, TNFα, IL-6 and other cytokines can be documented. In particular, high levels of acute phase reactants characterize these syndromes. Anti-TNFα therapies have no effect.

There have been 23 point mutations found in this gene (variably called NALP3/CIAS-1/PYPAF-1) but each is a single amino acid change in a protein called NACHT [32]. The protein NACHT (NAIP, CIITA, HET-E, TP1) is thought to function as an assembly platform, which facilitates the docking of other proteins, particularly proteins containing a motif called caspase recruitment domains (CARD). CARD-containing proteins participate in protein–protein interactions for ICE (caspase-1) activation. CARD motifs are also found in proteins called pyrins, and mutations in pyrins are associated with another periodic systemic inflammatory disease, familial mediterranean fever (FMF). In Muckle-Wells syndrome, the most common mutation in NACHT is R260W [26]. It is unclear how CARD motifs activate ICE. ICE is present as an inactive precursor in many cells but primarily in cells with phagocytic functions. The NALP3/CIAS-1/PYPAF-1 gene is expressed primarily in these same cells, as is ICE. It is now believed that proteins that comprise the NALP3 gene associate with other intracellular proteins containing the CARD motifs and activate ICE into an active enzyme for the cleavage of the inactive IL-1β precursor into a mature, active cytokine.

Another protein that is essential for activation of ICE is ASC, which contains a CARD motif in the pyrin domain on the protein. The ASC protein binds to inactive ICE. Together, the assembly is called the "inflammasome" [33], a large, signal-induced multiprotein complex. NALP3 and ASC associate with inactive ICE (pro-caspase-1) to form a complex [34]. The inactive ICE N-terminal also has a CARD motif, which binds to the CARD motif of ASC and results in activation of ICE and the secretion of active IL-1β [34]. It is known that the activation of ICE requires two cleavages of the proenzyme; the active form is a heterodimer (Fig. 1).

As shown in Figure 1, processing of the IL-1β (including IL-18) precursor requires several steps, beginning with the activation of ICE. First, the N-terminus of the proenzyme is cleaved. There is an ICE site itself at this location but it is also possible that caspase-5 may initiate the first step. Active ICE is a heterodimer and the internal cleavage site also contains an ICE site. As a result of the second cleavage, the heterodimer is active with the active cysteine on the longer peptide. In monocytes from human subjects, without mutations in NALP3 (wild-type), the activation of ICE is concomitant with the synthesis of the IL-1β precursor. Thus, agents such as Toll-like receptor ligand endotoxins or peptidoglycans initiate both the synthesis of the IL-1β precursor and the activation of ICE.

The finding that elevated levels of IL-1β are present in the sera of patients with Muckle-Wells syndrome, FCAS or CINA does not necessarily prove causation of the manifestations of the disease as due to IL-1β. However, patients with Muckle-Wells syndrome have a dramatic response to a single injection of IL-1Ra [35]. These patients have high levels of serum amyloid A protein, which results in the fatal nephropathy. IL-1Ra rapidly reduces serum amyloid A protein (from nearly 100 μg/dl to less than 100 ng/dl) within days and fever and related systemic symptoms are completely reduced 24 hours following IL-1Ra [35]. Although IL-1Ra blocks the IL-1 receptor triggered by either IL-1α or IL-1β, it seems highly likely that the dramatic efficacy of IL-1Ra in patients with Muckle-Wells syndrome is due to blockade of IL-1β triggered responses.

There are two mechanisms that may explain the link of point mutations in these patients with disease caused by IL-1β. The mutations in NACHT may result in overactive processing of the IL-1β precursor to an active cytokine, particularly during minor perturbations to monocytes. Thus, the assembly of the NALP3 scaffold of proteins participating in activation of ICE may be facilitated by the mutations. However, it seems more likely that the mutations in NACHT result in a failure of an endogenous inhibitor of ICE activation (inhibitor of NACHT assembly) to bind or function. Indeed, intracellular inhibitors of ICE have been reported. The periodic nature of the disease manifestations associated with these mutations would favor the concept of ineffective inhibitor function. It is presently unclear whether these

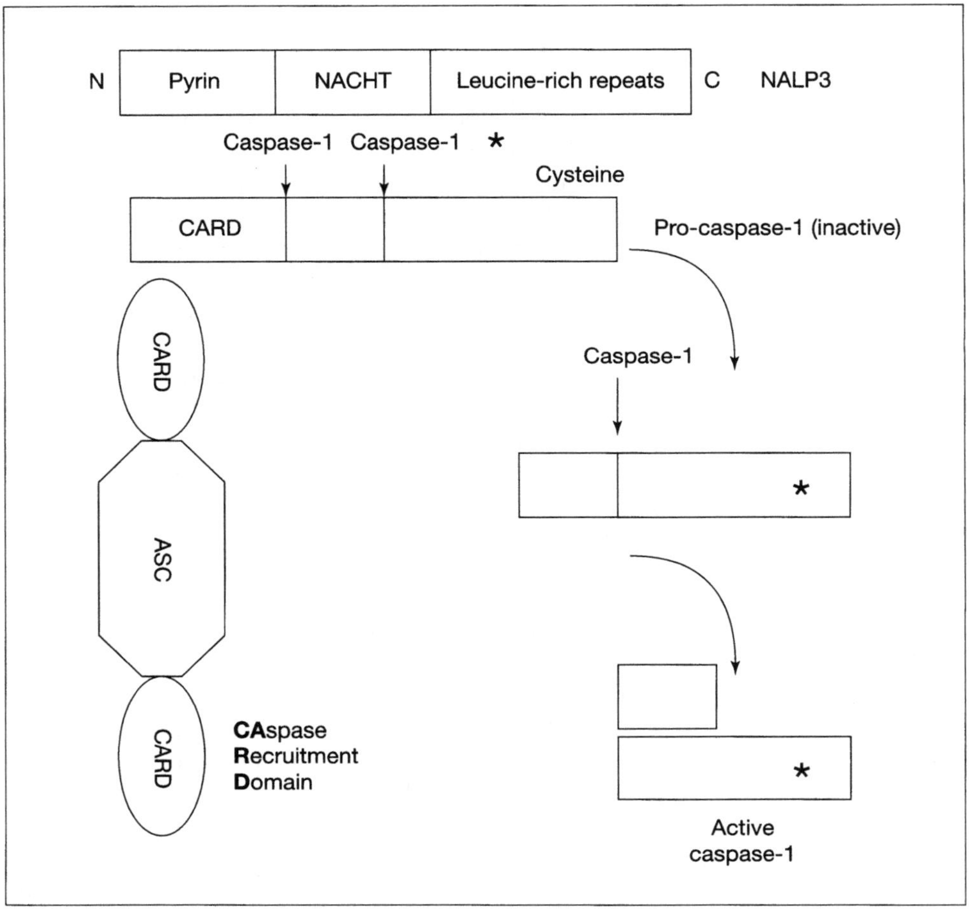

Figure 1
Processing of the IL-1β and IL-18 precursors into active molecules by the IL-1β converting enzyme (ICE). See test for details.

patients would respond to oral ICE inhibitors. However, from the dramatic biochemical and clinical responses, mutations in the NALP3/CIAS-1/PYPAF-1 protein result in an IL-1 disease with many of the characteristics of systemic and local inflammation. Because the ICE activation also results in secretion of active IL-18, it is unclear whether blocking IL-18 with the IL-18 binding protein (see below) would also result in the same reduction of disease activity.

Several periodic fever syndromes have been described, each with a different mutation in specific proteins. In addition to familial Mediterranean fever, Muckle-

Wells syndrome, familial cold urticaria and chronic infantile neurological cutaneous and articular syndrome, there are the TNF receptor 1A associated syndrome and hyperIgD syndrome. A website has been established (http://fmf.igh.cnrs.fr/infevers/) in order to provide updated data on mutations responsible for these syndromes [36]. These are distinct syndromes and not similarly treated with the same agents. For example, colchicine is the drug of choice for FMF but has no effect in Muckle-Wells syndrome [37]. The mutations in the Muckle-Wells syndrome gene appears to result in dysregulation of ICE activity, increased IL-1β processing and release and hence is treated with IL-1 receptor blockade [35]. The TNF receptor 1A associated syndrome is treated with soluble receptors to TNF.

Blocking IL-1 in disease

IL-1Ra

Historically, the late Phillipe Seckinger, working in the laboratory of Jean-Michel Dayer, demonstrated that the urinary inhibitor of IL-1 prevented IL-1 binding to cells [38]. A similar IL-1 inhibitor was induced by stimulating human blood monocytes with IgG [39]. This IL-1 inhibitor was purified to homogeneity and the amino acid sequence was used to clone the molecule from a monocyte cDNA library [40]. The recombinant IL-1 inhibitor was renamed the IL-1 receptor antagonist (IL-1Ra) and was shown to block the binding of IL-1 as a receptor antagonist. The surprise for the IL-1 field was the discovery that the amino acid sequence of IL-1Ra was more homologous to the amino acid sequence of IL-β than IL-1α was to IL-1β. There is only one IL-1Ra gene but several isoforms exist. Unlike IL-1β, IL-1Ra has a classic leader peptide and is readily secreted. The other forms of IL-1Ra, namely intracellular forms lacking a leader peptide but derived from the same gene, have been reported [41]. When expressed as a mature recombinant molecule, intracellular IL-1Ra blocks IL-1 binding as well as does the secreted form [42].

The crystal structural analysis of the IL-1RI/IL-1Ra complex reveals that IL-1Ra contacts all three domains of IL-1RI. IL-1β has two sites of binding to IL-1RI. There is a primary binding site located at the open top of its barrel shape and a second site on the reverse side of the IL-1β molecule. IL-1Ra also has two binding sites, which are similar to those of IL-1β. However, the reverse side site of IL-1Ra is more homologous to that of IL-1β than the primary binding site. Thus, the backside site of IL-1Ra binds to IL-1RI tightly and occupies the receptor. Because the second binding site is not available, IL-1Ra does not recruit the IL-1 receptor accessory protein (IL-1RacP) to form the heterocomplex required to trigger a signal. After binding of IL-1Ra to IL-1RI bearing cells, there was no phosphorylation of the epidermal growth factor receptor, a well-established and sensitive assessment of IL-1 signal transduction: Overwhelming evidence that IL-1Ra is a pure receptor antagonist.

The intravenous injection of IL-1Ra into healthy humans at doses 1,000,000-fold greater than that of IL-1α or IL-1β exhibits no agonist activity [43].

Studies in IL-1Ra-deficient mice

Mice deficient in IL-1Ra have low litter numbers and exhibit growth retardation in adult life [44]. These animals also have elevated basal concentrations of plasma IL-6 and exhibit higher levels of hepatic acute phase proteins compared to those of wild-type control mice. Endotoxemia or a turpentine abscess in IL-1Ra-deficient mice results in increased lethality. In a model of endotoxin-induced inflammation, IL-1Ra-deficient mice exhibit nearly two-fold higher serum levels of sphingomyelinase compared to wild-type mice. The most dramatic phenotype has been observed in IL-1Ra-deficient mice crossed from a C57BL/6 to a BALB/c genetic background [7]. In these IL-1Ra-deficient mice with a BALB/c background, a chronic inflammatory polyarthropathy develops spontaneously. The joints showed prominent synovial and periarticular infiltration of inflammatory cells, osteoclast activation and structural erosion associated with the presence of granulation tissue. Overall, the histological pattern appeared similar to that of humans with RA. Elevated levels of anti-IgG1, but not anti-IgM, were observed, as were elevated levels of rheumatoid factor and anti-double-stranded DNA antibodies. Steady-state levels of COX-2, IL-1RI, IL-1β, IL-6, and TNFα mRNA in the affected joints were also increased. Bone erosion with arthritis was present in the joints of IL-1Ra-deficient mice, which was consistent with the observation that mice deficient in IL-1RI exhibited no significant trabecular bone loss following ovariectomy compared to wild-type controls [45].

The finding that IL-1Ra-deficient mice spontaneously develop a destructive, inflammatory arthropathy strongly suggests that endogenous IL-1Ra functions to suppress inflammation in mice living in a normal environment. Furthermore, since IL-1Ra binds only IL-1 receptors, the results also implicate IL-1 as an essential cytokine in the pathological process. The onset of this autoimmune process requires a genetic background favoring the Th2 response, which produces antibodies rather than cytotoxic T cells in response to antigens. The immunologic stimulus likely occurs when either an endogenous antigen or an antigen from the intestinal flora triggers a Th2 response, which, in the absence of IL-1 blockade, is uncontrolled. IL-17 expression is elevated in IL-1Ra-deficient mice and the spontaneous development of arthritis does not take place in IL-1Ra-deficient mice that are also deficient in IL-17 [46]. IL-1 induces the expression of OX40 on $CD4^+$ T cells and activation of surface OX40 facilitates antigen recognition by T cells. Therefore, IL-17 production appears to be a consequence of IL-1-induced OX40, which results in heightened responses to antigens. Although the naturally occurring antigen(s) that triggers the development of a RA-like disease in IL-1Ra-deficient mice has not been identified, it appears that without endogenous control of IL-1 activity, the immune response to either foreign (floral) or self antigens results in an thoroughly pathologic process. Of considerable importance is that no other cytokine or cytokine receptor knockout mouse manifests a spontaneous arthritic phenotype. Knockout mice for other anti-inflammatory cytokines such as IL-10 or transforming growth factor-β develop inflammatory bowel disease and systemic inflammation, but not arthritis.

IL-1Ra deficient mice also develop a lethal arterial inflammation involving primary and secondary branch points of the aorta [8]. These are stress points in the vessel wall due to blood flow and are also the same locations at which atherosclerotic plaques are commonly found. The lesions are characterized by transmural infiltration of neutrophils, macrophages and $CD4^+$ T cells. Death is due to vessel wall collapse, stenosis and organ infarction. Heterozygotes, which have reduced but detectable levels of endogenous IL-1Ra compared to wild-type controls, do not die from this severe arteritis but do develop small arterial lesions.

Numerous studies have implicated IL-1 in the pathogenesis of RA using animal models of the disease [47]. Although the systemic administration of IL-1 or the instillation of IL-1 into a joint space provides supportive data, there are drawbacks to interpreting the effects of exogenously applied IL-1 as a preclinical model. The most convincing preclinical evidence for a pivotal role for IL-1 in RA is derived from experimental models in which specific blockade of IL-1 activity reduces one or more of the pathological processes that arise in the context of the naturally occurring disease. The administration of IL-1Ra reduces the inflammation as well as the loss of bone and cartilage in the rat adjuvant arthritis model and hence fulfills the criteria of reduction in disease severity using a complex model that mimics human RA pathology. Rats with developing adjuvant arthritis were treated with IL-1Ra by continuous infusion. The results showed modest but significant reductions in swelling of the ankle joints and in paw weights and histologic improvement of bone and cartilage lesions. However, marked inhibition (53%) of bone resorption was also observed, even at doses at which anti-inflammatory activity was not seen [48]. A well-studied model for mimicking human RA is collagen-induced arthritis in using immunization with type II collagen. In this model, high levels of exogenous IL-1Ra completely suppress the disease [48]. In the collagen-induced arthritis model in rats, methotrexate alone reduced bone erosions by 57%, but the combination of IL-1Ra plus methotrexate reduced bone erosions by 97%. In addition, IL-1Ra treatment of rats with established collagen-induced arthritis resulted in nearly complete suppression of all parameters of the disease [48]. Other methods to provide higher therapeutic levels of IL-1Ra have been investigated. For example, following transplantation of murine 3T3 fibroblasts transfected with the gene for human IL-1Ra, paws and knee joints were inspected to evaluate inflammation and cartilage destruction in a murine model of type II collagen-induced arthritis. The onset of collagen-induced arthritis was almost prevented in joints containing the transfected IL-1Ra-expressing cells, whereas joints containing cells transfected with the control (empty) vector showed severe inflammation and destruction of cartilage. In the paw ipsilateral to the IL-1Ra gene-expressing knees, reduced inflammation and joint destruction were observed [49]. Following instillation of endotoxin directly into the joints of rabbits, leukocyte infiltration and protein leak developed; however, administration of a neutralizing monoclonal antibody against rabbit IL-1Ra resulted in a 50% increase in the level of IL-1β and a 20–40% enhancement of leukocyte infiltration and protein leakage [50]. One can conclude that endogenous IL-1Ra functions as an anti-inflammatory cytokine in this model by limiting the production of IL-1β as well as the intensity of the inflammatory response.

Human chondrocytes can be cultured in the presence of a cartilaginous matrix and studied for the synthesis of proteoglycans. When synovial fibroblasts from patients with RA are added to this system, the cartilaginous matrix is destroyed in 10–18 days. The addition of IL-1Ra or anti–IL-1β monoclonal antibodies reduced the destruction of the matrix by 45% and 35%, respectively [51].

Reducing IL-1 activity with the soluble form of the IL-1 receptor type I

The extracellular domain of the type I receptor (also called soluble IL-1 receptor type I, IL-1sRI) has the highest binding affinities for IL-1Ra and IL-1α but the least affinity for IL-1β. Injection of murine IL-1sRI to mice increased the survival of heterotopic heart allografts and reduced the hyperplasic lymph node response to allogeneic cells [52]. In a rat model of antigen-induced arthritis, local instillation of the murine IL-1sRI reduced joint swelling and tissue destruction [53]. Even when instilled into the contralateral, unaffected joint, a reduction in tissue damage was also observed in the affected joint, suggesting that the amount of IL-1sRI injected into the contralateral joint was acting systemically. In a model of experimental autoimmune encephalitis, the IL-1sRI also reduced the severity of this disease [54]. Administration of IL-1sRI to animals has also been reported to reduce the physiologic response to LPS, acute lung injury and delayed-type hypersensitivity (reviewed in [53]).

However, there are also data suggesting that exogenous administration of IL-1sRI may act as carrier. In mice, an i.v. injection of IL-1sRI alone induced a rapid increase in circulating IL-1α, but not of TNFα or IL-1β [55]. The soluble receptor did not interfere with the IL-1α assay. This observation is consistent with the view that IL-1sRI acts as a carrier for IL-1α. Treatment of mice with IL-1sRI improved the survival during a lethal infection with *Candida albicans*.

Using the accelerated model of autoimmune diabetes induced by cyclophosphamide in the non-obese diabetic (NOD) mouse, repeated injections with IL-1sRI protected NOD mice from insulin-dependent diabetes mellitus in a dose-dependent fashion; the incidence of the diabetes was 53.3% in the mice treated with 0.2 mg/kg and only 6.7% in mice treated with 2 mg/kg. However, none of the doses of IL-1sRI reduced the extent of insulitis in NOD mice. Splenic lymphoid cells from NOD mice treated with 2 mg/kg IL-1sRI for five consecutive days showed a normal distribution of mononuclear cell subsets and maintained their capacity to secrete IFNγ and IL-2 [56].

Effect of IL-1sRI in humans

Recombinant human IL-1sRI has been administered intravenously to healthy humans in a Phase I trial without side effects or changes in physiologic, hematologic or endocrinologic parameters. Thus, similar to infusions of IL-1Ra, IL-1sRI appears safe and reinforces the conclusion that IL-1 does not have a role in homeostasis in humans.

Healthy volunteers have also been injected with LPS following pre-treatment with IL-1sRI [57]. Volunteers were also pre-treated with IL-1sRI or placebo and then challenged with endotoxin. There were no effects on fever or systemic symptoms. Although there was a decrease in the level of circulating IL-1β compared to placebo-treated volunteers, there was also a decrease in the level of circulating IL-1Ra ($p < 0.001$); this was due to complexing of the soluble receptor to endogenous IL-1Ra [57]. This was dose-dependent and resulted in a 43-fold decrease in endotoxin-induced IL-1Ra. High doses of IL-1sRI were also associated with higher levels of circulating TNFα and IL-8 as well as cell-associated IL-1β [57]. These results support the concept that IL-1sRI binds endogenous IL-1Ra and reduces its effectiveness in inhibiting IL-1. As discussed below, patients with RA treated with IL-1sRI do not exhibit improved clinical outcomes.

IL-1sRI was administered to 23 patients with active RA in a randomized, double-blind, two-center study. Patients received subcutaneous doses of the receptor at 25, 250, 500, or 1,000 $\mu g/m^2$/day or placebo for 28 consecutive days. Although four of eight patients receiving 1,000 $\mu g/m^2$/day showed improvement in at least one measure of disease activity, only one of these four patients exhibited clinical improvement [58]. Similar to the placebo-treated patients, lower doses of the receptor did not result in any improvement by acceptable criteria. Despite this lack of clinical or objective improvement in disease activity, cell surface monocyte IL-1α expression in all patients receiving IL-1sRI was significantly reduced. One explanation for the lack of clinical response despite efficacy in suppressing immune responses could be the inhibition of endogenous IL-1Ra. This was observed in volunteers receiving IL-1sRI before challenge by endotoxin [57].

A Phase I trial of IL-1sRI was also conducted in patients with relapsed and refractory acute myeloid leukemia. Circulating levels of IL-1sRI were elevated 360- and 25-fold after i.v. and s.c. administration, respectively. However, serum levels of IL-1β, IL-6 and TNFα did not change and there were no complete, partial or minor responses to treatment [59]. There have been no further trials of IL-1sRI in human disease.

Other strategies to reduce IL-1 activity in disease

The goal of any IL-1 based strategy is to prevent the formation of the heterodimer complex of surface receptors. Using IL-1sRI to block IL-1 activity in disease is similar to using neutralizing antibodies against IL-1 and distinct from using receptor blockade with IL-1Ra. Since the molar concentrations of circulating IL-1 in disease are relatively low, pharmacologic administration of IL-1sRI to reach a 100-fold molar excess of the soluble receptor over that of IL-1 is feasible. The human trial of IL-1sRI in delayed hypersensitivity reactions supports the notion that low doses (100 μg/patient) can have anti-inflammatory effects. The fusion of two chains of extracellular domains of the type IL-1RI to the Fc portion of immunoglobulin enhances the binding IL-1 over that of monomeric IL-1sRI [60] and may have a greater plasma half-life compared to that of the monomeric form. However, as shown in the study of IL-1sRI in RA [58], binding of the endogenous IL-1Ra worsened the disease. The goal of receptor antagonists requires occupancy of large numbers of IL-1 surface receptors with natural or synthetic antagonists [61] since triggering only a few evokes a response. Receptor blockade is a formidable task and the large amounts of IL-1Ra required to reduce disease activity contributes to this conclusion. The disadvantage of IL-1sRI therapy is that it is prolonging the clearance of IL-1 or reducing the function of the natural IL-1Ra.

The soluble form of IL-1RI and IL-1RII circulate in healthy humans at molar concentrations which are 10–50-fold greater than those of IL-1β measured in septic patients and 100-fold greater than the concentration of IL-1β following intravenously administration [62]. Why do humans have a systemic response to an infusion of IL-1α [2] or IL-1β? One can conclude that binding of IL-1 to the soluble forms of IL-1R types I and II exhibits a slow "on" rate compared to the cell-bound IL-1RI. Another interpretation is that the soluble forms of both the type I and type II receptors found in the circulation are already occupied.

Naturally occurring, neutralizing antibodies to IL-1α are present in many patients with autoimmune disease and even in health subjects and likely reduce the activity of IL-1α [63]. However, there have been few animal studies that support the concept that IL-1α in human disease is a sensible target. Despite the presence of soluble forms of both receptors, naturally occurring antibodies and circulating IL-1Ra, the injection of IL-1 as low as 1 ng/kg subcutaneously evokes a systemic response in humans [64]. These findings underscore the high functional level of only a few IL-1 type I receptors. They also imply that the post-receptor triggering events are greatly amplified. It seems reasonable to conclude that treating disease based on blocking IL-1 type I receptor needs to take into account the efficiency of so few type I receptors initiating a biological event. The use of soluble forms of the IL-1R type II, neutralizing antibodies to IL-1β, blocking antibodies to the IL-1R type I or blocking antibodies to the IL-1R accessory protein are each likely to be tested inn human disease.

The IL-1 trap

A cytokine trap is a construction of the extracellular domains of the two chains of a heterodimeric cytokine receptor complex [65]. The trap takes advantage that two separate receptor chains provide high affinity binding compared to single cytokine receptor chains. The cytokine trap provides both receptor chains in a dimeric configuration much like the extracellular cell surface receptor. The trap is, however, different in that each arm of the dimeric structure is identical and comprised of domains of both receptor chains. The two chains of the trap are linked by fusion of complement binding domain of IgG1. As such, the structure is not naturally occurring but is comprised of solely natural proteins.

Neutralization of cytokine activity is the therapeutic objective of soluble receptors. The IL-1 trap is comprised of the ligand binding receptor chain (IL-1 receptor type I) plus the IL-1 receptor accessory protein, the second chain of the IL-1 receptor complex. The two chains are expressed linearly and fused to the Fc domain of IgG1. The resulting affinity of the IL-1 trap has a dissociation constant of 1–2 pM whereas the affinity of the complete cell surface IL-1 receptor complex for IL-1β is 400 pM. Cytokine traps offer additional therapeutic benefits in that they have long plasma half-lives and can neutralize members of the same cytokine family binding to the same receptor chains. For example, the IL-1 trap also binds and neutralizes IL-1β as well as IL-1α activity. In contrast, two distinct monoclonal antibodies would be needed in order to reduce both IL-1α and IL-1β.

Although the IL-1 trap also binds the IL-1Ra, the affinity is 23-fold less than that for IL-1β. Hence, the advantage of the IL-1 trap is its preferential binding of IL-1β and not to IL-1Ra. Not unexpectedly, cytokine traps also work *in vivo*. A murine receptor component IL-1 trap reduced inflammatory parameters and joint bone erosions [65]. The IL-1 trap also inhibited IL-1-induced IL-6 production. In a Phase I trial of the IL-1 trap in patients with RA, there was a dose-dependent fall in serum C-reactive protein, and at a single dose of 100 mg/week, a 40% reduction was reported compared to placebo-treated patients.

Studies using IL-1Ra in animals

Because IL-1Ra exhibits no species specificity, a large body of data exists revealing a role for IL-1 in models of disease. As shown in Figure 2, when IL-1Ra binds to the type I IL-1R, there is no formation of the heterodimer with the IL-1R-AcP chain. In the presence of IL-1β, there is no signal transduction and hence there is classic competitive inhibition similar to other receptor antagonists. However, unlike small molecule receptor antagonists, IL-1Ra appears to be a pure receptor antagonist. IL-1Ra occupancy of the type I receptor also prevents the binding of IL-1α to the cell receptor. Since there is considerable evidence that the biological response of IL-1α is due

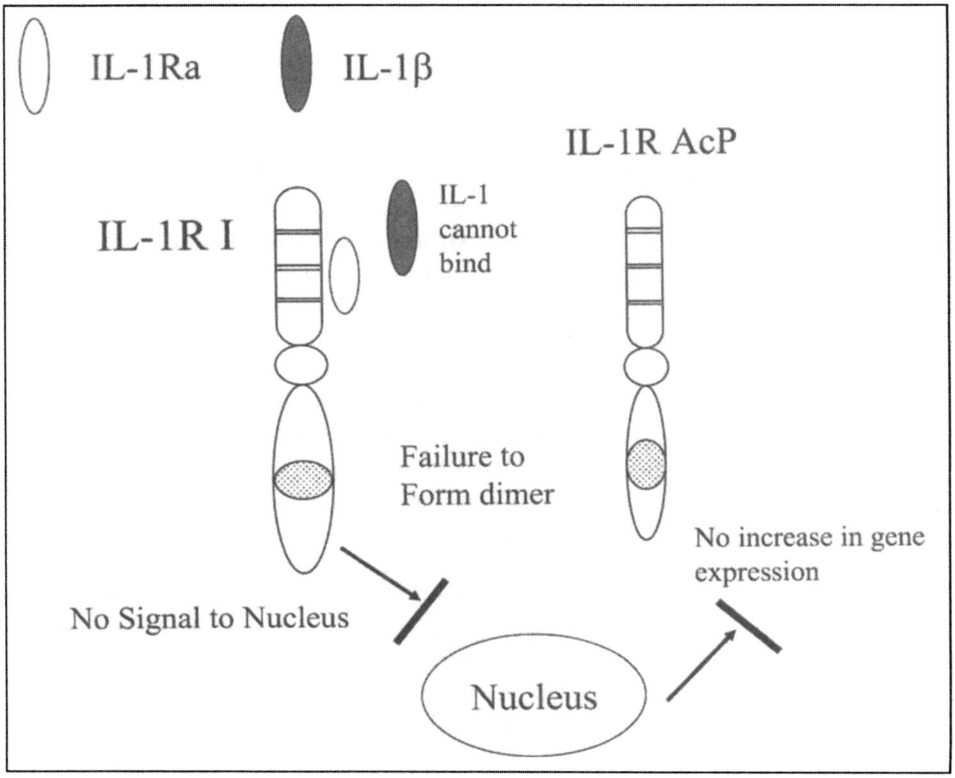

Figure 2
IL-1 receptor antagonist prevents the formation of the IL-1 receptor type I heterodimer with the IL-1 receptor accessory protein.

to its membrane form, nevertheless, IL-1Ra blocks the activities of membrane IL-1α. Before IL-1Ra was approved for treating RA in humans, there were extensive preclinical trials in various animal models, particularly in models of autoimmune diseases and model of RA. Table 1 lists the effects of blocking IL-1 in various animal models of disease.

IL-1Ra treatment in humans with RA

The recombinant form of IL-1Ra produced in *E. coli* is called by the generic name anakinra. In a randomized, double-blind trial, anakinra was administered to 175 patients at 21 sites in the United States [66]. During the first three weeks of the trial,

Table 1: Effects of IL-1Ra

Models of infection
Improved survival in LPS-induced shock in primates, mice, rats and rabbits
Improved survival in *Klebsiella pneumoniae* infection in newborn rats
Reduction in shock and mortality in rabbits and baboons from *E. coli* or *Staphylococcus epidermidis* bacteremia
Amelioration of shock and reduction in death after cecal ligation and puncture
Attenuation of LPS-induced lung nitric oxide activity
Decreased hypoglycemia, production of CSF and early tolerance in mice after administration of endotoxin
Reduction in LPS-induced hyperalgesia
Protection against TNF-induced lethality in D-galactosamine treated mice
Reduction in nematode-induced intestinal nerve dysfunction
Decreased circulating or cellular TNF production in models of sepsis
Decreased IL-6 production after LPS or enteric LPS administration
Protection from *Bacillus anthracis* toxin-induced lethality in mice
Decreased intestinal inflammation and bacterial invasion in shigellosis
Reversal of decreased survival by insulin-like growth factor-1 in sepsis
Decreased in live *E. coli*-induced thrombin, tissue plasminogen activator, plasminogen activator inhibitor and elastase elevations

Models of local inflammation
Decreased neutrophil accumulation in inflammatory peritonitis in mice
Reduction in immune complex-induced neutrophil infiltration, eicosanoid production and tissue necrosis in rabbit colitis
Reduction in acid-induced neutrophil infiltration and enterocolitis in rats
Decreased endotoxin-induced intestinal secretory diarrhea in mice
Inhibition of permanganate-induced granulomas in rats
Inhibition of LPS-induced intra-articular neutrophil infiltration
Decreased IL-1-induced synovitis and loss of cartilage proteoglycan
Reduced myocardial neutrophil accumulation after coronary occlusions in dogs
Reduced inflammation and mortality in acute pancreatitis
Decreased hepatic inflammation following hemorrhagic shock
Modest reduction in acetaminophen-induced liver damage
Decreased IL-8 and MCP-1 levels induced by intravitreal LPS
Reduction in hapten-induced intestinal motor dysfunction and intestinal myeloperoxidase levels
Decreased cerulean-induced pancreatic inflammation

Table 1 - continued

Models of acute or chronic lung injury
Decreased local LPS-induced neutrophil infiltration in rats
Inhibition of antigen-induced pulmonary eosinophil accumulation and airway hyperactivity in guinea pigs
Prevention of bleomycin or silica-induced pulmonary fibrosis
Reduction in hypoxia-induced pulmonary hypertension
Reduction in carrageenan-induced pleurisy in rats
Decreased intratracheal IL-1-induced fluid leak (systemic administration)
Decreased albumin leak after systemic LPS
Inhibition of antigen-induced eosinophil accumulation in guinea pigs
Models of central nervous system functions
Decreased stress-induced hypothalamic-pituitary axis
Decreased immobilization-induced stress on hypothalamic-pituitary axis
Reduced astrocytosis after spinal cord transection
Increased survival after heat stroke in rats
Decreased cerebral ischemia-induced edema and infarct size
Reduced astrocytosis-mediated wound closure after brain damage
Decreased LPS-induced brain monoamine levels
Decreased in number of necrotic neurons in cerebral artery occlusion
Models of metabolic dysfunction
Reduction in hepatocellular damage following ischemia-reperfusion
Improved survival after hemorrhagic shock in mice
Inhibition of SAA gene expression and synthesis in high dose IL-2 toxicity
Decreased muscle protein breakdown in rats with peritonitis due to cecal ligation
Reduced muscle protein breakdown in rats with chronic septic peritonitis
Inhibition of weight loss following muscle tissue injury
Decrease in bone loss in ovariectomized rats
Decreased multinucleated osteoclasts in ovariectomized rats
Reversal of LPS-induced CRF gene expression in the hypothalamus
Prevention of LPS-induced ACTH release
Decreased *E. coli*-induced chaperone protein-induced osteolytic activity in rat calvarium
Decreased LPS-induced bone resorption in mouse calvarium

Table 1 - continued

Models of autoimmune disease
Diminution of *Streptococcus* wall-induced arthritis in rats
Reduction in collagen-induced arthritis in rats and mice
Suppression of anti-basement membrane glomerulonephritis
Delayed hyperglycemia in the diabetic BB rat
Decreased hyperglycemia in non-obese diabetic mice
Reduction in autoimmune myelin basic protein-induced encephalomyelitis
Reduction in elevated cortisone levels in allergic encephalomyelitis
Models of immune-mediated disease
Prevention of graft *versus* host disease in mice
Prolongation of islet allograft survival
Reduction in skin contact hypersensitivity to haptens
Decrease in coronary artery fibronectin deposition in heterotopic cardiac transplant
Decreased adjuvant arthritis in rats
Reduction in streptozotocin-induced diabetes
Increased survival of corneal transplants
No effect on mitogen- or antigen-induced T cell proliferation
Decreased IL-2-induced LAK cell generation
Decreased in IL-2-induced LAK cell TNF production
Decreased phorbol ester-induced B cell proliferation and IgG synthesis
Models of malignant disease
Reduction in the number and size of metastatic melanoma
Reduction in growth of subcutaneous melanoma tumors
Reduced LPS-induced augmentation of metastatic melanoma
Reduction in tumor-mediated cachexia (intratumoral injection)
Reduced spontaneous proliferation, colony formation and cytokine production of human AML, CML (adult and juvenile) leukemia cells
Reduced spontaneous IL-6 and PGE2 production in multiple myeloma cells
Decreased spontaneous release of serotonin and histamine in rat basophilic leukemia cells
Decreased spontaneous blast proliferation and cytokine production
Effect of models of angiogenesis
Decreased new blood vessel growth following sciatic nerve injury in mice
Decreased vascularization of inflammatory polymers implanted subcutaneously in mice

Table 1 - continued

Impairment of host responses
Increased mortality to *Klebsiella pneumoniae* in newborn rats (high dose)
Increased mortality to *Listeria* infection
Enhanced growth of *Mycobacterium avium* in organs
Worsening of infectious arthritis (late administration)
Increased vascular leak in mice given high dose IL-2
Effects on production of cytokines and other molecules
Reduced LPS-induced monocyte production of IL-1β, IL-1α, IL-6
Decreased LPS-induced fibroblast production of IL-6
Decreased asbestos-stimulated IL-8 from mesothelial cells
Reduced spontaneous production of substance P by cultured neurons
Enhanced natural killer cell activity
Increased spontaneous PGE_2 production by cultured decidual cells
Increased smooth muscle cell proliferation
Other effects *in vitro*
Reduction in *Actinobacillus* LPS-induced bone resorption
Reduction in spleen cell colony forming units after radiation
Decreased LPS-induced nitric oxide synthase in glial cells
Decreased *Clostridia dificile*-toxin activity on hepatocytes
Reduced VCAM-1 and ICAM expression induced by Rickettsia LPS on HUVEC
Decreased TSH-stimulated cAMP release in thyroid cells
Decreased *Clostridia dificile*-toxin A-induced intestinal secretory factor release
Decreased NO release and insulin production in rat islets following TNF
Decreased PDGF receptor expression in rat lung myofibroblasts following exposure to particulates

patients received subcutaneous doses of 20, 70 or 200 mg of anakinra one, three, or seven times per week. To maintain the blindness of the study, patients received placebo injections on the days anakinra was not administered. After three weeks, a statistically significant reduction in the swollen joint count was observed in patients receiving 70 or 200 mg per day ($P < 0.01$). Daily dosing was more effective than dosing three times per week when assessed by the number of swollen joints, the investigator and patient assessments of disease activity, pain score and CRP levels. CRP levels fell from a mean baseline of 4.7 μg/dl to 2.6 μg/dl after three weeks of daily therapy of 70 mg/day during the treatment phase. Following the above study, anakinra was evaluated in a double-blind, placebo-controlled European multicenter study in 472 patients [3]. Patients with moderate-to-severe RA (disease duration between six months and eight years) and who had discontinued use of disease modifying drugs six weeks prior to entry comprised the study. Patients were randomly assigned to a 24-week course of therapy with placebo or one of three subcutaneous doses of anakinra: 30, 75 or 150 mg/day. After 24 weeks, 43% of the patients receiving 150 mg/day of anakinra exhibited a significant reduction in their disease compared to 27% of patients injected with placebo. In addition, there was a similar reduction in CRP levels erythrocyte sedimentation levels at all doses, whereas CRP and erythrocyte sedimentation levels did not change in the placebo group.

The rate of radiologic progression of joint destruction in the patients receiving anakinra was significantly less than in the placebo group at 24 weeks [67]. An additional evaluation of these patients revealed a decrease in the rate of progression in erosion and joint space narrowing compared to placebo. These clinical findings are consistent with anakinra blocking the osteoclast activating factor (OAF) property of IL-1 as has been reported using *in vitro* cultures [68]. In fact, when purified to homogeneity, the amino terminal sequence of osteoclast activating factor revealed that OAF was, in fact, IL-1β [69]. In patients who received anakinra, synovial biopsies before and after 24 weeks of daily treatment revealed evidence of decreased cellular infiltration and expression of the adhesion molecules E-selectin and VCAM-1 using immunohistochemical techniques [5]. In addition, the numbers of macrophages and lymphocytes in the subintimal tissue were reduced compared to those found in biopsies from patients treated with injections of placebo. The study also found evidence of a reduction or arrest in cellular markers of progressive joint disease in the synovial membrane.

In 2001, the United States Food and Drug Administration (FDA) approved anakinra for treating the signs and symptoms of RA; anakinra is also approved in Europe. The use of anakinra to reduce joint erosions in patients is approved in the United States. Following the monotherapy clinical trail of anakinra, other trials of the combination of anakinra plus a disease modifying drug have been reported, including the safety of anakinra [4, 6, 70, 71]. Unlike agents that neutralize TNFα, anakinra in combination with disease modifying drugs have not been associated with opportunistic infections [70].

IL-18

Interleukin-IL-18 (IL-18) is a member of the IL-1 family of ligands and IL-18 receptors, although distinct from those of IL-1, are also members of the IL-1 family of receptors. The IL-1 family is now expanded from the original two agonist members (IL-1α and IL-1β) to include the novel cytokine IL-18 and several newly described genes. This has resulted in a new nomenclature; IL-18 is also termed IL-1 Family 4 (IL-1F4). However, for this chapter, the term IL-18 will be used. IL-18 appears to have unique characteristics some of which are important for its role in RA. The most salient of these is the induction of IFNγ, particularly in the presence of IL-12. IL-18 is clearly a prominent IFNγ-inducing factor. In fact, the term "IFNγ-inducing factor" was the term used during the initial description of the cytokine by Haruki Okamura in 1989 [72].

In the years surrounding the 1989 report, the prominent IFNγ-inducing factor was IL-12 and therefore investigators concluded that the factor described by Okamura was IL-12. However, the putative IFNγ-inducing factor activity was purified from mouse livers and the amino acid sequence revealed a unique cytokine, not related to IL-12, but rather to IL-1 and particularly to IL-1β. The nomenclature of newly described cytokine genes provides an interleukin number if the recombinant cytokine is human and has a unique biological activity. Thus, with the molecular cloning of "IFNγ-inducing factor" in 1995 [73], the name was changed to IL-18.

Because IL-18 often functions to augment IFNγ production, the role of IL-18 in disease must consider the role of IFNγ itself. IFNγ is unlike other proinflammatory cytokines. IFNγ has been administered in hundreds of thousands of humans with a variety of diseases, including RA. In fact, IFNγ was approved for the treatment of RA in several European countries. The biological basis for its use in RA is, in part, due to its ability to suppress IL-1β-induced prostaglandin E_2 (PGE_2) [74, 75] and IL-1-induced collagenase activity [76]. In patients with atypical mycobacterial infections, chronic granulomatous disease of childhood or leprosy, IFNγ is used in conjunction with specific antibiotic therapies in order to improve intracellular killing of bacteria, particularly by mononuclear phagocytes [77]. There are also a large number of reports of IFNγ treatment in cancer. Overall, there has been no worsening of disease with the exception of exacerbation of central nervous system lesions in multiple sclerosis. Therefore, unlike the systemic inflammatory response of humans injected with IL-1 [78] or TNFα [79], IFNγ appears to be tolerated by humans and in some disease states, can be considered therapeutic.

As with any cytokine, its role in a particular disease process is best assessed during preclinical testing employing specific blockade or neutralization of the cytokine in a complex disease model such as adjuvant, collagen-induced arthritis (CIA) or streptococcal cell wall-induced arthritis (SCW). Although mice deficient in IL-18 have been generated and tested for the development of CIA [80], the observed reduction in disease severity may be due to a reduction in the immune response to the collagen because primary immunization with collagen in mice deficient in IL-18 or deficient in the IL-18 receptor result in low titers of anti-collagen antibodies and likely affect the disease model itself. In some cases, severe inflammatory arthritis develops in the total absence of TNFα with elevated $CD4^+$ T cells and lymphadenopathy [81] raising the issue of compensatory cytokines in gene-deficient studies. Nevertheless, IL-18 neutralization in wild-type mice is effective in reducing CIA [82] as well as SCW-induced arthritis [83]. The SCW-induced arthritis is of particular relevance since this is a model of cartilage loss due to decreased proteoglycan synthesis and is independent of IFNγ.

In addition to the disadvantage of an immune deficiency state in some gene deficient mice, no therapeutic agent can achieve the total absence of a particular cytokine in humans as takes place in gene deficient animals. On the other hand, if a single cytokine or receptor is blocked or neutralized in wild-type mice and there is a dramatic reduction in the severity of the model, then that cytokine or receptor likely plays a critical role in the disease being tested. The difficulty arises when one can show that blocking other cytokines in the same model, for example, blocking IL-1, IL-15, IL-17 or TNFα, results in a comparable reduction in severity of the arthritis. This certainly is the case with IL-18 in that one can observe similar reductions in the severity of CIA with blocking each of the cytokines stated above [82–85]. The dilemma arises as to which cytokine is best to block or neutralize clinically. In general, the cytokine to block or neutralize is the cytokine with the least role in normal cell function, repair or host defense.

In addition to local disease such as CIA and SCW arthritis, IL-18 neutralization, IL-18 deficient mice or mice deficient in IL-18 receptor-α chain reveal varying degrees of reduced severity in models of systemic disease. In fact, there is considerable overlap in blocking IL-1, IL-18, IFNγ and TNFα in models of endotoxemia. For targeting IL-18 in humans with RA, the therapeutic agents are presently neutralizing antibodies against IL-18, antibodies that block the IL-18 receptor α or β chains, the IL-18 binding protein (IL-18BP) and methods to reduce ICE activity.

Production and release of active IL-18

IL-18 is first synthesized as an inactive precursor, which requires processing to mature, active IL-18 by ICE. Oral ICE inhibitors are in clinical development in patients with RA and in some cases, Phase II trials. The therapeutic goal for inhibition of ICE activity is the reduction in the biological activity of both IL-1β and IL-18. Another member of the IL-1 family (IL-1F7) may also be affected by inhibition of ICE [86]. Although the role for IL-1F7 in RA is presently unknown, IL-1F7 binds to the IL-18BP [87]. IL-1F7 also binds to the IL-18 receptor α-chain but without a demonstrable biological effect [88].

Production of IL-18 from human dendritic cells

Dendritic cells play an important role in presentation of antigens to $CD4^+$ T cells. Like human blood monocytes [89], human dendritic cells constitutively produce IL-

18 [90]. For comparison, IL-1β is not expressed constitutively in these same cells. Dendritic cells readily interact with T cells, and this interaction contributes to the processing and secretion of mature IL-18. Dendritic cells incubated with all-specific T cells release mature IL-18 into the supernatant but not when incubated with autologous T cells. The stimulation and secretion of IL-18 is mediated *via* a CD40 signal since anti-CD40 agonistic antibody also stimulated secretion of IL-18 [90]. Similar to IL-1β, the IL-18 precursor is found diffusely in the cytosol but approximately 20% is co-localized to the lysosomal compartment. Following subcellular fractionations, the IL-18 precursor co-localizes with the endolysosomes [91]. The precursor of IL-1β also is found diffusely in the cytosol and co-localizes with these specialized endolysosomes [92]. Whereas stimulation with endotoxin results in release of mature IL-1β from monocytes *via* ICE cleavage, there is little if any secretion of mature IL-18 [89]. A similar lack of endotoxin-induced secretion of IL-18 is observed in dendritic cells [91]. However, upon contact with activated, alloreactive T cells, dendritic cells release the IL-18 precursor *via* the endolysosomal pathway. As a result, the levels of constitutive IL-18 precursor inside the cell rapidly decrease [91]. It is well established that exocytosis of these specialized lysosomes is a calcium-dependent mechanism. In fact, increasing the extracellular calcium flux results in release of the IL-18 precursor [91]. Interestingly, increasing the intracellular release of calcium affects the secretion of the IL-18 precursor. Rubartelli has proposed that the low level of recovery of mature IL-18 may be due to the rapid uptake by activated T cells [91].

Non-ICE processing of IL-18

In considering a therapeutic role for oral inhibitors of ICE, do other proteases convert the inactive precursors of IL-1β and IL-18 to active cytokines? Extracellular processing of the IL-1β precursor has been reported and is an ICE-independent mechanism. The enzyme that seems most associated with the processing of both the IL-1β as well as the IL-18 precursor is proteinase-3 (PR-3) [18]. Autoantibodies to PR-3 are pathological in Wegener's disease. The IL-18 precursor is constitutively expressed in primary human oral epithelial cells and several epithelial cell lines. When primed by IFNγ and then stimulated with PR-3 in the presence of endotoxin, these cells release active IL-18 into the supernatant [19]. Since lactate dehydrogenase activity is not released, the appearance of active IL-18 is not due to cell leakage or death. The release of active IL-18 was independent of ICE activity. Injecting mice with recombinant Fas-ligand results in hepatic damage, which is IL-18-dependent. However, the same results are observed in ICE-deficient mice [93]. Therefore, IL-18 and IL-1β precursor processing seems to take place in the absence of ICE. The implications for treating RA with oral inhibitors of ICE are presently unknown.

IL-18 as a mediator of systemic inflammation

Like all cytokine responses to infections, there are two sides to the coin. IL-18 functions to protect the host in that its ability to induce IFNγ and other immunostimulatory cytokines assists the immune system in a specific T and B cell mediated response. However, the other pathological consequences of infection are, in part, also mediated by IL-18 in somewhat the same fashion are mediated by IL-1 and TNFα. These include the increases in cell adhesion molecules, the production of chemokines, inflammatory mediators such as nitric oxide (NO) and neutrophil activation [94]. A role for IL-18 in the pathological processes of systemic inflammation is derived from animal studies in which specific blockade of IL-18 reduces the impact on organ damage or improves the survival of the host. The first experiments showed that mice deficient in ICE, failure to process the IL-18 and IL-1β precursors survived lethal endotoxemia [95, 96] but mice deficient in IL-1β died [97]. In fact, specific antibodies against mouse IL-18 also protected against the hepatic toxicity of endotoxemia [73, 98]. On the other hand, in naïve mice not preconditioned with a prior infection of *Propionibacterium acnes*, IL-18 neutralization also reduces lethal endotoxemia. Moreover, this protection is observed in mice deficient in IFNγ [99]. Thus, one may conclude that IL-18 is a proinflammatory cytokine independent of IFNγ.

Since IL-18 induces synthesis of the proinflammatory cytokines TNFα, IL-1β and the chemokines IL-8 and macrophage inflammatory protein-1α [100], neutralization of IL-18 would have a beneficial effect in lethal endotoxemia in naïve mice. Anti-IL-18 antibodies protected mice against a lethal injection of *E. coli* or *S. typhimurium* [99]. Anti-IL-18 also reduced myeloperoxidase levels in the liver and lungs [99]. An increased survival was accompanied by decreased levels of IFNγ and macrophage inflammatory protein-2 in anti-IL-18-treated animals challenged with *E. coli* endotoxin, whereas IFNγ and TNFα concentrations were decreased in treated mice challenged with *S. typhimurium*.

IL-18 binding protein

IL-18BP is a naturally occurring, secreted protein, which possesses a high affinity binding to IL-18 (dissociation constant of 400 pM) and therefore neutralizes the biological activity of IL-18 [101, 102]. IL-18BP is specific for mature IL-18 and does not bind the IL-18 precursor when assessed by ELISA [103] or BiaCore binding [102]. With the exception of IL-1F7 [87], IL-18BP does not bind to other members of the IL-1 family or several cytokines tested. IL-18BP is in clinical trials for a variety of autoimmune diseases including RA.

IL-18BP is not a soluble form of the membrane-bound IL-18 receptor, although it has many characteristics of a soluble receptor similar to the IL-1 type II receptor [104]. IL-18BP was discovered using ligand affinity chromatography, which had been used to isolate soluble (extracellular domain) receptors for other cytokines [105, 106], including TNF receptors p55 and p75 [107, 108]. However, IL-18BP is not the soluble receptor for IL-18 and is only distantly related structurally to IL-18Rα [102]. Unlike all members of the IL-1 receptor family, which have three Ig-like domains in the extracellular receptor segment, IL-18BP has only one Ig-like domain. It seems that the transmembrane and the first two extracellular domains of the ancestral IL-18 receptor were deleted during evolution. The only amino acid identity with the IL-18Rα chain and IL-18BP is found in the third Ig domain of the α chain [109]. There is limited amino acid homology between IL-18BP and the IL-1 receptor type II, also known as the IL-1 decoy receptor. In fact, IL-18 is similar biologically to the IL-1 decoy receptor in that its function is primarily to bind and neutralize the ligand rather than act as a ligand passer.

The human IL-18BP gene is located on chromosome 11q13. Using Northern Blot analysis, IL-18BP is highly expressed in spleen and the intestinal tract, both immunologically active tissues. There are four isotypes of human IL-18BP and two isotypes of murine IL-18BP [101]. These isotypes are formed by alternate mRNA splicing of the respective genes. A single copy of IL-18IL-18BP gene exists for humans, mice and rats [110]. Only those isoforms that retain the intact Ig domain are biologically functional by neutralizing IL-18 [102]. For example, human IL-18BP has four isotypes termed IL-18BPa, b, c and d. Only IL-18BPa and IL-18BPc have the intact Ig domain and neutralize IL-18 [102]. The other two isoforms, although they are produced in humans, do not bind and do not neutralize IL-18. However, the mRNA splicing that creates these isoforms is not a haphazard event in that the spliced mRNA has an open reading frame, which results in the same carboxyl terminal for all isoforms. It is possible that IL-18BP isoform b and d bind another member of the IL-1 family. The mouse has two isoforms, IL-18BPc and IL-18BPd. Murine IL-18BPc and IL-18BPd isoforms, possessing the identical Ig domain, also neutralize > 95% murine IL-18. However, murine IL-18BPd, which shares a common C-terminal motif with human IL-18BPa, also neutralizes human IL-18 [102].

The binding sites for IL-1 to the IL-1 receptor type I were used to model the binding of IL-18 to IL-18BP [102]. Modeling predicted a large mixed electrostatic and hydrophobic binding site in the Ig domain of IL-18BP, which could account for its high affinity binding to the ligand. In the site, the binding of IL-18 glutamic acid at position 35 and lysine at position 89 to oppositely charged amino acids in the IL-18BP were thought to participate. If correct, mutations in these amino acids would alter two properties of IL-18: binding to the IL-18Rα chain and binding to the IL-18BP. Therefore, biological activity, as well as the ability to be neutralized by IL-18BP, would be affected. In fact, mutational analysis established that these two amino acids are functionally important for biological activity [109, 111] as well as for neutralization [109].

A divalent fusion protein of human IL-18BP linked human IgG1 Fc (IL-18BP:Fc) binds and neutralizes human, mouse and rat IL-18 with a dissociation constant of

0.3–5 nM. Using *E. coli*-derived endotoxin with a lethal dose of 90%, IL-18BP:Fc administered 10 minutes prior to the endotoxin significantly reduced mortality [112]. IFNγ levels were also reduced in these mice. Because of the long plasma half-life of Fc fusion protein, IL-18BP:Fc reduced endotoxin-induced IFNγ when administered six days before the endotoxin challenge. IL-18BP:Fc reduced hepatic injury as well as expression of Fas [112]. IL-18BP:Fc also decreased granuloma formation, macrophage-inflammatory protein-1-α and macrophage-inflammatory protein-2 production. As shown previously using anti-mouse IL-18 [113], IL-18 mediates the hepatic damage caused by intravenously injected Concanavalin-A [112]. Fas ligand expression as well as liver damage induced by *Pseudomonas aeruginosa* exotoxin A or by anti-Fas agonistic antibody were also reduced by IL-18BP [112]. IL-18BP:Fc reduces the severity of CIA [84].

Viral IL-18BP: a natural experiment for the importance of IL-18 in inflammatory and immune responses

The human sequence for IL-18BP is found in various members of the Pox viruses. The greatest homology is found in *Molluscum contagiosum* [101], a common infection of the skin that is characterized by a large number of viral particles in the epithelial cells without a significant presence of inflammatory or immunologically active cells. The viral genes encoding for IL-18BP-like have been expressed and the recombinant proteins neutralize mammalian IL-18 activity [114]. The ability of viral IL-18BP to reduce the activity of mammalian IL-18 may explain the blandness of the inflammatory and immune response of the viral infection. The viral genes encoding for the IL-18BP-like proteins also exhibit signal peptides, N-glycosylation sites and cysteine residues similar to those in the human IL-18BP [115]. The ability of human IL-18BP to neutralize human and murine IL-18 was compared to the ability of pox viral IL-18BP to neutralize human and murine IL-18. The dissociation constants of the viral IL-18BP for murine IL-18 were 12- to 50-fold lower than that for human IL-18 [116].

IL-18BP levels in health and disease

The serum levels of IL-18BP isoform "a" in a cohort of healthy subjects as determined by a specific ELISA are 2.15 ± 0.15 ng/ml (range 0.5–7 ng/ml) [103]. In patients with sepsis and acute renal failure, the levels rose to 21.9 ± 1.44 ng/ml (range 4–132 ng/ml) but this rise was due to increased production and not to renal retention since there was no correlation with creatinine levels. Using the law of mass action and knowing the dissociation constant of IL-18BP to IL-18, total IL-18 and free IL-18 was calculated. Total IL-18 in healthy individuals was 64 ± 17 pg/ml and approximately 85% was in the free form [103]. Total IL-18 and IL-18BPa were both elevated in sepsis patients upon admission (1.5 ± 0.4 ng/ml and 28.6 ± 4.5 ng/ml, respectively). At these levels, most of the IL-18 is bound to IL-18BPa, however the remaining free IL-18 in sepsis patients is still higher than in healthy individuals. One can conclude from these studies that IL-18BPa considerably inhibits circulating IL-18 in sepsis. Nevertheless, exogenous administration of IL-18BP may further reduce circulating IL-18 activity.

The relative gene expression of the IL-18-neutralizing (a and c isoforms) and non-neutralizing (b and d isoforms) of IL-18BP was studied in Crohn's disease during active phases of the disease [117]. Intestinal endothelial cells and macrophages were the major source of IL-18BP within the submucosa. These findings were similar to those for IL-18BP in cultures human endothelial cells and peripheral blood monocytes. Gene expression as measured by steady state mRNA levels for IL-18BP as well as the IL-18BP protein were elevated in intestinal biopsies from patients with active disease [117]. A control group was assessed for comparison. Unbound IL-18BP isoforms a and c and inactive isoform d were present in specimens from patients with active disease as well as from tissues from control patients. The IL-18BP isoform b was not detected. Elevated IL-18BP has been described in several autoimmune diseases including RA, hepatitis C treated with IFNγ [118] and in patients with chronic liver diseases [119]. Serum IL-18 and IL-18BP are elevated in RA [120].

The IL-18BP promoter

The regulation of IL-18BP gene expression appears to be *via* IFNγ [121]. In a human colon carcinoma epithelial cell line, IFNγ induced gene expression and release of IL-18BPa. The increase in IL-18BP was also observed in a variety of intestinal cell lines and in the human keratinocyte cell line. The histone deacetylase inhibitor sodium butyrate suppressed IFNγ-induced IL-18BP gene and protein expression [121]. The promoter for IL-18BP has been described and various response elements, including two IFNγ response elements [122]. Thus, like other naturally occurring genes encoding proteins that block a specific cytokine (soluble receptors, receptors antagonists and binding proteins), the cytokine itself or a related cytokine induces its own negative regulator. IL-1 itself and TNFα are inducers of IL-1IL-1Ra.

Comparison of soluble IL-18 receptor α and β chains

Although the β chain of IL-18R is required for signaling, the soluble (extracellular) form does not bind IL-18, and its role in inhibiting IL-18 is unclear. Neutralization of IL-18 by soluble receptors was compared with that of IL-18BP. At an equimolar concentration IL-18BP to IL-18, inhibition of 90% of IL-18 activity was observed, whereas a four-fold molar excess of the soluble IL-18Rα had no effect [123]. A dimeric construct of soluble IL-18R linked to the Fc domain of IgG1 increased IL-18 activity 2.5-fold. In PBMC stimulated with endotoxin or in whole blood stimulated with *Staphylococcus epidermidis*, 3 nM IL-18BP reduced IFNγ by 80%, whereas IL-18Rα:Fc had no effect. A construct of the soluble IL-18Rα linked to Fc did not affect IL-18-induced IFNγ even at 80-fold molar excess of IL-18. However, the combination of both soluble receptors reduced IFNγ by 80%. In KG-1 macrophages like cells, a 50% reduction in IL-18 activity was observed using an 80-fold molar excess of soluble IL-18Rα:Fc but only in the presence of soluble IL-18Rα:Fc. These studies reveal that the combination of the ligand binding and the non-ligand binding extracellular domains of IL-18R are needed to inhibit IL-18, whereas IL-18BP neutralizes at an equimolar concentration.

Role of IL-18 in autoimmune disease

IL-18 in humans with RA

Synovial tissue from 29 patients with RA has been studied using specific staining for IL-18. IL-18 was detectable in 80% of the patients, in both the lining and sublining of synovial tissues from knees. There was a strong correlation of IL-18 with IL-1_ expression but less with TNFα staining [124]. Moreover, IL-18 expression correlated with macrophage infiltration and local inflammation scores. Using measurement of systemic inflammation by erythrocyte sedimentation rate, IL-18 staining also correlated with this biomarker. Of considerable interest was the observation that in patients with co-expression of IL-18 and IL-12, there were also greater levels of IL-17. Blocking of endogenous IL-17 with specific inhibitors resulted in a protective inhibition of bone destruction in this model [125]. Together with the data described below, these findings suggest that IL-18 likely plays a role in RA and perhaps regulates the activities of IL-1β, TNFα, IFNγ and IL-17. In a mouse model of endotoxemia, a neutralizing anti-murine IL-18 antibody resulted in near suppression of myocardial IL-1β levels [126]. In another study, circulating IL-18 and IL-18BP did not correlate significantly with measurements of clinical of disease activity or the response to treatment in patients with early RA [120]. However, in general, circulating cytokines do not reflect the level of the cytokine in the affected organ or tissue. For example, IL-18 bioactivity in affected joints of patients with RA has been reported to correlate with disease activity [127].

Models of arthritis

Initial studies were carried out using SCW-induced arthritis [83]. A neutralizing rabbit anti-murine IL-18 antibody was injected shortly before induction of arthritis by intra-articular injection of SCW fragments into the right knee joint. Significant (> 60%) suppression of joint swelling was noted on days 1 and 2 of SCW arthritis after blockade of endogenous IL-18 and joint TNFα and IL-1 levels were also decreased. Severe reduction of chondrocyte proteoglycan synthesis is a prominent component of SCW-induced arthritis but a near complete reversal of the depressed chondrocyte proteoglycan synthesis was observed in the anti-IL-18-treated animals. Although these studies clearly established the pathological role for endogenous IL-18 in this model, the effect of IL-18 is apparently independent of IFNγ since mice deficient in IFNγ showed similar results using anti-IL-18 antibodies [83].

IL-18 also plays a role in CIA. IL-18 was injected into DBA-1 mice immunized with collagen in incomplete Freund's adjuvant. There was an increase in the erosive and inflammatory component of the arthritis [128]. Using mice deficient in IL-18, CIA was less severe compared to wild-type controls [80]. Histological evidence of decreased joint inflammation and destruction were observed. Levels of bovine collagen-induced IFNγ, TNFα, IL-6 and IL-12 from spleen cell cultures were decreased in IL-18-deficient mice. However, there was a significant reduction in serum anti-collagen antibody levels in the IL-18-deficient mice, raising the perennial issue that gene deletions on immunologically active cytokines can obscure the role of a cytokine in CIA. Nevertheless, from these studies, there is likely a pathological role for IL-18 in CIA.

Other studies in CIA used wild-type DBA-1 mice treated with either neutralizing antibodies to IL-18 or the IL-18BP after clinical onset of disease. The therapeutic efficacy of neutralizing endogenous IL-18 was assessed using different pathological parameters of disease progression. The clinical severity in mice undergoing CIA was significantly reduced after treatment with either IL-18 neutralizing antibodies or IL-18BP [82]. Attenuation of the disease was associated with reduced cartilage erosion evident on histology. The decreased cartilage degradation was further documented by a significant reduction in the levels of circulating cartilage oligomeric matrix protein (an indicator of cartilage turnover). Both strategies efficiently slowed disease progression, but only anti-IL-18 antibody treatment significantly decreased an established synovitis. Serum levels of IL-6 were significantly reduced with both neutralizing strategies. *In vitro*, neutralizing IL-18 resulted in a significant inhibition of TNFα, IL-6, and IFNγ secretion by macrophages [82].

Mice with established CIA (21 days after the primary immunization with collagen) were treated for three weeks with either vehicle or with two doses of murine IL-18BP (0.5 and 3 mg/kg) as a fusion protein with the Fc portion of murine IgG1 [84]. Both the clinical disease activity scores and the histological scores of joint damage were reduced by 50% in mice treated with either dose of IL-18BP. Proliferation of collagen-stimulated spleen and lymph node cells as well as the change in serum levels of IgG1 and IgG2a antibodies to collagen between days 21 and 42 were decreased in mice treated with IL-18BP. The production of IFNγ, TNFα, and IL-1β in cultured spleen cells was reduced by *in vivo* treatment with low dose, but not high dose, IL-18BP. Cell sorting analysis showed a decrease in NK cells and an increase

Table 2 - Reduction in disease severity reported with neutralization of endogenous IL-18

(DSS) dextran sulfate sodium-induced colitis
TNBS colitis colitis-induced colitis
CD69 transfer-induced colitis
Streptococcal Wall-induced arthritis
Collagen-induced arthritis
Con-A-induced hepatitis
Pseudomonas exotoxin-A-induced hepatic damage
IL-12-induced intestinal inflammation
Endotoxin-induced hepatic necrosis
Endotoxin-induced lethality
Endotoxin-induced pulmonary neutrophils
Melanoma hepatic metastasis
Melanoma-induced endothelial VCAM-1 expression
Ischemia-induced acute renal failure
Ischemia-induced myocardial dysfunction
Endotoxin-induced myocardial dysfunction

in $CD4^+$ T cells in spleens of mice treated with IL-18BP. The steady state mRNA levels of IFNγ, TNFα, and IL-1β in isolated joints were each decreased in mice treated with either dose of IL-18BP. The mechanisms of IL-18BP inhibition of CIA include reductions in cell-mediated and humoral immunity to collagen as well as decreases in production of proinflammatory cytokines in the spleen and joints.

Employing an adenoviral vector containing the murine IL-18IL-18BP, intra-articular over expression of IL-18BP significantly reduced the incidence CIA in treated knee joints [85]. Affected knee joints of IL-18BP-treated mice showed less severe arthritis, with reduced cellular infiltration and less bone erosions. Loss of cartilage was also observed in treated mice. IgG1 anti-collagen type II antibodies were similar to those in the control vector group.

Table 2 summarizes the effects of blocking IL-18 in various models of disease.

Acknowledgments

Supported by NIH Grants AI-15614 and HL-68743

References

1 Voronov E, Shouval DS, Krelin Y, Cagnano E, Benharroch D, Iwakura Y, Dinarello CA, Apte RN (2003) IL-1 is required for tumor invasiveness and angiogenesis. *Proc Natl Acad Sci USA* 100: 2645–2650

2 Smith JW, Longo D, Alford WG, Janik JE, Sharfman WH, Gause BL, Curti BD, Creekmore SP, Holmlund JT, Fenton RG et al (1993) The effects of treatment with interleukin-1a on platelet recovery after high-dose carboplatin. *N Engl J Med* 328: 756–761

3 Bresnihan B, Alvaro-Gracia JM, Cobby M, Doherty M, Domljan Z, Emery P, Nuki G, Pavelka K, Rau R, Rozman B et al (1998) Treatment of rheumatoid arthritis with recombinant human interleukin-1 receptor antagonist. *Arthritis Rheum* 41: 2196–2204

4 Jiang Y, Genant HK, Watt I, Cobby M, Bresnihan B, Aitchison R, McCabe D (2000) A multicenter, double-blind, dose-ranging, randomized, placebo-controlled study of recombinant human interleukin-1 receptor antagonist in patients with rheumatoid arthritis: radiologic progression and correlation of Genant and Larsen scores. *Arthritis Rheum* 43: 1001–1009

5 Cunnane G, Madigan A, Murphy E, FitzGerald O, Bresnihan B (2001) The effects of treatment with interleukin-1 receptor antagonist on the inflamed synovial membrane in rheumatoid arthritis. *Rheumatology (Oxford)* 40: 62–69

6 Cohen SB, Woolley JM, Chan W (2003) Interleukin 1 receptor antagonist anakinra improves functional status in patients with rheumatoid arthritis. *J Rheumatol* 30: 225–231

7 Horai R, Saijo S, Tanioka H, Nakae S, Sudo K, Okahara A, Ikuse T, Asano M, Iwakura Y (2000) Development of chronic inflammatory arthropathy resembling rheumatoid arthritis in interleukin 1 receptor antagonist-deficient mice. *J Exp Med* 191: 313–320

8 Nicklin MJ, Hughes DE, Barton JL, Ure JM, Duff GW (2000) Arterial inflammation in mice lacking the interleukin 1 receptor antagonist gene. *J Exp Med* 191: 303–312

9 Kurt-Jones EA, Beller DI, Mizel SB, Unanue ER (1985) Identification of a membrane-associated interleukin-1 in macrophages. *Proc Natl Acad Sci USA* 82: 1204–1208

10 Kaplanski G, Farnarier C, Kaplanski S, Porat R, Shapiro L, Bongrand P, Dinarello CA (1994) Interleukin-1 induces interleukin-8 from endothelial cells by a juxacrine mechanism. *Blood* 84: 4242–4248

11 Stevenson FT, Bursten SL, Fanton C, Locksley RM, Lovett DH (1993) The 31-kDa precursor of interleukin-1α is myristoylated on specific lysines within the 16-kDa N-terminal propiece. *Proc Natl Acad Sci USA* 90: 7245–7249

12 Stevenson FT, Turck J, Locksley RM, Lovett DH (1997) The N-terminal propiece of interleukin 1α is a transforming nuclear oncoprotein. *Proc Natl Acad Sci USA* 94: 508–513

13 Miller AC, Schattenberg DG, Malkinson AM, Ross D (1994) Decreased content of the IL-1α processing enzyme calpain in murine bone marrow-derived macrophages after treatment with the bezene metabolite hydroquinone. *Tox Lett* 74: 177–184

14 Nakae S, Naruse-Nakajima C, Sudo K, Horai R, Asano M, Iwakura Y (2001) IL-1 alpha, but not IL-1 beta, is required for contact-allergen-specific T cell activation during the sensitization phase in contact hypersensitivity. *Int Immunol* 13: 1471–1478

15 Alnemri ES, Livingston DJ, Nicholson DW, Salvesen G, Thornberry NA, Wong WW, Yuan J (1996) Human ICE/CED-3 protease nomenclature. *Cell* 87: 123

16 Watanabe N, Kawaguchi M, Kobayashi Y (1998) Activation of interleukin-1b converting enzyme by nigericin is independent of apoptosis. *Cytokine* 10: 645–653
17 Thornberry NA, Bull HG, Calaycay JR, Chapman KT, Howard AD, Kostura MJ, Miller DK, Molineaux SM, Weidner JR, Aunins J et al (1992) A novel heterodimeric cysteine protease is required for interleukin-1 beta processing in monocytes. *Nature* 356: 768–774
18 Coeshott C, Ohnemus C, Pilyavskaya A, Ross S, Wieczorek M, Kroona H, Leimer AH, Cheronis J (1999) Converting enzyme-independent release of TNFα and IL-1β from a stimulated human monocytic cell line in the presence of activated neutrophils or purified proteinase-3. *Proc Natl Acad Sci USA* 96: 6261–6266
19 Sugawara S, Uehara A, Nochi T, Yamaguchi T, Ueda H, Sugiyama A, Hanzawa K, Kumagai K, Okamura H, Takada H (2001) Neutrophil proteinase 3-mediated induction of bioactive IL-18 secretion by human oral epithelial cells. *J Immunol* 167: 6568–6575
20 Laliberte R, Perregaux D, Svensson L, Pazoles CJ, Gabel CA (1994) Tenidap modulates cytoplasmic pH and inhibits anion transport *in vitro*. II. Inhibition of IL-1β production from ATP-treated monocytes and macrophages. *J Immunol* 153: 2168–2179
21 Ferrari D, Chiozzi P, Falzoni S, Dal Susino M, Melchiorri L, Baricordi OR, Di Virgilio F (1997) Extracellular ATP triggers IL-1 beta release by activating the purinergic P2Z receptor of human macrophages. *J Immunol* 159: 1451–1458
22 Buell G, Chessell IP, Michel AD, Collo G, Salazzo M, Herren S, Gretener D, Grahames C, Kaur R, Kosco-Vilbois MH et al (1998) Blockade of human P2X7 receptor function with a monoclonal antibody. *Blood* 92: 3521–3528
23 Perregaux DG, Gabel CA (1998) Post-translational processing of murine IL-1: evidence that ATP-induced release of IL-1 alpha and IL-1 beta occurs *via* a similar mechanism. *J Immunol* 160: 2469–2477
24 Perregaux DG, McNiff P, Laliberte R, Conklyn M, Gabel CA (2000) ATP acts as an agonist to promote stimulus-induced secretion of IL-1 beta and IL-18 in human blood. *J Immunol* 165: 4615–4623
25 Solle M, Labasi J, Perregaux DG, Stam E, Petrushova N, Koller BH, Griffiths RJ, Gabel CA (2001) Altered cytokine production in mice lacking P2X(7) receptors. *J Biol Chem* 276: 125–132
26 Aganna E, Martinon F, Hawkins PN, Ross JB, Swan DC, Booth DR, Lachmann HJ, Bybee A, Gaudet R, Woo P et al (2002) Association of mutations in the NALP3/CIAS1/PYPAF1 gene with a broad phenotype including recurrent fever, cold sensitivity, sensorineural deafness, and AA amyloidosis. *Arthritis Rheum* 46: 2445–2452
27 Muckle TJ (1979) The 'Muckle-Wells' syndrome. *Br J Dermatol* 100: 87–92
28 Watts RA, Nicholls A, Scott DG (1994) The arthropathy of the Muckle-Wells syndrome. *Br J Rheumatol* 33: 1184–1187
29 Prost A, Barriere H, Legent F, Cottin S, Wallez B (1976) [Intermittent rheumatism revealing a familial syndrome. Arthritis-urticarian eruptions-deafness: Muckle-Wells syndrome without kidney amylosis]. *Rev Rhum Mal Osteoartic* 43: 201–208
30 Hoffman HM, Mueller JL, Broide DH, Wanderer AA, Kolodner RD (2001) Mutation

of a new gene encoding a putative pyrin-like protein causes familial cold autoinflammatory syndrome and Muckle-Wells syndrome. *Nat Genet* 29: 301–305
31 Aksentijevich I, Nowak M, Mallah M, Chae JJ, Watford WT, Hofmann SR, Stein L, Russo R, Goldsmith D, Dent P et al (2002) *De novo* CIAS1 mutations, cytokine activation, and evidence for genetic heterogeneity in patients with neonatal-onset multisystem inflammatory disease (NOMID): a new member of the expanding family of pyrin-associated autoinflammatory diseases. *Arthritis Rheum* 46: 3340–3348
32 McDermott MF (2002) Genetic clues to understanding periodic fevers, and possible therapies. *Trends Mol Med* 8: 550–554
33 Tschopp J, Martinon F, Burns K (2003) NALPs: a novel protein family involved in inflammation. *Nat Rev Mol Cell Biol* 4: 95–104
34 Srinivasula SM, Poyet JL, Razmara M, Datta P, Zhang Z, Alnemri ES (2002) The PYRIN-CARD protein ASC is an activating adaptor for caspase-1. *J Biol Chem* 277: 21119–21122
35 Hawkins PN, Lachmann HJ, McDermott MF (2003) Interleukin-1 receptor antagonist in the Muckle-Wells syndrome. *N Engl J Med* 348: 2583–2584
36 Sarrauste de Menthiere C, Terriere S, Pugnere D, Ruiz M, Demaille J, Touitou I (2003) INFEVERS: the Registry for FMF and hereditary inflammatory disorders mutations. *Nucleic Acids Res* 31: 282–285
37 Bustorff MM, Oliveria JP, Moura C, Carvalho E, Faria V, Guerra L (1995) Muckle-Wells syndrome nephropathy: lack of response to colchicine therapy. *Nephrol Dial Transplant* 10: 709–710
38 Seckinger P, Lowenthal JW, Williamson K, Dayer JM, MacDonald HR (1987) A urine inhibitor of interleukin-1 activity that blocks ligand binding. *J Immunol* 139: 1546–1549
39 Arend WP, Joslin FG, Massoni RJ (1985) Effects of immune complexes on production by human monocytes of interleukin 1 or an interleukin 1 inhibitor. *J Immunol* 134: 3868–387
40 Eisenberg SP, Evans RJ, Arend WP, Verderber E, Brewer MT, Hannum CH, Thompson RC (1990) Primary structure and functional expression from complementary DNA of a human interleukin-1 receptor antagonist. *Nature* 343: 341–346
41 Arend WP, Malyak M, Guthridge CJ, Gabay C (1998) Interleukin-1 receptor antagonist: role in biology. *Ann Rev Immunol* 16: 27–55
42 Haskill S, Martin M, VanLe L, Morris J, Peace A, Bigler CF, Jaffe GJ, Sporn SA, Fong S, Arend WP et al (1991) cDNA cloning of a novel form of the interleukin-1 receptor antagonist associated with epithelium. *Proc Natl Acad Sci USA* 88: 3681–3685
43 Granowitz EV, Porat R, Mier JW, Pribble JP, Stiles DM, Bloedow DC, Catalano MA, Wolff SM, Dinarello CA (1992) Pharmacokinetics, saftey, and immunomodulatory effects of human recombinant interleukin-1 receptor antagonist in healthy humans. *Cytokine* 4: 353–360
44 Hirsch E, Irikura VM, Paul SM, Hirsh D (1996) Functions of interleukin 1 receptor

antagonist in gene knockout and overproducing mice. *Proc Natl Acad Sci USA* 93: 11008–11013
45 Lorenzo JE, Naprta A, Rao Y, Alander C, Glaccum M, Widmer M, Gronowicz G, Kalinowski J, Pilbeam CC (1998) Mice lacking the type I interleukin-1 receptor do not lose bone mass after ovariectomy. *Endocrinol* 139: 3022–3025
46 Nakae S, Saijo S, Horai R, Sudo K, Mori S, Iwakura Y (2003) IL-17 production from activated T cells is required for the spontaneous development of destructive arthritis in mice deficient in IL-1 receptor antagonist. *Proc Natl Acad Sci USA* 100: 5986–5990
47 van den Berg WB (2001) Uncoupling of inflammatory and destructive mechanisms in arthritis. Semin *Arthritis Rheum* 30: 7–16
48 Bendele A, McAbee T, Sennello G, Frazier J, Chlipala E, McCabe D (1999) Efficacy of sustained blood levels of interleukin-1 receptor antagonist in animal models of arthritis: comparison of efficacy in animal models with human clinical data. *Arthritis Rheum* 42: 498–506
49 Bakker AC, Joosten LA, Arntz OJ, Helsen MM, Bendele AM, van de Loo FA, van den Berg WB (1997) Prevention of murine collagen-induced arthritis in the knee and ipsilateral paw by local expression of human interleukin-1 receptor antagonist protein in the knee. *Arthritis Rheum* 40: 893–900
50 Fukumoto T, Matsukawa A, Ohkawara S, Takagi K, Yoshinaga M (1996) Administration of neutralizing antibody against rabbit IL-1 receptor antagonist exacerbates lipopolysaccharide-induced arthritis in rabbits. *Inflamm Res* 45: 479–485
51 Neidhart M, Gay RE, Gay S (2000) Anti-interleukin-1 and anti-CD44 interventions producing significant inhibition of cartilage destruction in an *in vitro* model of cartilage invasion by rheumatoid arthritis synovial fibroblasts. *Arthritis Rheum* 43: 1719–1728
52 Fanslow WC, Sims JE, Sassenfeld H, Morrissey PJ, Gillis S, Dower SK, Widmer MB (1990) Regulation of alloreactivity *in vivo* by a soluble form of the interleukin-1 receptor. *Science* 248: 739–742
53 Dower SK, Fanslow W, Jacobs C, Waugh S, Sims JE, Widmer MB (1994) Interleukin-1 antagonists. *Therapeutic Immunol* 1: 113–122
54 Jacobs CA, Baker PE, Roux ER, Picha KS, Toivola B, Waugh S, Kennedy MK (1991) Experimental autoimmune encephalomyelitis is exacerbated by IL-1α and suppressed by soluble IL-1 receptor. *J Immunol* 146: 2983–2989
55 Netea MG, Kullberg BJ, Boerman OC, Verschueren I, Dinarello CA, Van der Meer JW (1999) Soluble murine IL-1 receptor type I induces release of constitutive IL-1 alpha. *J Immunol* 162: 4876–4881
56 Nicoletti F, Di Marco R, Barcellini W, Magro G, Schorlemmer HU, Kurrle R, Lunetta M, Grasso S, Zaccone P, Meroni P (1994) Protection from experimental autoimmune diabetes in the non-obese diabetic mouse with soluble interleukin-1 receptor. *Eur J Immunol* 24: 1843–1847
57 Preas HL, 2nd, Reda D, Tropea M, Vandivier RW, Banks SM, Agosti JM, Suffredini AF (1996) Effects of recombinant soluble type I interleukin-1 receptor on human inflammatory responses to endotoxin. *Blood* 88: 2465–2472

58 Drevlow BE, Lovis R, Haag MA, Sinacore JM, Jacobs C, Blosche C, Landay A, Moreland LW, Pope RM (1996) Recombinant human interleukin-1 receptor type I in the treatment of patients with active rheumatoid arthritis. *Arthritis Rheum* 39: 257–265
59 Bernstein SH, Fay J, Frankel S, Christiansen N, Baer MR, Jacobs C, Blosch C, Hanna R, Herzig G (1999) A phase I study of recombinant human soluble interleukin-1 receptor (rhu IL-1R) in patients with relapsed and refractory acute myeloid leukemia. *Cancer Chemother Pharmacol* 43: 141–144
60 Pitti RM, Marsters SA, Haak-Frendscho M, Osaka GC, Mordenti J, Chamow SM, Ashkenazi A (1994) Molecular and biological properties of an interleukin-1 receptor immunoadhesin. *Mol Immunol* 31: 1345–1351
61 Palaszynski EW (1987) Synthetic C-terminal peptide of IL-1 functions as a binding domain as well as an antagonist for the IL-1 receptor. *Biochem Biophy Res Comm* 147: 204–211
62 Crown J, Jakubowski A, Kemeny N, Gordon M, Gasparetto C, Wong G, Toner G, Meisenberg B, Botet J, Applewhite J et al (1991) A phase I trial of recombinant human interleukin-1b alone and in combination with myelosuppressive doses of 5-fluoruracil in patients with gastrointestinal cancer. *Blood* 78: 1420–1427
63 Bendtzen K, Svenson M, Jonsson V, Hippe E (1990) Autoantibodies to cytokines – friends or foes? *Immunol Today* 11: 167–169
64 Tewari A, Buhles WC Jr, Starnes HF Jr (1990) Preliminary report: effects of interleukin-1 on platelet counts. *Lancet* 336: 712–714
65 Economides AN, Carpenter LR, Rudge JS, Wong V, Koehler-Stec EM, Hartnett C, Pyles EA, Xu X, Daly TJ, Young MR et al (2003) Cytokine traps: multi-component, high-affinity blockers of cytokine action. *Nat Med* 9: 47–52
66 Campion GV, Lebsack ME, Lookabaugh J, Gordon G, Catalano M (1996) Dose-range and dose-frequency study of recombinant human interleukin-1 receptor antagonist in patients with rheumatoid arthritis. *Arthritis Rheum* 39: 1092–1101
67 Genant HK, Bresnihan B, Ng E, Robbins S, Newmark RD, McCabe D (2001) Treatment with anakinra reduces the rate of joint destruction and shows accelerated benefit in the secon 6 months of treatment for patients with rheumatoid arthritis. *Ann Rheumat Dis* 40 (Suppl 1): 169 (abs)
68 Torcia M, Lucibello M, Vannier E, Fabiani S, Miliani A, Guidi G, Spada O, Dower SK, Sims JE, Shaw AR et al (1996) Modulation of osteoclast-activating factor activity of multiple myeloma bone marrow cells by different interleukin-1 inhibitors. *Exp Hematol* 24: 868–874
69 Dewhirst FE, Stashenko PP, Mole JE, Tsurumachi T (1985) Purification and partial sequence of human osteoclast-activating factor: identity with interleukin 1 beta. *J Immunol* 135: 2562–2568
70 Fleischmann RM, Schechtman J, Bennett R, Handel ML, Burmester GR, Tesser J, Modafferi D, Poulakos J, Sun G (2003) Anakinra, a recombinant human interleukin-1

receptor antagonist (r-metHuIL-1ra), in patients with rheumatoid arthritis: A large, international, multicenter, placebo-controlled trial. *Arthritis Rheum* 48: 927–934
71 Bresnihan B, Cobby M (2003) Clinical and radiological effects of anakinra in patients with rheumatoid arthritis. *Rheumatology (Oxford)* 42 (Suppl 2): ii22–ii28
72 Nakamura K, Okamura H, Wada M, Nagata K, Tamura T (1989) Endotoxin-induced serum factor that stimulates gamma inteferon production. *Infect Immun* 57: 590–595
73 Okamura H, Tsutsui H, Komatsu T, Yutsudo M, Hakura A, Tanimoto T, Torigoe K, Okura T, Nukada Y, Hattori K et al (1995) Cloning of a new cytokine that induces interferon-γ. *Nature* 378: 88–91
74 Browning JL, Ribolini A (1987) Interferon blocks interleukin 1-induced prostaglandin release from human peripheral monocytes. *J Immunol* 138: 2857–2863
75 Reznikov LL, Kim SH, Westcott JY, Frishman J, Fantuzzi G, Novick D, Rubinstein M, Dinarello CA (2000) IL-18 binding protein increases spontaneous and IL-1-induced prostaglandin production *via* inhibition of IFN-gamma. *Proc Natl Acad Sci USA* 97: 2174–2179
76 Andrews HJ, Bunning RA, Plumpton TA, Clark IM, Russell RG, Cawston TE (1990) Inhibition of interleukin-1-induced collagenase production in human articular chondrocytes *in vitro* by recombinant human interferon-gamma. *Arthritis Rheum* 33: 1733–1738
77 Gallin JI, Farber JM, Holland SM, Nutman TB (1995) Interferon-gamma in the management of infectious diseases. *Ann Intern Med* 123: 216–224
78 Dinarello CA (1996) Biological basis for interleukin-1 in disease. *Blood* 87: 2095–2147
79 Van der Poll T, Romijn JA, Endert E, Borm JJ, Buller HR, Sauerwein HP (1991) Tumor necrosis factor mimics the metabolic response to acute infection in healthy humans. *Am J Physiol* 261: E457–465
80 Wei XQ, Leung BP, Arthur HM, McInnes IB, Liew FY (2001) Reduced incidence and severity of collagen-induced arthritis in mice lacking IL-18. *J Immunol* 166: 517–521
81 Campbell IK, O'Donnell K, Lawlor KE, Wicks IP (2001) Severe inflammatory arthritis and lymphadenopathy in the absence of TNF. *J Clin Invest* 107: 1519–1527
82 Plater-Zyberk C, Joosten LA, Helsen MM, Sattonnet-Roche P, Siegfried C, Alouani S, van De Loo FA, Graber P, Aloni S, Cirillo R et al (2001) Therapeutic effect of neutralizing endogenous IL-18 activity in the collagen-induced model of arthritis. *J Clin Invest* 108: 1825–1832
83 Joosten LA, van De Loo FA, Lubberts E, Helsen MM, Netea MG, van Der Meer JW, Dinarello CA, van Den Berg WB (2000) An IFN-gamma-independent proinflammatory role of IL-18 in murine streptococcal cell wall arthritis. *J Immunol* 165: 6553–6558
84 Banda NK, Vondracek A, Kraus D, Dinarello CA, Kim SH, Bendele A, Senaldi G, Arend WP (2003) Mechanisms of inhibition of collagen-induced arthritis by murine IL-18 binding protein. *J Immunol* 170: 2100–2105
85 Smeets RL, van de Loo FA, Arntz OJ, Bennink MB, Joosten LA, van den Berg WB

(2003) Adenoviral delivery of IL-18 binding protein C ameliorates collagen-induced arthritis in mice. *Gene Ther* 10: 1004–1011

86 Kumar S, McDonnell PC, Lehr R, Tierney L, Tzimas MN, Griswold DE, Capper EA, Tal-Singer R, Wells GI, Doyle ML et al (2000) Identification and initial characterization of four novel members of the interleukin-1 family. *J Biol Chem* 275: 10308–10314

87 Bufler P, Azam T, Gamboni-Robertson F, Reznikov LL, Kumar S, Dinarello CA, Kim SH (2002) A complex of the IL-1 homologue IL-1F7b and IL-18-binding protein reduces IL-18 activity. *Proc Natl Acad Sci USA* 99: 13723–13728

88 Pan G, Risser P, Mao W, Baldwin DT, Zhong AW, Filvaroff E, Yansura D, Lewis L, Eigenbrot C, Henzel WJ et al (2001) IL-1H, an interleukin 1-related protein that binds IL-18 receptor/IL-1Rrp. *Cytokine* 13: 1–7

89 Puren AJ, Fantuzzi G, Dinarello CA (1999) Gene expression, synthesis and secretion of IL-1b and IL-18 are differentially regulated in human blood mononuclear cells and mouse spleen cells. *Proc Natl Acad Sci USA* 96: 2256–2261

90 Gardella S, Andrei C, Costigliolo S, Poggi A, Zocchi MR, Rubartelli A (1999) Interleukin-18 synthesis and secretion by dendritic cells are modulated by interaction with antigen-specific T cells. *J Leukoc Biol* 66: 237–241

91 Gardella S, Andrei C, Poggi A, Zocchi MR, Rubartelli A (2000) Control of interleukin-18 secretion by dendritic cells: role of calcium influxes. *FEBS Lett* 481: 245–248

92 Andrei C, Dazzi C, Lotti L, Torrisi MR, Chimini G, Rubartelli A (1999) The secretory route of the leaderless protein interleukin 1beta involves exocytosis of endolysosome-related vesicles. *Mol Biol Cell* 10: 1463–1475

93 Tsutsui H, Kayagaki N, Kuida K, Nakano H, Hayashi N, Takeda K, Matsui K, Kashiwamura S, Hada T, Akira S et al (1999) Caspase-1-independent, Fas/Fas ligand-mediated IL-18 secretion from macrophages causes acute liver injury in mice. *Immunity* 11: 359–367

94 Wyman TH, Dinarello CA, Banerjee A, Gamboni-Robertson F, Hiester AA, England KM, Kelher M, Silliman CC (2002) Physiological levels of interleukin-18 stimulate multiple neutrophil functions through p38 MAP kinase activation. *J Leukoc Biol* 72: 401–409

95 Gu Y, Kuida K, Tsutsui H, Ku G, Hsiao K, Fleming MA, Hayashi N, Higashino K, Okamura H, Nakanishi K et al (1997) Activation of interferon-g inducing factor mediated by interleukin-1β converting enzyme. *Science* 275: 206–209

96 Ghayur T, Banerjee S, Hugunin M, Butler D, Herzog L, Carter A, Quintal L, Sekut L, Talanian R, Paskind M et al (1997) Caspase-1 processes IFN-gamma-inducing factor and regulates LPS-induced IFN-gamma production. *Nature* 386: 619–623

97 Fantuzzi G, Zheng H, Faggioni R, Benigni F, Ghezzi P, Sipe JD, Shaw AR, Dinarello CA (1996) Effect of endotoxin in IL-1β-deficient mice. *J Immunol* 157: 291–296

98 Okamura H, Nagata K, Komatsu T, Tanimoto T, Nukata Y, Tanabe F, Akita K, Torigoe K, Okura T, Fukuda S et al (1995) A novel costimulatory factor for gamma interferon induction found in the livers of mice causes endotoxic shock. *Infect Immun* 63: 3966–3972

99 Netea MG, Fantuzzi G, Kullberg BJ, Stuyt RJ, Pulido EJ, McIntyre RC Jr, Joosten LA, Van der Meer JW, Dinarello CA (2000) Neutralization of IL-18 reduces neutrophil tissue accumulation and protects mice against lethal *Escherichia coli* and *Salmonella typhimurium* endotoxemia. *J Immunol* 164: 2644–2649

100 Puren AJ, Fantuzzi G, Gu Y, Su MS-S, Dinarello CA (1998) Interleukin-18 (IFN-γ-inducing factor) induces IL-1β and IL-8 *via* TNFα production from non-CD14$^+$ human blood mononuclear cells. *J Clin Invest* 101: 711–724

101 Novick D, Kim S-H, Fantuzzi G, Reznikov L, Dinarello CA, Rubinstein M (1999) Interleukin-18 binding protein: a novel modulator of the Th1 cytokine response. *Immunity* 10: 127–136

102 Kim S-H, Eisenstein M, Reznikov L, Fantuzzi G, Novick D, Rubinstein M, Dinarello CA (2000) Structural requirements of six naturally occurring isoforms of the interleukin-18 binding protein to inhibit interleukin-18. *Proc Natl Acad Sci USA* 97: 1190–1195

103 Novick D, Schwartsburd B, Pinkus R, Suissa D, Belzer I, Sthoeger Z, Keane WF, Chvatchko Y, Kim SH, Fantuzzi G et al (2001) A novel IL-18BP ELISA shows elevated serum il-18BP in sepsis and extensive decrease of free IL-18. *Cytokine* 14: 334–342

104 Colotta F, Dower SK, Sims JE, Mantovani A (1994) The type II "decoy" receptor: a novel regulatory pathway for interleukin-1. *Immunol Today* 15: 562–566

105 Novick D, Engelmann H, Wallach D, Leitner O, Revel M, Rubinstein M (1990) Purification of soluble cytokine receptors from normal human urine by ligand-affinity and immunoaffinity chromatography. *J Chromatogr* 510: 331–337

106 Novick D, Engelmann H, Wallach D, Rubinstein M (1989) Soluble cytokine receptors are present in normal human urine. *J Exp Med* 170: 1409–1414

107 Engelmann H, Aderka D, Rubinstein M, Rotman D, Wallach D (1989) A tumor necrosis factor-binding protein purified to homogeneity from human urine protects cells from tumor necrosis factor toxicity. *J Biol Chem* 264: 11974–11980

108 Engelmann H, Novick D, Wallach D (1990) Two tumor necrosis factor-binding proteins purified from human urine. Evidence for immunological cross-reactivity with cell surface tumor necrosis factor receptors. *J Biol Chem* 265: 1531–1536

109 Kim SH, Azam T, Novick D, Yoon DY, Reznikov LL, Bufler P, Rubinstein M, Dinarello CA (2002) Identification of amino acid residues critical for biological activity in human interleukin-18. *J Biol Chem* 14: 14

110 Im SH, Kim SH, Azam T, Venkatesh N, Dinarello CA, Fuchs S, Souroujon MC (2002) Rat interleukin-18 binding protein: cloning, expression, and characterization. *J Interferon Cytokine Res* 22: 321–328

111 Kim SH, Azam T, Yoon DY, Reznikov LL, Novick D, Rubinstein M, Dinarello CA (2001) Site-specific mutations in the mature form of human IL-18 with enhanced biological activity and decreased neutralization by IL-18 binding protein. *Proc Natl Acad Sci USA* 98: 3304–3309

112 Faggioni R, Cattley RC, Guo J, Flores S, Brown H, Qi M, Yin S, Hill D, Scully S, Chen C et al (2001) IL-18-binding protein protects against lipopolysaccharide-induced lethal-

ity and prevents the development of Fas/Fas ligand-mediated models of liver disease in mice. *J Immunol* 167: 5913–5920

113 Faggioni R, Jones-Carson J, Reed DA, Dinarello CA, Feingold KR, Grunfeld C, Fantuzzi G (2000) Leptin-deficient (ob/ob) mice are protected from T cell-mediated hepatotoxicity: role of tumor necrosis factor alpha and IL-18. *Proc Natl Acad Sci USA* 97: 2367–2372

114 Xiang Y, Moss B (1999) IL-18 binding and inhibition of interferon gamma induction by human poxvirus-encoded proteins. *Proc Natl Acad Sci USA* 96: 11537–11542

115 Xiang Y, Moss B (1999) Identification of human and mouse homologs of the MC51L-53L-54L family of secreted glycoproteins encoded by the Molluscum contagiosum poxvirus. *Virology* 257: 297–302

116 Calderara S, Xiang Y, Moss B (2001) Orthopoxvirus IL-18 binding proteins: affinities and antagonist activities. *Virology* 279: 22–26

117 Corbaz A, ten Hove T, Herren S, Graber P, Schwartsburd B, Belzer I, Harrison J, Plitz T, Kosco-Vilbois MH, Kim SH et al (2002) IL-18-binding protein expression by endothelial cells and macrophages is upregulated during active Crohn's disease. *J Immunol* 168: 3608–3616

118 Kaser A, Novick D, Rubinstein M, Siegmund B, Enrich B, Koch RO, Vogel W, Kim SH, Dinarello CA, Tilg H (2002) Interferon-alpha induces interleukin-18 binding protein in chronic hepatitis C patients. *Clin Exp Immunol* 129: 332–338

119 Ludwiczek O, Kaser A, Novick D, Dinarello CA, Rubinstein M, Vogel W, Tilg H (2002) Plasma levels of interleukin-18 and interleukin-18 binding protein are elevated in patients with chronic liver disease. *J Clin Immunol* 22: 331–337

120 Bresnihan B, Roux-Lombard P, Murphy E, Kane D, FitzGerald O, Dayer JM (2002) Serum interleukin 18 and interleukin 18 binding protein in rheumatoid arthritis. *Ann Rheum Dis* 61: 726–729

121 Paulukat J, Bosmann M, Nold M, Garkisch S, Kampfer H, Frank S, Raedle J, Zeuzem S, Pfeilschifter J, Muhl H (2001) Expression and release of IL-18 binding protein in response to IFN-gamma. *J Immunol* 167: 7038–7043

122 Hurgin V, Novick D, Rubinstein M (2002) The promoter of IL-18 binding protein: activation by an IFN-gamma-induced complex of IFN regulatory factor 1 and CCAAT/enhancer binding protein beta. *Proc Natl Acad Sci USA* 99: 16957–16962

123 Reznikov LL, Kim SH, Zhou L, Bufler P, Goncharov I, Tsang M, Dinarello CA (2002) The combination of soluble IL-18Ra and IL-18Rb chains inhibits IL-18-induced IFN-γ. *J Interferon Cytokine Res* 22: 593–601

124 Joosten LA, Radstake TR, Lubberts E, van den Bersselaar LA, van Riel PL, van Lent PL, Barrera P, van den Berg WB (2003) Association of interleukin-18 expression with enhanced levels of both interleukin-1beta and tumor necrosis factor alpha in knee synovial tissue of patients with rheumatoid arthritis. *Arthritis Rheum* 48: 339–347

125 Chabaud M, Lubberts E, Joosten L, van Den Berg W, Miossec P (2001) IL-17 derived from juxta-articular bone and synovium contributes to joint degradation in rheumatoid arthritis. *Arthritis Res* 3: 168–177

126 Raeburn CD, Dinarello CA, Zimmerman MA, Calkins CM, Pomerantz BJ, McIntyre RC Jr, Harken AH, Meng X (2002) Neutralization of IL-18 attenuates lipopolysaccharide-induced myocardial dysfunction. *Am J Physiol* 283: H650–657
127 Yamamura M, Kawashima M, Taniai M, Yamauchi H, Tanimoto T, Kurimoto M, Morita Y, Ohmoto Y, Makino H (2001) Interferon-gamma-inducing activity of interleukin-18 in the joint with rheumatoid arthritis. *Arthritis Rheum* 44: 275–285
128 Gracie JA, Forsey RJ, Chan WL, Gilmour A, Leung BP, Greer MR, Kennedy K, Carter R, Wei XQ, Xu D et al (1999) A proinflammatory role for IL-18 in rheumatoid arthritis. *J Clin Invest* 104: 1393–1401

Targeted IL-1 inhibition

Cem Gabay

Division of Rheumatology, University Hospital of Geneva & Department of Pathology-Immunology, University of Geneva School of Medicine, 26 avenue Beau-Séjour, 1211 Geneva 14, Switzerland

Interleukin-1 and its receptors

Interleukin (IL)-1α and IL-1β are produced by two different genes and share 26% amino acid homology. Both forms are synthesized as 31 kDa precursor peptides (pro-IL-1α and pro-IL-1β), which are specifically cleaved to generate 17 kDa proteins for mature IL-1α and IL-1β. IL-1β is primarily produced by macrophages and is primarily secreted after cleavage of its proform by IL-1β-convertase enzyme (ICE, also termed caspase 1). In addition, other enzymes in the absence of ICE can process pro-IL-1β [1, 2]. IL-1β is produced without a hydrophobic leader peptide and thus does not follow the typical secretion pathway. Recently it has become apparent that IL-1β can be secreted either through the exocytosis of endoplasmic vesicles or can be rapidly released by microvesicles budding off the cell membrane [3, 4]. In contrast, IL-1α is expressed as a 31 kDa intracellular or membrane-bound protein in the human. An important distinction between these two molecules is that pro-IL-1β is biologically inactive, whereas both pro-IL-1α and mature IL-1α exhibit full receptor binding activity. In addition, to its biologic effects following receptor binding, pro-IL-1α exerts other functions inside cells [5–7].

Besides IL-1α and IL-1β, the IL-1 family of cytokines includes IL-1 receptor antagonist (IL-1Ra), IL-18, and six novel members. The recently proposed nomenclature for the IL-1 family is IL-1F1 (IL-1α), IL-1F2 (IL-1β), IL-1F3 (IL-1Ra), IL-1F4 (IL-18), IL-1F5 (Il-1H3, IL-1Hy1, FIL1δ, IL-1RP3, IL-1L1, IL-1δ), IL-1F6 (FIL1ε), IL-1F7 (IL-1H4, FIL1ζ, IL-1RP1, IL-1H), IL-1F8 (IL-1H2 and FILη) IL-1F9 (IL-1H1, IL-1RP2, IL-1ε), and IL-1F10 (IL-1Hy2 and FKSG75). The genes encoding for IL-1α, IL-1β, IL-1Ra and the six novel cytokines form a cluster on the long arm of chromosome 2. IL-18 and its binding protein (IL-18BP) are located on chromosome 11 (reviewed in [8]). Little information is available on the function of the recent IL-1 homologs *in vivo*.

Three types of IL-1 receptors (IL-1R) exist as both membrane-bound and soluble forms including IL-1 receptor type I (IL-1RI), IL-1RII, and IL-1R accessory protein (IL-1RAcP). The extracellular domains of the three types of IL-1 receptors belong to the immunoglobulin superfamily and share some amino-acid sequence

Cytokines and Joint Injury, edited by Wim B. van den Berg and Pierre Miossec

homology. In addition, they also belong to the Toll-like receptor superfamily, which has been shown to play a crucial role in inflammation and host response to infectious agents. IL-1 treatment of cells induces the formation of a complex of high affinity containing IL-1RI and IL-1RAcP. Cells that express IL-1RI but lack IL-1RAcP are not responsive to IL-1, indicating that the IL-1 signal transduction machinery is dependent on the presence of IL-1RAcP [9]. IL-1 activates cells through the NF-κB, c-Jun N-terminal kinase (JNK)/AP-1, p38 MAPK, and ERK1/ERK2 transduction pathways.

In addition to IL-1R and IL-18 receptors, five other members of the IL-1R family are known, including T1/ST2, IL-1R rp2, APL, TIGIRR (or APL-2) and SIGIRR. Limited information is available regarding the function and the putative ligands of these IL-1R homologs. Recent findings suggest that receptor cytoplasmic domains of T1/ST2 and IL-1R rp2 can complement with either IL-1RAcP or IL-18R AP cytoplasmic domains to induce NF-κB activation. In Jurkat cells transfected with IL-1R rp2, IL-1F9 can activate NF-κB and this effect is inhibited by IL-1F5 [10]. T1/ST2 is expressed on T-helper type-2 (Th2) lymphocytes and some evidence indicates that T1/ST2 plays an important role in Th2 responses [11].

Effects of IL-1 in arthritis

IL-1 is an important proinflammatory cytokine in the pathophysiology of various inflammatory conditions. IL-1 induces several systemic manifestations that can be present in RA patients such as fever, anemia, anorexia, bone loss and the increased production of acute-phase proteins by the liver [12]. However, the local effects of IL-1 may be more important in the pathophysiology of RA (Table 1). The results of many studies, including those derived from experimental models of arthritis (reviewed by W. van den Berg), strongly suggest that IL-1 is involved in the pathophysiology of arthritis both in early events as well as in subsequent tissue damage. IL-1 induces the chemotaxis of neutrophils, lymphocytes and monocytes by increasing the expression of both chemokines and adhesion molecules, enhances the proliferation of fibroblasts leading to the pannus formation, and stimulates the production of prostaglandin E_2 (PGE_2). IL-1 also contributes to the destruction of cartilage, bone and periarticular tissues through effects on both synovial fibroblasts and chondrocytes. The effects of IL-1 on cartilage include an increase in proteoglycan degradation, through inducing the production of neutral matrix metalloproteinases (MMPs) such as collagenase and stromelysin, and a decrease in proteoglycan synthesis by chondrocytes. IL-1 also alters the production of collagen by chondrocytes, decreasing the production of collagen type II, the main constituent of cartilage, and increasing the production of collagen type I. IL-1 stimulates the formation of new blood vessels and hence contributes to the development of pannus mass through the release of vascular endothelial growth factor by synovial cells [13].

Table 1 - Effects of IL-1 in arthritis

Attraction of inflammatory cells
Stimulates expression of adhesion molecules
Induces production of chemokines
Tissue damage
Stimulates pannus formation
Synovial fibroblast proliferation
Production of VEGF
Enhances the release of MMP
Stimulates production of PGE2
Stimulates directly and indirectly the effects of RANKL
Inhibits tissue repair
Decreased production of type 2 collagen
Decreased production of proteoglycan

Periarticular bone loss and bony erosions are typical features in patients with RA. IL-1 has a catabolic effect on bone primarily through the maturation and activation of osteoclasts. This effect may be mediated in part by upregulating the expression of receptor activator of nuclear factor κB ligand (RANKL), an essential cytokine for osteoclast maturation and activation [14]. The biological effect of RANKL is controlled by osteoprotegerin (OPG), a soluble receptor that binds RANKL and prevents its interaction with its cell surface receptor RANK. In cultured human vascular endothelial cells, IL-1 stimulated the expression of both RANKL and OPG. However, OPG levels peak early and then fall, whereas the production of RANKL increases slowly but is sustained [15]. IL-1 has also been shown to enhance the stimulatory effect of activated T lymphocytes on osteoclasts [16]. In addition, IL-1 induces osteoclast activation through a RANKL independent pathway. Altogether, these results indicate that IL-1 directly and indirectly is implicated in the mechanisms of bone erosions.

Endogenous IL-1 inhibitors

IL-1 Receptor antagonist

IL-1Ra binds to cell-surface IL-1 receptors with the same affinity as IL-1 but does not stimulate any intracellular response. IL-1Ra competitively prevents the interactions between IL-1 and its target cells. However, a large amount of IL-1Ra is necessary to block the biological effects of IL-1 since binding of only a few molecules of IL-1 per cell suffices to stimulate a full biologic response. IL-1Ra is produced as four different isoforms. One isoform has a hydrophobic leader peptide and is secreted (sIL-1Ra), whereas the three others are intracellular (icIL-1Ra1, 2, 3). These different IL-1Ra peptides are produced from the same gene by the use of alternate first exons, alternative mRNA splicing, or alternative translation initiation (reviewed in [17]).

The results of many studies indicate that endogenous IL-1Ra is produced in human forms of arthritis and may represent an important natural anti-inflammatory mechanism. Elevated synovial fluid levels of IL-1Ra were found primarily in RA [18]. Neutrophils may be the major source of IL-1Ra in synovial fluids, although these cells produced relatively less IL-1Ra and more IL-1β than did peripheral blood neutrophils [19, 20]. Synovial fluid mononuclear cells, as well as synovial tissue macrophages are also potential sources of IL-1Ra [21]. An important anti-inflammatory role for endogenous IL-1Ra was suggested by a study that compared the clinical course of knee arthritis in patients with Lyme disease. Patients with high concentrations of synovial fluid IL-1Ra and low concentrations of IL-1β had rapid resolution of acute attacks of arthritis, whereas patients with the reverse pattern of cytokine concentrations had a more protracted course [22].

In the rheumatoid synovium, IL-1Ra mRNA and protein were found primarily in the sublining and perivascular areas and were present at lower levels in the intimal lining layer [23, 24]. Freshly isolated synovial macrophages produced primarily sIL-1Ra [21]. However, the production of IL-1Ra *in vivo* is relatively low in comparison to that of IL-1. Up to 90% of the cells at the cartilage–pannus interface in the rheumatoid synovium stained for IL-1α, but fewer than 10% of these cells expressed IL-1Ra protein [23].

The production of IL-1Ra was also investigated in articular chondrocytes and synovial fibroblasts in culture. Chondrocytes produced low amounts of sIL-1Ra in response to IL-1. The addition of IL-6 enhanced the stimulatory effects of IL-1 [25]. Interferon-β had a similar effect as IL-6 on IL-1Ra production by human articular chondrocytes [23b]. Although relatively weak, the synthesis of IL-1Ra by chondrocytes may be sufficient to exert a protective effect against IL-1-mediated cartilage lesions that occur in RA. Synovial fibroblasts produced both sIL-1Ra and intracellular species of IL-1Ra in response to various cytokines present in rheumatoid joints including IL-1, TNFα, and IFN-γ [26–28].

Elevated serum levels of IL-1Ra were found both in adults with RA and in children with juvenile RA. Circulating levels of IL-1Ra were higher in patients with active RA than in those with inactive disease [29]. Results from *in vitro* and *in vivo* studies suggest that serum IL-1Ra may be derived primarily from hepatocytes as an acute-phase protein in response to stimulation with IL-1β and IL-6 [30, 31]. Interestingly, rheumatoid arthritis patients exhibited a lower ratio of IL-1Ra to IL-1β in plasma at both baseline and after surgery in comparison to patients with osteoarthritis or osteomyelitis [32]. The results of these studies suggest that a relatively deficient local and systemic production of IL-1Ra in comparison to that of IL-1 may predispose to the perpetuation of inflammatory mechanisms in the rheumatoid joint, leading to progressive tissue destruction.

Disease modifying anti-rheumatic drugs (DMARDs) used in the treatment of RA can influence the ratio of IL-1 to IL-1Ra. PBMC from RA patients treated with methotrexate produced less IL-1 *in vitro*; this decrease was correlated with clinical improvement [29]. Gold sodium thiomalate or auranofin inhibited the production of IL-1 by LPS-stimulated PBMC and increased the production of IL-1Ra [33]. Leflunomide enhanced the stimulatory effect of IL-1 on production of IL-1Ra by both articular chondrocytes and synovial fibroblasts (Palmer et al., *Arthritis Res Ther; in press*).

Soluble IL-1 receptor type II

The most striking structural difference between IL-1RI and IL-1RII is the short cytoplasmic domain of IL-1RII (29 amino acids), whereas IL-1RI possesses a cytoplasmic tail of 213 residues. After ligand binding, intracellular signaling occurs only through IL-1RI. IL-1RI binds to IL-1α with a highest affinity, and IL-1RII binds to IL-1β with highest affinity. IL-1 receptor antagonist binds to IL-1R with the same affinity as IL-1α, thus when both receptors are expressed at the cell surface, IL-1Ra will preferentially bind to IL-1RI.

IL-1RII may exist on the cell surface or in a soluble form only as a decoy molecule [34]. Soluble IL-1RII is found in biological fluids in a variety of pathophysiological conditions [18]. Glucocorticoids, IL-4, LPS, and TNFα stimulate expression or release of IL-1RII. Soluble IL-1RII can be generated by proteolysis of the extracellular domain by matrix metalloproteinases [35] or by alternative splicing of a primary transcript leading to a secreted protein [36]. Soluble IL-1RII binds IL-1β with high avidity and prevents its interaction with cell surface IL-1RI. In contrast, soluble IL-1RII binds IL-1Ra with a much lower affinity than IL-1β and may supplement the anti-inflammatory effects of IL-1Ra [37]. In addition, membrane-bound IL-1RII has also an inhibitory effect on IL-1 responses, which is probably related to a ligand sink function [38].

Other cytokines

IL-1 production may be decreased both *in vitro* and *in vivo* by other cytokines. IL-4, IL-10, and IL-13 are primarily Th2 cytokines and they promote the expansion of antibody producing B cells, suppress the Th1-mediated responses such as delayed hypersensitivity. $CD4^+$ Th2 lymphocytes are the major, but not only, source of these cytokines. Indeed, monocytes, B cells, keratinocytes also synthesize IL-10, and mast cells and basophils produce IL-4. IL-4, IL-10, and IL-13 downregulated the production of IL-1β by cultured PBMC and rheumatoid synovial explants [39]. In addition, these cytokines enhanced the production of IL-1Ra by cultured PBMC and rheumatoid synovial explants [40, 41]. IL-4 and IL-13 increased the stimulatory effect of IL-1β on production of IL-1Ra by cultured human primary hepatocytes [42]. Administration of IL-10 reduced the levels of LPS-induced IL-1β in the circulation of human volunteers [43]. Transforming growth factor (TGF)β is another potent immunosuppressive cytokine. TGFβ decreased the production of IL-1β and other proinflammatory cytokines *in vitro* while enhancing the production of IL-1Ra [39].

Clinical trials with IL-1 inhibitors

IL-1 has been demonstrated to play an important role in the pathophysiology of arthritis in various experimental models. Thus, different strategies to inhibit IL-1 have been developed. Some are currently used in clinical practice, while others are in clinical trials or tested in experimental models (Table 2).

IL-1Ra

Clinical trials with IL-1Ra alone

Recombinant human sIL-1Ra, also known as anakinra, is produced by Amgen (USA). Anakinra has been recently approved for the treatment of RA in the USA and in Europe, and is now commercially available as Kineret®.

The results of clinical trials with anakinra administered alone or in combination with other DMARDs are listed in Table 3.

An initial randomized, double-blind study was performed with 172 patients with RA divided in nine different treatment groups. The patients received recombinant IL-1Ra 20, 70 or 200 mg once, three times, or seven times each week for three weeks, followed by a four-week maintenance phase of once weekly injections [44]. Because of the multiple small treatment groups and the lack of placebo control, it was impossible to draw any firm conclusion from this first trial. However, the patients receiving the daily injections appeared to exhibit some clinical improvement that was associated with decreased serum levels of C-reactive protein.

A subsequent randomized double-blind, placebo-controlled multicenter trial was performed. 472 patients with RA were randomized in four different groups to receive daily subcutaneous injections of placebo or three different doses of IL-1Ra (30, 75, 150 mg) for 24 weeks [45]. After 24 weeks, the American College of Rheumatology (ACR) 20% (ACR20) response was achieved by 43% of the patients receiving the largest dose (150 mg/injection) of IL-1Ra in comparison with 27% of those in the placebo group. The clinical responses in the 150 mg/day group were

Table 2 - Current and future IL-1 inhibitors

Block ligand-receptor interaction
IL-1Ra
Peptides with IL-1Ra activity
Block extracellular IL-1
IL-1 trap
Soluble IL-1 receptor type II
ICE inhibitors
Pralnacasan
Block IL-1 intracellular signaling (not specific for IL-1)
IKK2 inhibitors
JNK inhibitors
P38 MAPK inhibitors

superior to other treatment groups and were significantly better than patients receiving placebo. In addition, the radiological evaluation of the hands demonstrated a significant decrease in the severity of radiological signs of joint damage among patients receiving the highest dose of IL-1Ra as compared with the placebo group [45]. A subsequent evaluation was performed using the Genant score that distinguishes the joint space narrowing and the presence of bony erosions. The results of this study showed greater reduction in joint space narrowing (58% compared with placebo) than in erosion (38% *versus* placebo), raising the possibility that IL-1 inhibition provides a greater protection for cartilage than for subchondral bone [46].

Recombinant human IL-1Ra was well tolerated besides the occurrence of injection site reactions causing 5% of discontinuation in the 150 mg/day group as compared to 2% in the patients receiving the placebo. These reactions were usually mild, appeared within the first four weeks of treatment, and tended to regress with time. A 24-week extension of this study was carried out where IL-1Ra was also given to the placebo group. The results of this extension showed that administration of IL-1Ra to the placebo patients resulted in new clinical improvement and that the patients who previously received IL-1Ra maintained the improvement seen in the first 24-week trial [47].

Serial synovial biopsies were performed in the 24-week randomized trial and in the extension study. The 12 patients included were divided in three groups: placebo (n = 3), 30 mg/day (n = 6), and 150 mg IL-1Ra (n = 3). Cellular infiltration and expression of adhesion molecules was examined before and after 24 weeks of IL-

Table 3 - Results of clinical trials with anakinra (IL-1Ra)

Study	Type of study	Patients	Results
Monotherapy	RCT	172	Clinical response with highest dose of anakinra, no placebo group
Monotherapy	RDBPCT	453	Significant improvement according to ACR responses Decreased X-ray progression
With MTX	RDBPCT	419	Significant improvement according to ACR responses
With DMARDS	RPCT	1414	Anakinra in combination with DMARDs is safe
With etanercept	open	58	Some improvement but one-third withdrawal, 11/58 side-effects, serious infections
With etanercept	RDBCT	244	Combination is not more efficacious than etanercept alone. Increased rate of infections

RDBPCT: randomized, double-blind, placebo-controlled trial

1Ra treatment. The results showed a reduction in intimal macrophages and subintimal macrophages and lymphocytes, and a downregulation of E-selectin and vascular cell adhesion molecule-1 in patients receiving 150 mg/day IL-1Ra. The absence of progression in radiological signs of joint damage seen in some patients correlated with the decrease in intimal macrophages [48]. Although this study included only a limited number of patients, the results further demonstrated the therapeutic effect of IL-1Ra in the rheumatoid synovium with possible consequences on joint damage.

Combination of IL-1Ra with other DMARDs

There has been a growing interest over the past decade in the combined use of several anti-rheumatic drugs in the treatment of RA. The effect of IL-1Ra in combination with methotrexate has been studied recently in a randomized, double-blind, placebo-controlled, multicenter trial. 419 patients with active RA despite being treated with methotrexate for six consecutive months with stable doses for ≥ 3 months were randomized into six groups including placebo or 0.04, 0.1, 0.4, 1.2 mg/kg IL-1Ra administered in a single, daily, subcutaneous injection. The primary endpoint was the proportion of subjects who met the ACR20 improvement criteria after 24 weeks. The mean methotrexate dose ranged from 16.3 to 17.6 mg/week among the six groups. The results demonstrated that the ACR20 rates were significantly higher in patients who received the combination of methotrexate and 1 mg/kg IL-1Ra than those who received methotrexate and the placebo only (42% *versus* 23%). Subjects with IL-1Ra 0.4 and 2 mg/kg also had better ACR20 responses (36 and 35%, respectively), although not significantly different than the placebo group. The effect of IL-1Ra, according to onset of ACR20 response, was already significant at week four. The percentage of patients who achieved more stringent response criteria such as ACR50 and ACR70 was also significantly higher among those treated with 1 mg/kg and 2 mg/kg IL-1Ra than those who received placebo. The patients treated with the combination of methotrexate and IL-1Ra, particularly those receiving the highest dosage of IL-1Ra (1 and 2 mg/kg) exhibited a significant improvement in their functional status as assessed by the health assessment questionnaire disability index. This improvement was already present at four weeks of treatment. The percentage of patients reporting no functional impairment at the end of the trial was significantly higher in those who received 1 mg/kg IL-1Ra in combination with methotrexate than in those treated with methotrexate alone. The characterization of the functional status showed that largest improvement were seen in grip, activities, grooming and arising, whereas walking was associated with the smallest improvement [49].

Most withdrawal in the IL-1Ra 1 mg/kg and 2 mg/kg groups were due to injection site reactions (6.8 and 9.7%, respectively). Five patients withdrew from the study because of leucopenia. White blood cell count returned to normal values after discontinuation of IL-1Ra treatment. Leucopenia was not associated with episodes of infection. No serious infections or deaths were noted during this study. The results of this clinical trial indicate that IL-1Ra is safe and provides a beneficial effect in RA patients with incomplete response to methotrexate [50].

Safety of IL-1Ra was also assessed in a large placebo-controlled population of RA patients seen in clinical practice. 1,414 patients were randomized to receive either a combination of DMARDs and 100 mg/day IL-1Ra (n = 1116) or DMARDs and placebo (n = 283). The rate of serious infections was slightly higher in the IL-1Ra than in the placebo group (2.1 *versus* 0.4%). However, none of these serious infections resulted in deaths and none were attributed to opportunistic microorganisms and tuberculosis. These results from a large population indicate that IL-1Ra in association with one DMARD or a combination of DMARDs is safe [51].

Combination of IL-1Ra with TNF inhibitors resulted in additional improvement in experimental models of arthritis indicating that this approach may prove useful in RA patients. A Phase II open trial was conducted in 58 RA patients with active disease despite treatment with etanercept (a fusion protein containing the extracellular domain of TNF receptor p75 and the Fc region of human IgG class 1). These patients received IL-1Ra at a dosage of 1 mg/kg/day. After 24 weeks, different clinical and biological parameters of disease activity were ameliorated by the combination of IL-1Ra and etanercept. However, approximately one-third of the patients discontinued this treatment. 11/58 patients had serious side effects, including four cases with severe infection [52]. Of note, no cases of tuberculosis or opportunistic infections were reported. Recently, an eight-week safety Phase I double-blind, placebo-controlled, dose escalation clinical trial with IL-1Ra and different dose of pegsunercept (pegylated extracellular portion of TNF receptor p55) was conducted in patients with RA. A total of 16 patients received a combination of 100 mg/day IL-1Ra and placebo or different doses of pegsunercept. 5/16 patients with the combination regimen experienced mild infectious episodes and one patient in the control group reported a mild infection. No serious infectious adverse events were reported [53]. The efficacy of the combination of IL-1Ra with etanercept was recently examined in a six month, randomized, double-blind, controlled study in methotrexate partial responders. RA patients receiving a background methotrexate therapy were randomized in three groups to receive either etanercept 25 mg twice weekly plus anakinra placebo once daily, 100 mg/day anakinra plus etanercept 25 mg once weekly, or 100 mg/day IL-1Ra plus etanercept 25 mg twice weekly. The primary efficacy outcome was the percentage of patients achieving the ACR50 response at 24 weeks. The results did not show any advantage of the combination over etanercept alone. The proportion of infectious adverse events was significantly higher in those treated with the combination of IL-1 and TNFα inhibitors [53b].

Gene therapy

The use of IL-1Ra may be limited by the fact that high amounts of IL-1Ra are necessary to block the effects of IL-1. Thus, different authors have considered the possibility to deliver IL-1Ra by local gene therapy in arthritic joints. The use of several experimental models of arthritis led to encouraging results [54–56]. In addition, anti-arthritic effects of gene transfer was also seen in contralateral untreated joints indicating that local gene therapy can be used in the treatment of a polyarticular and systemic disease [57]. The results of clinical trials of gene therapy with IL-1Ra in patients with RA demonstrated the successful transfer of *ex vivo* transduced synoviocytes, resulting in intra-articular expression of IL-1Ra [58]. However, two important issues including the safety and transient nature of cDNA expression in transduced cells need to be solved prior to gene therapy being considered in the treatment of RA.

Soluble IL-1 receptors

The administration of human soluble IL-1R type I exhibited some beneficial effects in antigen-induced arthritis in mice [59]. The efficacy of recombinant human IL-1R type I in patients with RA was recently examined in a randomized, double-blind, placebo-controlled trial. No significant improvement was observed at four weeks after treatment with soluble IL-1R type I administered either by intra-articular or subcutaneous injection [60]. This result may be due, at least in part, to the fact that IL-1R type I binds much more avidly to IL-1Ra than to IL-1β or IL-1α [18]. The administration of soluble IL-1R type I may have inhibited binding of IL-1Ra to cell surface IL-1 receptors, thus further enhancing the inflammatory effects of IL-1 on target cells.

In contrast, soluble IL-1R type II binds to IL-1Ra with much lower affinity than IL-1β and may supplement the anti-inflammatory effects of IL-1Ra [37]. The administration of soluble IL-1R type II in experimental models resulted in a marked inhibition of joint swelling and joint damage [61, 62]. In addition, recent data indicates that soluble IL-1R type II exerts a chondroprotective effect *in vitro* [63]. Thus, the administration of soluble IL-1R type II may be an interesting approach in the treatment of RA and other IL-1-mediated diseases.

IL-1 trap

IL-1 trap is a novel approach in the treatment of RA. It is a fusion protein containing some of the extracellular binding motifs of IL-1RI and IL-1RAcP coupled to the Fc fraction of the human immunoglobulin IgG. IL-1 trap is synthesized as a recombinant protein by cultured eukaryotic cells. It is a fusion protein in which the extracellular binding domains of IL-1RI and IL-1RAcP have been engineered into a single chain with two of these chains coupled to the Fc portion of a human IgG to form a dimeric molecule. The presence of the Fc fraction of human immunoglobulin enhances the circulating half-life of IL-1 trap.

IL-1 trap binds IL-1β and IL-1α with high affinity (Kd = 1.5 and 3 pM, respectively) [64]. Once IL-1 is bound to the trap, it cannot interact with its cell surface receptors and loses its biological effects. The inhibitory effect of IL-1 trap on IL-1-induced IL-6 production by the fibroblastic cell line MRC5 was significantly stronger than that of soluble IL-1R I and II, and of IL-1Ra. This result is consistent with the weaker binding affinity of IL-1RI and IL-1RII to IL-1 (1 to 3 nM and 500 pM, respectively). Most importantly, although soluble IL-1RI alone binds IL-1Ra more tightly than IL-1β, IL-1 trap binds IL-1β with a much stronger affinity than IL-1Ra. Thus, administration of IL-1 trap should not affect the anti-inflammatory effect of endogenous IL-1Ra [64].

The results of *in vivo* studies indicate that IL-1 trap inhibited the biological effects of IL-1. Administration of IL-1 trap at a dose of 20 mg/kg was effective in inhibiting IL-1-induced IL-6 release in mice. A murine form of IL-1 trap at 20 mg/kg almost completely blocked the development of collagen-induced arthritis with substantial reduction of bone erosions [64].

The pharmacological studies performed in patients with RA showed that pharmacokinetics of IL-1 trap are dose-dependent. T_{max} values ranged from 90 to 120 h and C_{max} ranged from 218 to 1,896 ng/ml following subcutaneous injections of 50 to 400 μg/kg. Terminal half-life varied from 128 to 182 h (unpublished data). The safety and efficacy of IL-1 trap was assessed in a Phase Ib randomized, dose-escalating, double-blind, placebo-controlled trial including four groups of 15–20 patients with active RA. Each group received six weekly doses of 0, 200, 400, 800 μg/kg IL-1 trap. After six weeks, an average ACR20 was achieved by 74% of patients in the 800 μg/kg group as compared with 36% of placebo-treated patients. ACR50 was achieved by 49% of the 800 μg/kg group as compared with 16% in the placebo group. These results were statistically significant. After six weeks, a median improvement in ACR-N score of approximately 26% was achieved in the highest dose group as compared with 6% in the placebo group. Circulating levels of CRP decreased in a dose-dependent manner in IL-1 trap-treated patients (Presented at the 2002 ACR annual meeting). A multicenter, randomized, placebo-controlled, double blind Phase II trial is currently in progress.

ICE inhibitor

Another potential therapeutic approach to RA is to reduce the production of IL-1 by interfering with its processing and secretion. IL-1β is synthesized in the cytoplasm as an inactive precursor pro-IL-1β that lacks a typical leader sequence. Secretion of IL-1β, by an as yet unclear mechanism, is accompanied by processing by ICE, a cysteine protease that cleaves pro-IL-1β to generate the mature 17 kDa form. Mice lacking the ICE gene do not produce IL-1β in response to LPS, and, paradoxically, also are deficient in production of IL-1α [65]. In addition, ICE knockout mice are not susceptible to collagen-induced arthritis [39]. Specific small molecule inhibitors of ICE have been synthesized.

The administration of ICE inhibitors blocked the progression of collagen-induced arthritis in mice [66]. In a Phase I trial, pralnacasan, an ICE inhibitor, was well tolerated. A 12-week Phase II placebo-controlled multicenter study was performed in 285 RA patients receiving concurrent DMARDs. Pralnacasan produced a trend toward a dose-dependent protection, with ACR20 responses just failing to achieve significance. This was accompanied by a significant reduction in inflammatory serum markers (CRP and serum amyloid A) as well as MMP-1 and TIMP-1 [67]. This disappointing result can be explained by the fact that besides ICE, other enzymes are also able to cleave pro-IL-1β and thus, contribute to the release of biologically active IL-1β. In addition, ICE does not influence the biological effects of IL-1α.

Future targets and conclusion

IL-1 stimulates different post-receptor signaling pathways that play an important role in cell activation. Thus, it is conceivable that strategies aimed to interfere with these intracellular signaling pathways should be beneficial for the treatment of arthritis. Recently, small molecules with inhibitory effects on intracellular signaling have been used with success in experimental models of arthritis. In addition, in contrast to current anti-cytokine therapies, some of these small molecules can be administered orally.

NF-κB is one of the potential targets for therapy. The oral use of proteasome inhibitors, which block the degradation of IκB and the translocation of NF-κB in the nucleus, reduced the severity of peptidoglycan polysaccharide-induced arthritis in female Lewis rats [68]. Recently, the oral administration of IKK2 inhibitors decreased the severity of arthritis in two different experimental models [69, 70]. IL-1 induces also the activation of p38 MAPK and JNK. Accordingly, the administration of p38 MAPK inhibitors decreased the severity of adjuvant-induced arthritis in the rat [71]. Most interestingly, inhibition of JNK was devoid of any effect on joint swelling but inhibited the expression of MMP-3 and prevented the development of joint damage in rats with adjuvant-induced arthritis [72].

Collectively, the results of experimental studies and of clinical trials with IL-1 inhibitors clearly demonstrated that IL-1 is an important target for the treatment of RA. However, the clinical trials showed that the effect of IL-1Ra on ACR response rates were weaker than those observed with different TNF inhibitors. These data contrast with the marked positive effect of IL-1 inhibitors in several animal models of arthritis and suggest either that IL-1 has a less predominant role in human RA or that IL-1Ra does not block all the effects of IL-1 *in vivo*. In this latter case, IL-1 trap may possess an advantage over IL-1Ra because it inhibits the biological effect of IL-1 with a much stronger affinity than IL-1Ra. In addition, IL-1 trap has a longer circulating half-life than IL-1Ra and can be administered once a week. The results of future clinical trials with IL-1 trap will provide important information regarding the efficacy of IL-1 targeting in RA. The efficacy and safety of combination of IL-1 inhibitors with other biological agents should be further tested in clinical trials. The combination of IL-1Ra with potent anti-inflammatory agents may also be of potential value. In this approach, IL-1Ra may act rather as a chondroprotective agent and prevent long-term joint damage. In addition, the development of other therapeutic agents administered orally including intracellular signaling inhibitors and small peptides blocking specifically the binding of IL-1 to its cell surface receptors may prove of potential value.

References

1 Fantuzzi G, Ku G, Harding MW, Livingston DJ, Sipe JD, Kuida K, Flavell RA, Dinarello CA (1997) Response to local inflammation of IL-1β-converting enzyme-deficient mice. *J Immunol* 158: 1818–1824

2 Schonbeck U, Mach F, Libby P (1998) Generation of biologically active IL-1β by matrix

metalloproteinases: a novel caspase-1-independent pathway of IL-1β processing. *J Immunol* 161: 3340–3346
3 Andrei C, Dazzi C, Lotti L, Torrisi MR, Chimini G, Rubartelli A (1999) The secretory route of the leaderless protein interleukin-1β involves exocytosis of endolysosome-related vesicles. *Mol Biol Cell* 10: 1463–1475
4 McKenzi A, Wilson HL, Kiss-Toth E, Dower SK, North RA, Surprenant A (2001) Rapid secretion of interleukin-1β by microvesicle shedding. *Immunity* 8: 825–835
5 Stevenson FT, Turck J, Locksley RM, Lovett DH (1997) The N-terminal propiece of interleukin 1α is a transforming nuclear oncoprotein. *Proc Natl Acad Sci USA* 94: 508–513
6 McMahon GA, Garfinkel S, Prudovsky I, Hu X, Maciag T (1997) Intracellular precursor interleukin (IL)-1α, but not mature IL-1α, is able to regulate human endothelial cell migration *in vitro*. *J Biol Chem* 272: 28202–28205
7 Pollock AS, Turck J, Lovett DH (2003) The prodomain of interleukin 1α interacts with elements of the RNA processing apparatus and induces apoptosis in malignant cells. *FASEB J* 17: 203–213
8 Dunn E, Sims JE, Nicklin MJH, O'Neill LAJ (2001) Annotating genes with potential roles inthe immune system: six new members of the IL-1 family. *Trends Immunol* 22: 533–537
9 Huang J, Gao X, Cao Z (1997) Recruitemont of IRAK to the interleukin 1 receptor complex requires interleukin 1 receptor accessory protein. *Proc Natl Acad Sci USA* 94: 12829–12832
10 Debets R, Timans JC, Homey B, Zurawski S, Sana TR, Lo S, Wagner J, Edwards G, Clifford T, Menon S et al (2001) Two novel IL-1 family members, IL-1 delta and IL-1 epsilon, function as an antagonist and agonist of NF-kappa B activation through the orphan IL-1 receptor-related protein 2. *J Immunol* 167: 1440–1446
11 Townsend MJ, Fallon PG, Matthews DJ, Jollin HE, McKenzy AN (2000) T1/ST2-deficient mice demonstrate the importance of T1/ST2 in developing primary T helper cell type 2 responses. *J Exp Med* 191: 1069–1076
12 Gabay C, Kushner I (1999) The acute-phase proteins and other systemic responses to inflammation. *N Engl J Med* 340: 448–454
13 Jackson JR, Minton JA, Ho ML, Wei N, Winkler JD (1997) Expression of vascular endothelial growth factor in synovial fibroblasts is induced by hypoxia and interleukin 1β. *J Rheumatol* 24: 1253–1259
14 Hofbauer LC, Heufelder AE (2001) The role of osteoprotegerin and receptor activator of nuclear factor κB ligand in the pathogenesis and treatment of rheumatoid arthritis. *Arthritis Rheum* 44: 253–259
15 Collin-Osdoby P, Rothe L, Anderson F, Nelson M, Maloney W, Osdoby P (2001) Receptor activator of the NF-κB and osteoprotegerin expression by human microvascular endothelial cells, regulation by inflammatory cytokines, and role in human osteoclastogenesis. *J Biol Chem* 276: 20659–20672
16 Horwood NJ, Kartsogiannis V, Quinn JMW, Romas E, Martin TJ, Gillespie MT (1999)

Activated T cells support osteoclast formation *in vitro*. *Biochem Biophys Res Commun* 265: 144–150
17 Arend WP, Malyak M, Guthridge CJ, Gabay C (1998) Interleukin-1 receptor antagonist: role in biology. *Annu Rev Immunol* 16: 27–55
18 Arend WP, Malyak M, Smith Jr MF, Whisenand TD, Slack JL, Sims JE, Giri JG, Dower SK (1994) Binding of IL-1α, IL-1β, and IL-1 receptor antagonist by soluble IL-1 receptors and levels of soluble IL-1 receptors in synovial fluids. *J Immunol* 153: 4766–4774
19 Malyak M, Swaney RE, Arend WP (1993) Levels of synovial fluid interleukin-1 receptor antagonist in rheumatoid arthritis and other arthropathies. Potential contribution from synovial fluid neutrophils. *Arthritis Rheum* 36: 781–789
20 Malyak M, Smith Jr MF, Abel AA, Arend WP (1994) Peripheral blood neutrophil production of interleukin-1 receptor antagonist and interleukin-1β. *J Clin Immunol* 14: 20–30
21 Firestein GS, Boyle DL, Yu C, Paine MM, Whisenand TD, Zvaifler NJ, Arend WP (1994) Synovial interleukin-1 receptor antagonist and interleukin-1 balance in rheumatoid arthritis. *Arthritis Rheum* 37: 644–652
22 Miller LC, Lynch EA, Isa S, Logan JW, Dinarello CA, Steere AC (1993) Balance of synovial fluid IL-1β and IL-1 receptor antagonist and recovery from Lyme arthritis. *Lancet* 341: 146–148
23 Deleuran BW, Chu CQ, Field M, Brennan FM, Katsikis P, Feldmann M, Maini RN (1992) Localization of interleukin-1 alpha, type 1 interleukin-1 receptor and interleukin-1 receptor antagonist in the synovial membrane and cartilage/pannus junction in rheumatoid arthritis. *Br J Rheumatol* 31: 801–809
23b Palmer G, Mezin F, Juge-Aubry CE, Plater-Zyberk C, Gabay C, Guerne PA (2004) Interferon β stimulates interleukin 1 receptor antagonist production in human articular chondrocytes and synovial fibroblasts. *Ann Rheum Dis* 63: 43–49
24 Firestein GS, Berger AE, Tracey DE, Chosay JG, Chapman DL, Paine MM, Yu C, Zvaifler NJ (1992) IL-1 receptor antagonist protein production and gene expression in rheumatoid arthritis and osteoarthritis synovium. *J Immunol* 149: 1054–1062
25 Palmer G, Guerne P-A, Mezin F, Maret M, Guicheux J, Gabay C (2002) Production of interleukin-1 receptor antagonist by human articular chondrocytes. *Arthritis Res* 4: 226–231
26 Martel-Pelletier J, McCollum R, Pelletier J-P (1993) The synthesis of IL-1 receptor antagonist (IL-1 ra) by synovial fibroblasts is markedly increased by cytokines TNF-alpha and IL-1. *Biochim Biophys Acta* 1175: 302–305
27 Krzesicki RF, Hatfield CA, Bienkowski MJ, Mcguire JC, Winterrowd GE, Chapman DL, Berger AE, Mcewan RN, Carter DB, Chosay JG (1993) Regulation of expression of IL-1 receptor antagonist protein in human synovial and dermal fibroblasts. *J Immunol* 150: 4008–4018
28 Seitz M, Loetscher P, Dewald B, Towbin H, Ceska M, Baggiolini M (1994) Production of interleukin-1 receptor antagonist, inflammatory chemotactic proteins, and prosta-

glandin E by rheumatoid and osteoarthritic synoviocytes. Regulation by IFN-γ and IL-4. *J Immunol* 152: 2060–2065
29 Seitz M, Loetscher P, Dewald B, Towbin H, Rordorf C, Gallati H, Gerber NJ (1996) Interleukin 1 (IL-1) receptor antagonist, soluble tumor necrosis factor receptors, IL-1β, and IL-8 markers of remission in rheumatoid arthritis during treatment with methotrexate. *J Rheumatol* 23: 1512–1516
30 Gabay C, Smith Jr. MF, Eidlen D, Arend WP (1997) Interleukin-1 receptor antagonist is an acute-phase protein. *J Clin Invest* 99: 2930–2940
31 Gabay C, Gigley JP, Sipe J, Arend WP, Fantuzzi G (2001) Production of interleukin-1 receptor antagonist by hepatocytes is regulated as an acute-phase protein *in vivo*. *Eur J Immunol* 31: 490–499
32 Chikanza IC, Roux-Lombard P, Dayer J-M, Panayi G (1995) Dysregulation of the *in vivo* production of interleukin-1 receptor antagonist in patients with rheumatoid arthritis. *Arthritis Rheum* 38: 642–648
33 Shingu M, Fujikawa Y, Wada T, Nonaka S, Nobunaga M (1995) Increased IL-1 receptor antagonist (IL-1ra) production and decreased IL-1β/IL-1ra ratio in mononuclear cells from rheumatoid arthritis patients. *Br J Rheumatol* 34: 24–30
34 Colotta F, Dower SK, Sims JE, Mantovani A (1994) The type II "decoy" receptor: a novel regulatory pathway for interleukin 1. *Immunol Today* 15: 562–566
35 Orlando S, Sironi M, Bianchi G, Drummond AH, Boraschi D, Yabes D, Mantovani A (1997) Role of metalloproteases in the release of the IL-1 type II decoy receptor. *J Biol Chem* 272: 31764–31769
36 Liu C, Hart RP, Liu XJ, Clevenger W, Maki RA, De Souza EB (1996) Cloning and characterization of an alternatively processed human type II interleukin-1 receptor mRNA. *J Biol Chem* 271: 20965–20972
37 Burger D, Chicheportiche R, Giri JG, Dayer J-M (1995) The inhibitory activity of human interleukin-1 receptor antagonist is enhanced by type II interleukin-1 soluble receptor and hindered by type I interleukin-1 soluble receptor. *J Clin Invest* 96: 38–41
38 Neumann D, Kollewe C, Martin MU, Boraschi D (2000) The membrane form of the type II IL-1 receptor accounts for inhibitory function. *J Immunol* 165: 3350–3357
39 Dinarello CA (1996) Biologic basis for interleukin-1 in disease. *Blood* 87: 2095–2147
40 Arend WP (1993) Interleukin-1 receptor antagonist. *Adv Immunol* 54: 167–227
41 De Waal Malefyt R, Figdor CG, Huijbens R, Mohan-Peterson S, Bennett B, Culpepper J, Dang W, Zurawski G, De Vries JE (1993) Effects of IL-13 on phenotype, cytokine production, and cytotoxic function of human monocytes. Comparison with IL-4 and modulation by IFNγ or IL-10. *J Immunol* 151: 6370–6381
42 Gabay C, Porter B, Guenette D, Bilir B, Arend WP (1999) IL-4 and IL-13 enhance the effect of interleukin-1 on IL-1 receptor antagonist production by human primary hepatocytes and HepG2 cells. Differential regulation for C-reactive protein. *Blood* 93: 1299–1307
43 Chernoff AE, Granowitz EV, Shapiro L, Vannier E, Lonnemann G, Angel JB, Kennedy Y JS, Rabson AR, Wolff SM, Dinarello CA (1995) A randomized, controlled trial of

interleukin-10 in humans: inhibition of inflammatory cytokine production and immune responses. *J Immunol* 154: 5492–5499
44 Campion GV, Lebsack ME, Lookabaugh J, Gordon G, Catalano MA (1996) Dose-range and dose-frequency study of recombinant human interleukin-1 receptor antagonist in patients with rheumatoid arthritis. *Arthritis Rheum* 39: 1092–1101
45 Bresnihan B, Alvaro-Gracia JM, Cobby M, Doherty M, Domljan Z, Emery P, Nuki G, Pavelka K, Rau R, Rozman B et al (1998) Treatment of rheumatoid arthritis with recombinant human interleukin-1 receptor antagonist. *Arthritis Rheum* 41: 2196–2204
46 Jiang Y, Genant HK, Watt I, Cobby M, Bresnihan B, Aitchison R, Mccabe D (2000) A multicenter, double-blind, dose-randing, randomized, placebo-controlled study of recombinant human interleukin-1 receptor antagonist in patients with rheumatoid arthritis. *Arthritis Rheum* 43: 1001–1009
47 Nuki G, Bresnihan B, Bear MB, McCabe D and the European Group of Clinical Investigators (2002) Long-term safety and maintenance of clinical improvement following treatment with anakinra (recombinant human interleukin-1 receptor antagonist) in patients with rheumatoid arthritis: Extension phase of a randomized, double-blind, placebo-controlled trial. *Arthritis Rheum* 46: 2838–2846
48 Cunnane G, Madigan A, Murphy E, FitzGerald O, Bresnihan B (2001) The effects of treatment with interleukin-1 receptor antagonist on the inflamed synovial membrane in rheumatoid arthritis. *Rheumatology* 40: 62–69
49 Cohen SB, Wooley JM, Chan WW (2003) Interleukin 1 receptor antagonist anakinra improves functional status in patients with rheumatoid arthritis. *J Rheumatol* 30: 225–231
50 Cohen S, Hurd E, Cush JJ, Schiff M, Weinblatt ME, Moreland LW, Kremer J, Bear MB, Rich WJ, Mccabe D (2002) Treatment of rheumatoid arthritis with anakinra a recombinant human interleukin-1 receptor antagonist, in combination with methotrexate. Results of a twenty-four-week, multicenter, randomized, double-bind, placebo-controlled trial. *Arthritis Rheum* 46: 614–624
51 Fleischmann RM, Schlechtman J, Bennett R, Handel ML, Burmester GR, Tesser J, Modafferi D, Poulakos J, Sun G (2003) Anakinra, a recombinant human interleukin-1 receptor antagonist (r-metHuIL-1Ra), in patients with rheumatoid arthritis: A large international, multicenter, placebo-controlled trial. *Arthritis Rheum* 48: 927–934
52 Schiff MH, Bulpitt K, Weaver AA, Genovese MC, Cohen S, Furst D, Weinblatt ME, Martin RW, Moreland L, Kazazi F et al (2001) Safety of combination therapy with anakinra and etanercept in patients with rheumatoid arthritis. *Arthritis Rheum* 44: S79 (Abstract)
53 Caldwell JR, Offenberg H, Furst D, Lam NP, Jayne G, Joh T, Davis M, Edwards CK III (2002) A Phase I safety study of combination treatment with pegylated soluble tumor necrosis factor receptor type I (PEG sTNF-RI) and anakinra (interleukin-1 receptor antagonist, IL-1Ra) in patients with rheumatoid arthritis. *Ann Rheum Dis* 61 (suppl I): 186 (Abstract)
53b Genovese MC, Cohen SB, Moreland LW, Lium D, Liu T, Newmark R, Bekker P (2003)

A randomized doouble-blind controlled study evaluating the safety and efficacy of etanercept versus etanercept plus analinra in patients with rheumatoid arthritis. *Ann Rheum Dis* 62 (Suppl I): 66
54 Bandara G, Mueller GM, Galea-Lauri J, Tindal MH, Georgescu HI, Suchanek MK, Hung GL, Glorioso JC, Robbins PD, Evans CH (1993) Intra-articular expression of biologically active interleukin 1 receptor antagonist protein by *ex vivo* gene transfer. *Proc Natl Acad Sci USA* 90: 10764–10768
55 Makarov SS, Olsen JC, Johnston WN, Anderle SK, Brown RR, Baldwin AS, Haskill JS, Schwab JH (1996) Suppression of experimental arthritis by gene transfer of interleukin 1 receptor antagonist cDNA. *Proc Natl Acad Sci USA* 93: 402–406
56 Otani K, Nita I, Macaulay W, Georgescu HI, Robbins PD, Evans CH (1996) Suppression of antigen-induced arthritis in rabbits by *ex vivo* gene therapy. *J Immunol* 156: 3558–3362
57 Bakker AC, Joosten LAB, Arntz OJ, Helsen MM, Bendele AM, Van de Loo FA, Van den Berg WB (1997) Prevention of murine collagen-induced arthritis in the knee and ipsilateral paw by local expression of human interleukin-1 receptor antagonist protein in the knee. *Arthritis Rheum* 40: 893–900
58 Ghivizzani SC, Kang R, Muzzonigro T, Whalen J, Watkins SC, Herndon JH, Robbins PD, Evans CH (1997) Gene therapy for arthritis – treatment of the first three patients. *Arthritis Rheum* 40: S223 (Abstract)
59 Dower SK, Fanslow W, Jacobs C, Waugh S, Sims JE, Widmer MB (1994) Interleukin-1 antagonists. *Therapeutic Immunol* 1: 113–122
60 Drevlow BE, Lovis R, Haag MA, Sinacore JM, Jacobs C, Blosche C, Landay A, Moreland LW, Pope RM (1996) Recombinant human interleukin-1 receptor type I in the treatment of patients with active rheumatoid arthritis. *Arthritis Rheum* 39: 257–265
61 Dawson J, Engelhardt P, Kastelic T, Cheneval D, Mackenzie A, Ramage P (1999) Effects of soluble interleukin-1 type II receptor on rabbit antigen-induced arthritis: clinical, biochemical and histological assessment. *Rheumatology* 38: 401–406
62 Bessis N, Guery L, Mantovani A, Vecchi A, Sims JE, Fradelizi D, Boissier MC (2000) The type II receptor of interleukin-1 inhibits murine collagen-induced arthritis. Eur *J Immunol* 30: 867–875
63 Attur MG, Dave MN, Leung MY, Cipolletta C, Meseck M, Woo SL, Amin AR (2002) Functional genomic analysis of type II IL-1β decoy receptor: Potential for gene therapy in human arthritis and inflammation. *J Immunol* 168: 2001–2010
64 Economides AN, Rocco Carpenter L, Rudge JS, Wong V, Koehler-Stec EM, Hartinett C, Pyles EA, Xu X, Daly TJ, Young MR et al (2003) Cytokine traps: multi-component, high-affinity blockers of cytokine action. *Nature Med* 9: 47–52
65 Li P, Allen H, Banerjee S, Franklin S, Herzog L, Johnston C, Mcdowell J, Paskind M, Rodman L, Salfeld J (1995) Mice deficient in IL-1β-converting enzyme are defective in production of mature IL-1β and resistant to endotoxic shock. *Cell* 80: 401
66 Ku G, Faust T, Lauffer LL, Livingston DJ, Harding MW (1996) Interleukin-1β convert-

ing enzyme inhibition blocks progression of type II collagen-induced arthritis in mice. *Cytokine* 8: 377–386

67 Pavelka K, Kuba V, Rasmussen MJ, Mikkelsen K, Tamasi L, Vitek P, Rozman B (2002) Clinical effects of pralnacasan (PRAL), an orally-active interleukin-1β converting enzyme (ICE) inhibitor, in a 285 patient Phase II trial in rheumatoid arthritis. *Arthritis Rheum* 46: LB02 (Abstract)

68 Conner EM, Fuseler JW, Davis JM, Laroux S, Palombella VJ, Brand S, Wolf RE, Grisham MB (1997) Selective proteasome inhibition attenuates experimental polyarthritis via inhibition of nuclear transcription factor κB (NFκB) activation. *Arthritis Rheum* 40: S322 (Abstract)

69 Sagot Y, Sattonet-Roche P, Bhagwatt SS, Grimshaw CE, Dreano M, Plater-Zyberk C (2001) Two IKK2 inhibitors are orally active small molecules decreasing severity of collagen-induced arthritis in DBA/1 mice. *Arthritis Rheum* 44: S368 (Abstract)

70 Bhagwat SS, Bennett BL, Satoh Y, O'Leary EC, Leisten J, Firestein GS, Boyle DL, Dreano M, Anderson DW, Grimshaw CE (2001) The small molecule IKK2 inhibitor SPC 839 is efficacious in an animal model of arthritis. *Arthritis Rheum* 44: S213 (Abstract)

71 Badger AM, Griswold DE, Kapadia R, Blake S, Swift BA, Hoffman SJ, Stroup GB, Webb E, Rieman DJ, Gowen M et al (2000) Disease-modifying activity of SB 242235, a selective inhibitor of p38 mitogen-activated protein kinase, in rat adjuvant-induced arthritis. *Arthritis Rheum* 43: 175–183

72 Han Z, Boyle DL, Chang L, Bennett B, Karin M, Yang L, Manning AM, Firestein GS (2001) c-Jun N-terminal kinase is required for metalloproteinase expression and joint destruction in inflammatory arthritis. *J Clin Invest* 108: 73–81

Is IL-6 a therapeutic target?

Norihiro Nishimoto and Tadamitsu Kishimoto

Laboratory of Immune Regulation, Graduate School of Frontier Biosciences, Osaka University, 1-3 Yamada-oka, Suita-City, Osaka 565-0871, Japan

Introduction

Rheumatoid arthritis (RA) is a chronic inflammatory disease characterized by persistent synovitis and progressive destruction of cartilage and bone in the multiple joints [1]. The pathological features of the affected joints consist of hyperplasia of inflamed synovium and erosion of cartilage and bone where the pannus tissue invades. There are numerous immunocompetent cells such as T cells, macrophages, dendritic cells and plasma cells infiltrating in the synovium [2]. Many small blood vessels are also observed and thought to be necessary to oxygenate the tissue [3]. In addition to local inflammation of the joints, patients are associated with systemic inflammatory manifestations such as fever, fatigue and anemia, and laboratory findings such as elevated erythrocyte sedimentation rate (ESR) and C-reactive protein (CRP), hyper-γ-globulinemia, and thrombocytosis.

The etiological causes of RA are not fully understood. Autoimmune mechanisms have been thought to exist in the pathogenesis of the disease [4]. Autoreactive T cells may recognize self-antigens due to shortcomings in the correct distinction of non-self-antigens from self-antigen or disruption of self-tolerance. Autoreactive T cells induce the proliferation and differentiation of immunocompetent cells and stimulate B cells to produce autoreactive antibodies such as rheumatoid factors. They also activate the tissues expressing self-antigens. In these processes, various mediators including both pro-inflammatory and anti-inflammatory cytokines are released. Pro-inflammatory cytokines such as tumor necrosis factor-α (TNFα), interleukin-1 (IL-1), IL-6, IL-8, granulocyte-macrophage colony-stimulating factor (GM-CSF) are abundant compared with anti-inflammatory cytokines or cytokine inhibitors such as IL-4, IL-10, IL-1 receptor antagonist (IL-1Ra), and soluble form of TNF receptor (TNF-R). The chronically deviated cytokine milieu is responsible for the appearance of various clinical abnormalities observed in RA [5]. It also generates the pathological changes such as immunocompetent cell activation and infiltration to synovium, synovial cell proliferation, osteoclast activation and angiogenesis in the inflamed joint. Therefore, the pro-inflammatory cytokines are potential therapeutic targets to block their actions for the treatment of RA.

Cytokines and Joint Injury, edited by Wim B. van den Berg and Pierre Miossec

Biological therapeutic agents targeting TNFα and IL-1 have been used successfully to treat patients with RA [6–9]. However, about 30% of the patients fail to respond to these agents. Therefore, we still need another tactic for the treatment of this refractory disease. IL-6 is a target molecule distinct from TNFα or IL-1 whose actions need to be blocked for the treatment of RA.

Rationale for anti-IL-6 therapy

Pleiotropic function of IL-6 and abnormalities in RA

IL-6 has been shown to be produced by various types of cells such as T cells, B cells, monocytes, fibroblasts, endothelial cells and several kinds of tumor cells. In the affected joints of patients with RA, IL-6 is produced by T cells, B cells [10] as well as synovial cells [11, 12]. This IL-6 production from synovial cells is augmented by IL-1β and TNFα [13].

IL-6 is a pleiotropic cytokine with a wide range of biological activities on various target cells. Therefore, continuous overproduction of IL-6 may result in the appearance of various clinical symptoms and abnormal laboratory findings observed in RA as shown in Figure 1.

IL-6 is an important regulator of immune response. It was originally identified as B cell differentiation factor that induces activated B cells to produce immunoglobulin [14]. Therefore, overproduction of IL-6 may cause hyper-γ-globulinemia and the increase in autoantibodies, including rheumatoid factors. IL-6 activates T cells through induction of IL-2 receptor expression [15, 16]. Overproduction of IL-6 results in the increase in autoreactive T cells. IL-6 may also regulate the interaction of immunocompetent cells through induction of adhesion molecules [17, 18]. Therefore, IL-6 may enhance autoimmune reaction.

IL-6 is also a pro-inflammatory mediator that promotes acute phase reaction. It induces fever and leukocytosis when it is administered *in vivo* [19, 20]. IL-6, as a hepatocyte stimulating factor, induces production of acute phase proteins such as C-reactive protein (CRP), fibrinogen (Fib), α1-antitrypsin and serum amyloid A (SAA) from hepatocytes, while it simultaneously suppresses production of albumin [21–24]. Increase in CRP, Fib, SAA levels, acceleration of erythrocyte sedimentation rate (ESR) and decrease in serum albumin may be explained by continuous overproduction of IL-6. Overproduction of SAA is responsible for the secondary amyloidosis that is an important complication in oriental patients, because only a therapy that successfully reduces the supply of amyloid fibril protein precursors is followed by substantial regression of amyloid [25].

IL-6, as a megakaryocyte differentiating factor, stimulates production of platelets [26] and over-expression of IL-6 may be responsible for the thrombocytosis found *in vivo* in RA patients [27].

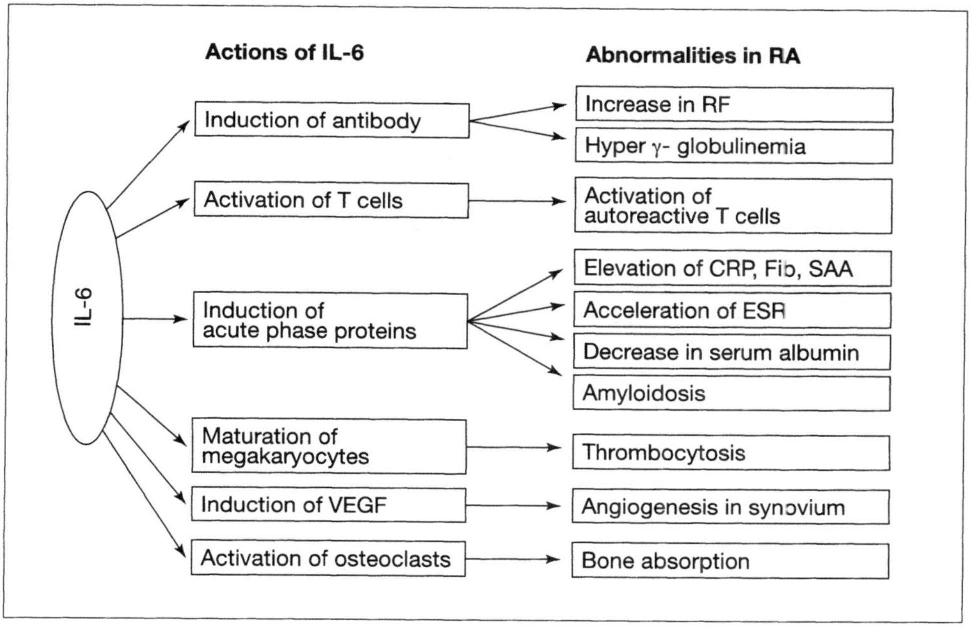

Figure 1
IL-6 actions and abnormalities in RA

IL-6 is a pivotal cytokine that induces vascular endothelial growth factor (VEGF) production synergistically with IL-1β or TNFα [28]. VEGF is a potent inducer of angiogenesis which is necessary to oxygenate the hyperplastic synovial tissues in the affected joints [29, 30].

In the bone metabolism, IL-6, in the presence of soluble IL-6 receptor (sIL-6R), induces osteoclast precursor cells to differentiate to authentic osteoclasts characterized by the presence of tartrate-resistant acid phosphatase (TRAP) activity, calcitonin receptors and pit formation in dentine slices [31]. Therefore, IL-6 may be responsible for bone and cartilage destruction and osteoporosis associated with RA.

In fact, elevation of IL-6 levels has been observed in both serum and synovial fluid in the patients with RA, and there have been correlations between serum IL-6 levels and clinical and laboratory indices of RA [10, 32–34]. These lead to the idea that interference with IL-6 actions could be a therapeutic approach for RA.

IL-6 receptor system

IL-6 signal transduction to exert the pleiotropic actions is mediated by a unique IL-6 receptor (IL-6R) system. The IL-6R system consists of two functional membrane proteins: an 80 kDa ligand-binding chain (IL-6R, IL-6R α-chain, CD126) [35] and a 130 kDa non-ligand-binding but signal-transducing chain (gp130, IL-6R β-chain, CD130) [36]. IL-6 binds to cell surface IL-6R, and the IL-6/IL-6R complex induces the homodimerization of gp130 molecule, followed by the formation of a high-affinity functional receptor complex of IL-6, IL-6R, and gp130 [37].

A soluble form of IL-6R (sIL-6R), lacking the intracytoplasmic portion of IL-6R, is also capable of transducing the signal as a ligand-binding receptor into the cells that express insufficient amount of cell surface IL-6R [36, 37]. Thus, IL-6 binding to either a membrane-anchored or soluble form of IL-6R can mediate IL-6 signal into cells as long as they express gp130. The sIL-6R, observed both in the serum and in the synovial fluids from RA patients, appears to play pathological roles in RA [38, 39].

Based on an understanding of this unique receptor system of IL-6, a strategy has been established to inhibit IL-6 functions. Several therapeutic approaches are proposed to block the signal of IL-6:

1. Inhibition of IL-6 production
2. Neutralization of IL-6
3. Blockade of IL-6 binding onto IL-6R
4. Blockade of IL-6/IL-6R complex binding to gp130
5. Suppression of IL-6R and gp130 expression.
6. Blockade of the intracytoplasmic signal from gp130.

Wendling et al. utilized mouse monoclonal anti-IL-6 antibody to neutralize IL-6 and found improvement in the disease activity of RA patients [40]. Therefore, interference with IL-6 action may constitute a new therapeutic strategy for RA. We have attempted to block IL-6 actions using a monoclonal antibody specific to IL-6R. The antibody recognizes the IL-6 binding site of both membranous and soluble forms of IL-6R and, therefore, is capable of inhibiting IL-6 signals mediated by both membranous and soluble forms of IL-6R (Fig. 2).

Humanized anti-human IL-6R antibody, MRA

Mouse antibody is highly antigenic in humans and causes emergence of human neutralizing antibodies against mouse antibody, so that the therapeutic value of mouse antibody in human chronic diseases remains limited. In order to administer the monoclonal antibody to patients in repeated doses, the antibody must be engineered to look like human antibody. Sato et al. achieved humanization of the anti-human IL-6R antibody by grafting the complementarity determining regions from mouse anti-IL-6R antibody to human IgG1 [41]. This humanized anti-IL-6R monoclonal antibody is now called MRA. It is equivalent to both original mouse and chimeric

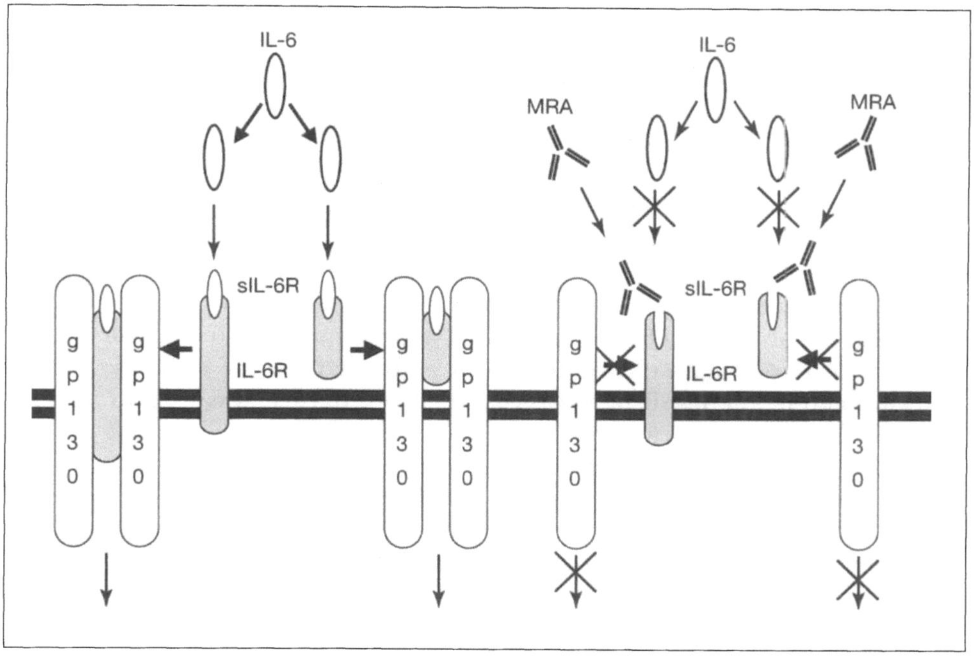

Figure 2
IL-6R system
IL-6 binds to both membrane and soluble forms of IL-6R followed by homodimerization of gp130 which mediates the signal into the cells. MRA blocks the binding of IL-6 to both membrane and soluble forms of IL-6R.

anti-IL-6R antibody in terms of antigen binding and blocking IL-6 actions. Furthermore, it is less immunogenic for human and shows longer half-life when administered *in vivo* in humans. We attempted to block IL-6 signal transduction by utilizing this humanized anti-IL-6R monoclonal antibody, MRA, for the treatment of RA.

Effect of anti-IL-6R antibody in animal models

Collagen induced arthritis (CIA) as a model for RA

Animal models for arthritis have been used to study the pathological significance of IL-6 as well as the therapeutic effects of agents in blocking IL-6 actions. Collagen induced arthritis (CIA) in DBA/1J mice is an experimental model widely used for studying disease processes in, and potential therapies for, human RA. CIA develops in rodents after immunization with bovine type II collagen with adjuvant. Excess production of IL-6 has been observed in sera from animals with CIA, suggesting that IL-6 is involved in the pathogenesis of CIA [42, 43].

Alonzi et al. showed the IL-6 gene knockout mice obtained a constitution completely resistant to CIA [44]. The finding suggests that IL-6 is essential for the development of CIA. Later on, Sasai et al. reported that IL-6 gene knockout delayed the onset of CIA and also reduced its severity [45]. Although the susceptibility of the mice to CIA differed between the two reports, both studies indicate that IL-6 plays an important role in CIA. However, inactivation of IL-6 gene affected the total immune response in mice, and therefore makes it difficult to identify the actual mechanism for decreasing susceptibility to and/or the severity of CIA.

We have analyzed the effect of anti-IL-6R antibody on CIA to establish whether blockade of IL-6 signal utilizing anti-IL-6R antibody could be a therapy for RA [46]. Rat anti-mouse IL-6R monoclonal antibody, MR16-1, was used to treat the DBA/1J mice immunized twice at a three-week interval with bovine type II collagen emulsified with complete adjuvant. In the mice, excess production of IL-6 in sera was observed within 24 hours after the first immunization but rapidly decreased (Fig. 3A). Serum IL-6 increased again 14 days after the first immunization in conjunction with the onset of arthritis, as previously reported [42, 43]. When MR16-1 was administered immediately after the first immunization, it inhibited the development of arthritis while isotype-matched control antibody KH-5 could not (Fig. 3B).

Humanized anti-human IL-6R monoclonal antibody, MRA, is specific to human IL-6R and cannot recognize mouse or rat IL-6R. However, it does cross-reacts with monkey IL-6R. So, the effect of MRA was examined for CIA in cynomolgus monkeys [47]. Similarly to MR16-1 treatment for CIA in DBA/1J mice, MRA treatment decreased the incidence and the severity of CIA in the monkeys. Histological examination also demonstrated the effect of MRA. In a non-treated control, affected joints showed synovial proliferation, pannus formation, infiltration of neutrophils, angiogenesis and destruction of cartilage and bone. On the contrary, in the monkey treated with MRA, these pathological changes were not observed.

These findings indicate that blocking IL-6 actions with anti-IL-6R antibodies could prevent the development of CIA.

Figure 3
Anti-IL-6R antibody treatment for collagen induced arthritis in mice
Collagen induced arthritis (CIA) in DBA/1J mice was used to test the efficacy of anti-IL-6R antibody.
A. Changes in serum IL-6 levels after immunization with bovine type II collagen with adjuvant
B. Rat anti-IL-6R antibody, MR16-1, prevented the development of CIA.
C. Early phase blockade of IL-6 actions is necessary to inhibit the development of CIA. (Modified figures from Takagi et al. [46])

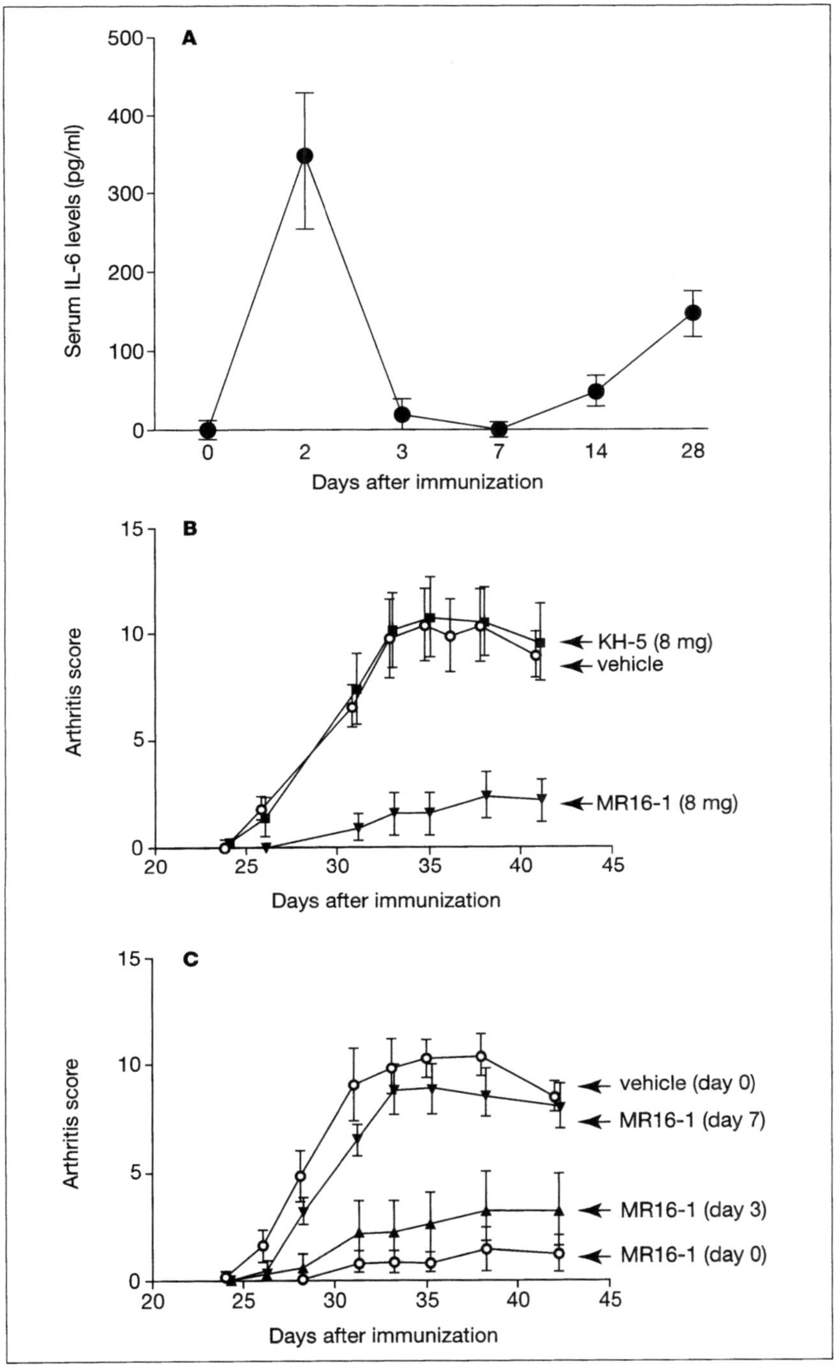
A
Serum IL-6 levels (pg/ml)
500
400
300
200
100
0
0
2
3
7
14
28
Days after immunization
B
Arthritis score
15
10
5
0
20
25
30
35
40
45
KH-5 (8 mg)
vehicle
MR16-1 (8 mg)
Days after immunization
C
Arthritis score
15
10
5
0
20
25
30
35
40
45
vehicle (day 0)
MR16-1 (day 7)
MR16-1 (day 3)
MR16-1 (day 0)
Days after immunization

Mechanisms of action of anti-IL-6R antibody in preventing CIA

What are the mechanisms for anti-IL-6R antibodies to prevent the development of arthritis induced by the immunization with type II collagen? Interestingly, the suppressive effect of MR16-1 for CIA was observed only when MR16-1 was injected within three days after the first immunization of bovine type II collagen, but not when MR16-1 was injected seven days after immunization or later (Fig. 3C). Furthermore, the production of IgG anti-type II collagen antibodies and the proliferation of splenic lymphocytes in response to type II collagen were suppressed in the mice whose CIA was prevented by the MR16-1 treatment [46]. Therefore, anti-IL-6R antibody MR16-1 acts mainly in a very early phase before appearance of arthritis symptoms by suppressing immunization of bovine type II collagen. In contrast, anti-TNFα and anti-IL-1 antibodies have shown to ameliorate the established arthritis in CIA models [48, 49]. This evidence indicates the difference in the mechanisms between the anti-IL-6 therapy and anti-TNFα or anti-IL-1 therapies for CIA.

So far we do not know the pathological significance of the second increase in serum IL-6 levels in this CIA model. IL-6 may play different roles in each phase of the development of CIA.

Next, we need to address a question as to whether or not anti-IL-6R antibody is indeed therapeutically effective in human RA because the arthritis has already been established in the patients who visit a clinic for treatment.

Severe combined immunodeficiency (SCID) mice transplanted with RA joint tissue

Severe combined immunodeficiency (SCID) mice transplanted with joint tissue from human RA can maintain histological features identical to human RA in the transplants [50]. Since the transplanted tissue is from human, therapeutic agents such as monoclonal antibodies, which react specifically to human antigens, can be examined for their efficacy in this mouse model. The model was used to examine the therapeutic usefulness of humanized anti-human IL-6R monoclonal antibody, MRA, for RA. Intraperitoneal administration of MRA (100 μg/body, weekly) for a period of one month significantly decreased the volume of implanted tissue. The treatment also reduced the number of infiltrating inflammatory cells, matrix metalloproteinase (MMP)-9 positive cells, and TRAP-positive cells in the implanted synovium. Therefore, IL-6 blockade could eliminate the inflammation of RA synovium and prevent joint destruction by decreasing MMP production and the number of osteoclasts. The findings suggest that MRA may be effective for the treatment of RA patients even though their arthritis has already been established.

In patients with RA, the repetition of amelioration and exacerbation is observed and suggests that the antigen stimulation for autoimmune response continues intermittently *in vivo* in patients, where IL-6 may regulate the immune reaction. Therefore, MRA can ameliorate the activity of arthritis by suppressing such autoimmune response. The SCID mice transferred with CD45Rb+CD4+ T cells from BALB/c mice develop Type 1 T helper cell-mediated murine colitis like Crohn's disease. IL-6 presumably contributes to the increased resistance of mucosal T cells against apoptosis. MR16-1 suppressed established experimental colitis by inducing apoptosis of lamina propria T cells [18, 51]. Therefore, blockade of IL-6 may induce apoptosis of autoreactive T cells in RA, too. In addition, MR16-1 treatment suppressed serum TNFα and IL-1β levels in the mice [18] suggesting the existence of regulatory mechanism for inflammatory cytokine production by IL-6. Blockade of IL-6 may downregulate the production of these cytokines.

Clinical studies with anti-IL-6 therapy

Pharmacokinetic of MRA

In the Phase I/II study of MRA for RA in Japan, pharmacokinetic, safety and efficacy was assessed [52]. In an open-labeled multi-dose trial, 15 RA patients received an i.v. infusion with three doses; 2, 4 or 8 mg of MRA/kg body weight every two weeks. The serum concentration of MRA decreased in a nonlinear manner with the dose range from 2–8 mg/kg (Fig. 4). In the 2 mg/kg group, serum MRA was detectable in four out of five patients, while in the other one, MRA was not detectable in the trough level. In the 4 mg/kg group, two patients could not maintain the serum MRA levels in the trough level. In the 8 mg/kg group, MRA was always detectable during the treatment period; and this trough level of MRA appeared important to block IL-6 actions assessed by serum CRP levels. The T1/2 increased as the dose increased and also as repeatedly administered. Multiple infusions also prolonged the T1/2 and that after the third dose in the 8 mg/kg group reached 241.8 ± 71.4 hours. The mean area-under-the-curve (AUC) increased as the dose increased and the values (mean ± SD) were 3.44 ± 8.22, 4.66 ± 2.18 and 10.66 ± 4.07 mg*hour/ml in the 2-, 4- and 8 mg/kg groups, respectively [52].

Clinical benefits

Wendling et al. reported clinical benefits of IL-6 blockade for the treatment of RA. They administered mouse anti-human IL-6 monoclonal antibody every day for total of two weeks to five patients with RA and found improvement in symptoms and laboratory findings [40]. Although the antibody that they used in the trial was derived from mouse, and therefore it was difficult to continue the treatment, they suggested that blockade of IL-6 actions can be a therapy for RA.

We conducted an early pilot study of humanized anti-IL-6R antibody, MRA, for patients with RA in 1995 to1997. We treated 11 patients with refractory RA with

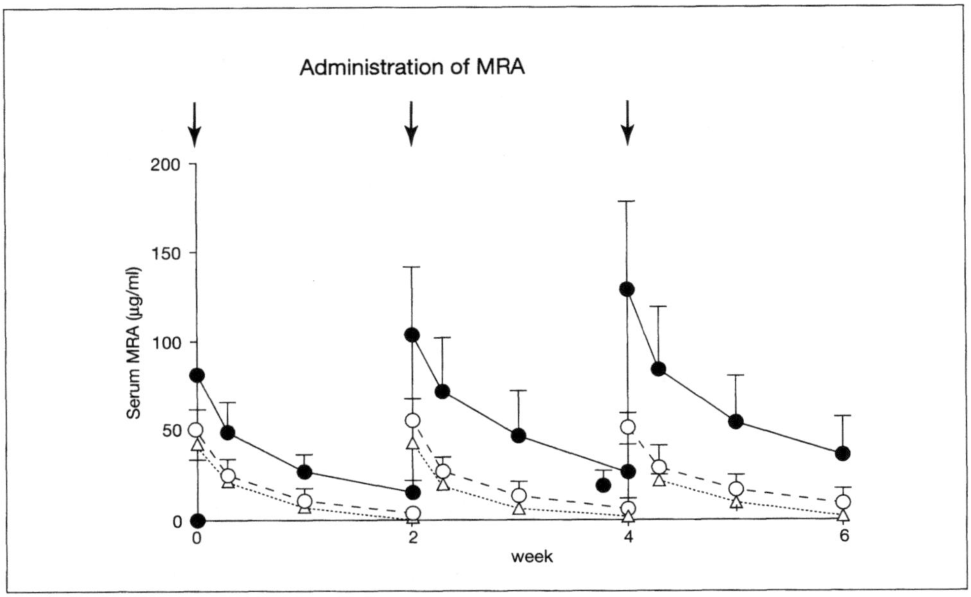

Figure 4
Serum MRA concentrations after i.v. administration of MRA
Serum levels of MRA were monitored after the administration of MRA. Note that 8 mg/kg every two weeks resulted in the continuous increase in the trough levels of MRA. 8 mg/kg: closed circle; 4 mg/kg: open circle; 2 mg/kg: open triangle.
(Modified figure from Nishimoto et al. [52])

MRA after obtaining the permission of the ethical committee of Osaka University [53]. All the patients had active disease resistant to conventional therapy using various disease modifying anti-rheumatic drugs (DMARDs) including methotrexate and low-dose corticosteroids. The patients were treated with MRA in a regimen to maintain their serum MRA levels (5~10 µg/ml in the trough levels) because the concentration of MRA had been proven to inhibit the proliferation of IL-6-dependent myeloma cells *in vitro*. Such concentrations of serum MRA were achieved in the patients by the administration of 50 mg of MRA twice a week or 100 mg of MRA once a week. The treatment rapidly improved CRP and ESR levels in all the patients followed by the reduction of swollen joint counts and pain/tender joint counts. The therapy was relatively well tolerated but appearance of anti-idiotypic antibody was observed in one case, who was therefore withdrawn. Two additional patients were withdrawn: one because of the occurrence of an attack of angina, the relation of which to MRA administration could not be classified, and the other for personal reasons. A decrease in neutrophil counts was observed in most of the cases on the day after MRA administration but mostly within the normal range. In the eight patients who received MRA treatment for more than eight weeks, both clinical and laboratory findings improved during treatment. At two months, clinical response was 88% assessed by ACR20 criteria, and 50% by ACR50, although this is an open trial. The response to MRA did not change over six months. These results suggested that MRA is relatively safe and useful in the treatment of RA. On the basis of our data, Phase I clinical trials for RA patients in the United Kingdom [54] and for healthy individuals in Japan were started in 1997. Later, Phase I/II studies in Japan were started in 1999 [52].

In the Phase I clinical trials in the United Kingdom, single doses of 0.1, 1, 5 and 10 mg of MRA per kg or placebo was administered to patients and the efficacy was assessed two weeks after the administration of study drug according to ACR20 criteria. In the 5 mg/kg group, 55.6% of the patients achieved ACR20, while none of the placebo group achieved ACR20. In the other dosages, however, no statistically significant difference was observed in comparison with placebo group. CRP was normalized at two weeks in 5 mg/kg and 10 mg/kg groups. In addition, no serious adverse reaction related to MRA treatment was observed. The study suggested the clinical benefit of MRA for the treatment of RA [54].

In the Phase I/II study of MRA for RA in Japan, 15 RA patients received an i.v. infusion with three doses of 2, 4 or 8 mg of MRA per kg body weight every two weeks [52]. Each cohort consists of five patients and the safety was evaluated after six weeks of treatment. Once the safety and tolerance was satisfactory, escalation to the next dose was examined with a group of newly recruited patients. The patients were allowed to stay on MRA treatment for 24 weeks and further assessed for the safety and efficacy. Extension study was also prepared for further assessment of safety and efficacy. A total of 70 adverse reactions were reported but none of them were serious. Interestingly, increase in total cholesterol was detected as an MRA-related reaction in 10 out of 15 patients. No antinuclear antibody or anti-DNA antibody newly appeared and no anti-MRA antibody was detected without the use of immunosuppressive agents such as methotrexate. Figure 5A shows the decrease in the objective inflammatory markers by MRA treatment. The decrease in CRP and SAA was rapid and normalized within six weeks in most of the patients. However, three patients did not become seronegative for CRP. In these three patients, MRA was undetectable in the trough level. When the injection dose was increased to 8 mg/kg bi-weekly, serum MRA levels were maintained and they also became negative for CRP. Similarly to CRP, SAA levels were also normalized. Tender joint counts and swollen joint counts also gradually decreased by MRA treatment (Fig. 5B). Response rate in the ACR criteria is shown in Figure 5C. At six weeks, 60% of the patients achieved ACR20, and after 24 weeks about 90% achieved ACR20 and 40% achieved ACR50. Although the optimal dosing schedule was not defined in this small study, maintenance of serum MRA concentration seemed important to achieve efficacy.

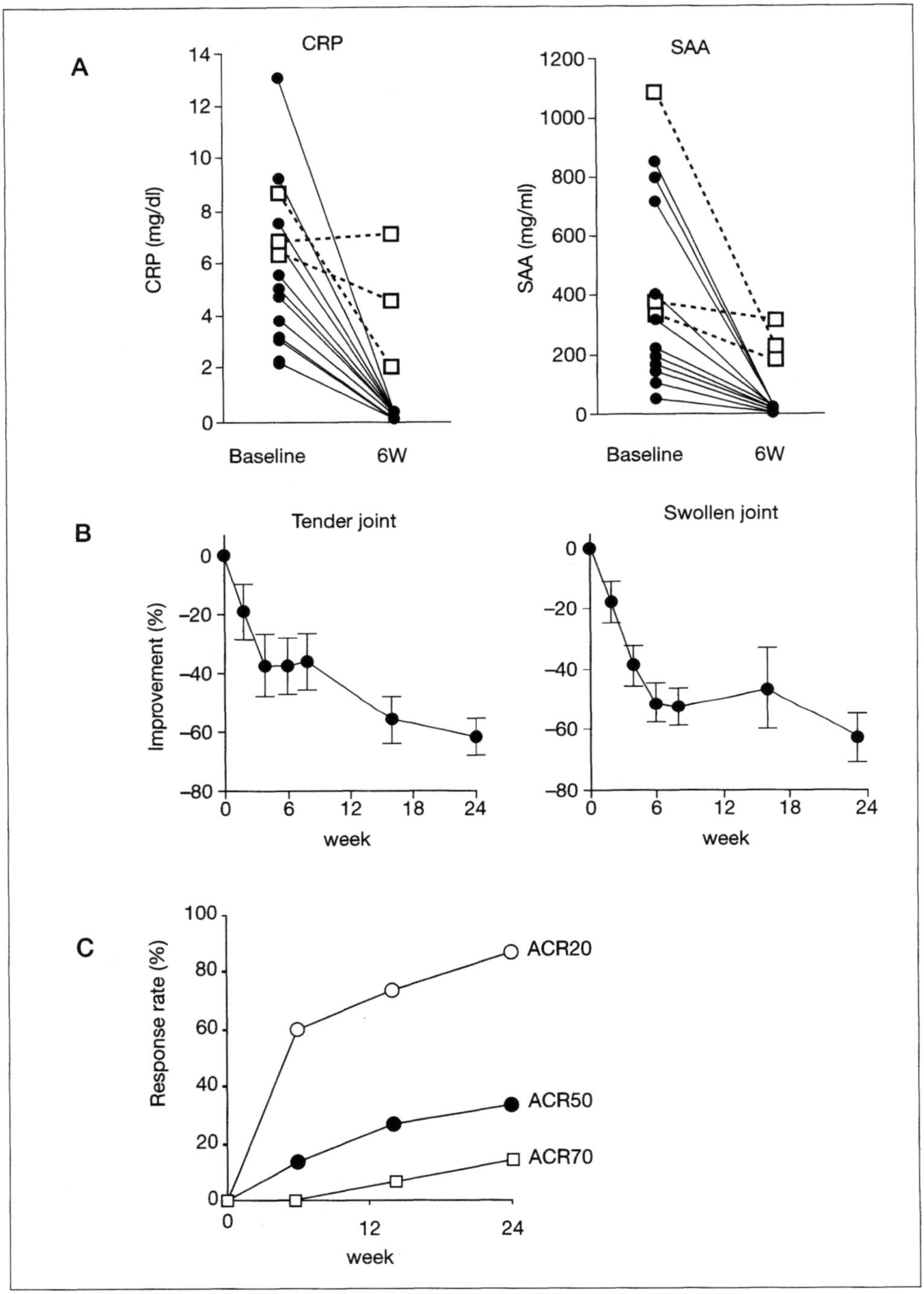
A
CRP
CRP (mg/dl)
Baseline
6W
SAA
SAA (mg/ml)
B
Tender joint
Swollen joint
Improvement (%)
week
C
Response rate (%)
ACR20
ACR50
ACR70
week

To confirm the result of these early studies, double-blind randomized placebo-controlled Phase II studies for RA were conducted both in Japan and in Europe from 2001 to 2002. In the Japanese Phase II study, 164 patients with refractory RA were randomized to receive i.v. treatment with placebo, 4 or 8 mg of MRA per kg body weight every four weeks for a total of three months and the clinical responses were measured using the ACR criteria [55]. At three months, 78% of the patients in the 8 mg group and 57% in the 4 mg group met the ACR20 criteria while only 11% in the placebo group achieved it. In the European Phase II study, 359 patients with active RA with an inadequate response to methotrexate were randomized to receive placebo, 2, 4 or 8 mg of MRA per kg every four weeks either as monotherapy or in combination with methotrexate [56]. Eight mg of MRA per kg most effectively decreased the disease activity. Efficacy was also seen with 2 and 4 mg/kg particularly in combination with methotrexate. Now Phase III studies are being conducted in Japan.

Conclusion

Clinical trials of MRA for RA have shown the clinical benefit of MRA for improving signs and symptoms of RA. However, it remains to be proven whether or not MRA can prevent joint destruction in patients with RA. It is also required to know the long term safety and efficacy of MRA. Since IL-6 is essential for the host defense, the clinical trials need to examine the incidence of serious infections and malignant diseases. An increase in total cholesterol was observed in the clinical trials. Since HDL cholesterol was also increased in association with LDL cholesterols, it does not always mean the increase in the risk of cardiovascular diseases. However, long term follow up of patients will be required to define the safety of MRA treatment.

The mechanism of the action by which MRA exerts its therapeutic effect is still not fully understood. Recently, Nakahara et al. reported that blockade of IL-6 inhibits the production of VEGF from synovial fibroblasts and also decreases serum VEGF levels in RA patients. It might be one of the mechanisms how MRA shows the efficacy because VEGF is a potent inducer of angiogenesis essential for the syn-

Figure 5
Efficacy of MRA for RA
A. Decrease in acute phase proteins by MRA treatment. Closed circles show the patients whose trough MRA levels were detectable. Open square show those whose trough MRA levels were undetectable.
B. Changes in tender and swollen joint counts by MRA treatment.
C. Changes in response rate to MRA treatment assessed by ACR criteria.
(Figure modified after Nishimoto et al. [52])

ovial hyperplasia. To establish the anti-IL-6 therapy for RA, it is necessary to elucidate the exact mechanisms for MRA to show the therapeutic efficacy.

IL-6 is also involved in various kinds of diseases such as Castleman's disease, Crohn's disease, systemic-onset juvenile idiopathic arthritis, multiple myeloma and systemic lupus erythematosus in addition to RA. Those are the possible targets of MRA in the future.

References

1 Harris ED Jr (1990) Rheumatoid arthritis: pathophysiology and implications for therapy. *N Engl J Med* 322: 1277–1289

2 Cush JJ, Lipsky PE (1988) Phenotypic analysis of synovial tissue and peripheral blood lymphocytes isolated from patients with rheumatoid arthritis. *Arthritis Rheum* 31: 1230–1238

3 Colville-Nash PR, Scott DL (1992) Angiogenesis and rheumatoid arthritis: pathogenic and therapeutic implications. *Ann Rheum Dis* 51: 919–925

4 Todd JA, Acha-Orbea H, Bell JI, Chao N, Fronek Z, Jacob CO, McDermott M, Sinha AA, Timmerman L, Steinman L, McDevitt HO (1988) A molecular basis for MHC class II associated autoimmunity. *Science* 240: 1003–1009

5 Feldmann M, Elliott MJ, Woody JN, Maini RN (1997) Anti-tumor necrosis factor-α therapy of rheumatoid arthritis. *Adv Immunol* 64: 283–350

6 Maini R, St Clair EW, Breedveld F, Furst D, Kalden J, Weisman M, Smolen J, Emery P, Harriman G, Feldmann M, Lipsky P (1999) Infliximab (chimeric anti-tumour necrosis factor α monoclonal antibody) *versus* placebo in rheumatoid arthritis patients receiving concomitant methotrexate: a randomised Phase III trial. *Lancet* 354: 1932–1939

7 Moreland LW, Schiff MH, Baumgartner SW, Tindall EA, Fleischmann RM, Bulpitt KJ, Weaver AL, Keystone EC, Furst DE, Mease PJ et al (1999) Etanercept therapy in rheumatoid arthritis. *Ann Intern Med* 130: 478–486

8 Lipsky PE, van der Heijde DM, St Clair EW, Furst DE, Breedveld FC, Kalden JR, Smolen JS, Weisman M, Emery P, Feldmann M et al (2000) Infliximab and methotrexate in the treatment of rheumatoid arthritis. *N Engl J Med* 343: 1594–1602

9 Fleischmann RM, Schechtman J, Bennett R, Handel ML, Burmester GR, Tesser J, Modafferi D, Poulakos J, Sun G (2003) Anakinra, a recombinant human interleukin-1 receptor antagonist (r-metHuIL-1ra), in patients with rheumatoid arthritis: A large, international, multicenter, placebo-controlled trial. *Arthritis Rheum* 48: 927–934

10 Hirano T, Matsuda T, Turner M, Miyasaka N, Buchan G, Tang B, Sato K, Shimizu M, Maini R, Feldmann M et al (1988) Excessive production of interleukin 6/B cell stimulatory factor-2 in rheumatoid arthritis. *Eur J Immunol* 18: 1797–1801

11 Miyasaka N, Sato K, Hashimoto J, Kohsaka H, Yamamoto K, Goto M, Inoue K, Matsuda T, Hirano T, Kishimoto T et al (1989) Constitutive production of interleukin 6/B

cell stimulatory factor-2 from inflammatory synovium. *Clin Immunol Immunopathol* 52: 238–247
12 Guerne P-A, Zuraw BL, Vaughan JH, Carson DA, Lotz M (1988) Synovium as a source of interleukin 6 *in vitro*. Contribution to local and systemic manifestations of arthritis. *J Clin Invest* 83: 585–592
13 Harigai M, Hara M, Kitani A, Norioka K, Hirose T, Hirose W, Suzuki K, Kawakami M, Masuda K, Shinmei M et al (1991) Interleukin 1 and tumor necrosis factor-alpha synergistically increase the production of interleukin 6 in human synovial fibroblast. *J Clin Lab Immunol* 34: 107–113
14 Hirano T, Yasukawa K, Harada H, Taga T, Watanabe Y, Matsuda T, Kashiwamura S, Nakajima K, Koyama K, Iwamatsu A et al (1986) Complementary DNA for a novel human interleukin (BSF-2) that induces B lymphocytes to produce immunoglobulin. *Nature* 324: 73–76
15 Noma T, Mizuta T, Rosen A, Hirano T, Kishimoto T, Honjo T (1987) Enhancement of the interleukin-2 receptor expression on T cells by multiple B-lymphotropic lymphokines. *Immunol Lett* 15: 249–253
16 Okada M, Kitahara M, Kishimoto S, Matsuda T, Hirano T, Kishimoto T (1988) IL-6/BSF-2 functions as killer helper factor in the *in vitro* induction of cytotoxic T cells. *J Immunol* 141: 1543–1549
17 Lindsley HB, Smith DD, Cohick CB, Koch AE, Davis LS (1993) Proinflammatory cytokines enhance human synoviocyte expression of functional intercellular adhesion molecule-1 (ICAM-1). *Clin Immunol Immunopathol* 68: 311–320
18 Yamamoto M, Yoshizaki K, Kishimoto T, Ito H (2000) IL-6 is required for the development of Th1 cell-mediated murine colitis. *J Immunol* 164: 4878–4882
19 Ulich TR, del Castillo J, Guo KZ (1989) *In vivo* hematologic effects of recombinant interleukin-6 on hematopoiesis and circulating numbers of RBCs and WBCs. *Blood* 73: 108–110
20 van Gameren MM, Willemse PH, Mulder NH, Limburg PC, Groen HJ, Vellenga E, de Vries EG (1994) Effects of recombinant human interleukin-6 in cancer patients: a Phase I-II study. *Blood* 84: 1434–1441
21 Gauldie J, Richards C, Harnish D, Landsdorp P, Baumann H (1987) Interferon-β2/B cell-stimulatory factor type 2 shares identity with monocyte-derived hepatocyte stimulating factor and regulates the major acute phase protein response in liver cells. *Proc Natl Acad Sci USA* 84: 7251–7255
22 Andus T, Geiger T, Hirano T, Northoff H, Ganter U, Bauer J, Kishimoto T, Heinrich PC (1987) Recombinant human B cell stimulatory factor 2 (BSF-2/INFβ2) regulates β-fibrinogen and albumin mRNA levels in Fao-9 cell. *FEBS Lett* 221: 18–22
23 Castell JV, Gomez-Lechon MJ, David M, Hirano T, Kishimoto T, Heinrich PC (1988) Recombinant human interleukin-6(IL-6/BSF-2/HSF) regulates the synthesis of acute phase proteins in human hepatocytes. *FEBS Lett* 232: 347–350
24 Isshiki H, Akira S, Sugita T, Nishio Y, Hashimoto S, Pawlowski T, Suematsu S, Kishi-

moto T (1991) Reciprocal expression of NF-IL6 and C/EBP in hepatocytes: possible involvement of NF-IL6 in acute phase protein gene expression. *New Biol* 3: 63–70
25 Gillmore JD, Lovat LB, Persey MR, Pepys MB, Hawkins PN (2001) Amyloid load and clinical outcome in AA amyloidosis in relation to circulating concentration of serum amyloid A protein. *Lancet* 358: 4–5
26 Ishibashi T, Kimura H, Uchida T, Kariyone S, Friese P, Burstein SA (1989) Human interleukin 6 is a direct promoter of maturation of megakaryocytes *in vitro*. *Proc Natl Acad Sci USA* 86: 5953–5957
27 Ishibashi T, Kimura H, Shikama Y, Uchida T, Kariyone S, Hirano T, Kishimoto T, Takatsuki F, Akiyama Y (1989) Interleukin-6 is a potent thrombopoietic factor *in vivo* in mice. Blood 74: 1241–1244
28 Nakahara H, Song J, Sugimoto M, Hagihara K, Kishimoto T, Yoshizaki K, Nishimoto N (2003) Anti-interleukin-6 receptor antibody therapy reduces vascular endothelial growth factor production in rheumatoid arthritis. *Arthritis Rheum* 48: 1521–1529
29 Colville-Nash PR, Scott DL (1992) Angiogenesis and rheumatoid arthritis: pathogenic and therapeutic implications. *Ann Rheum Dis* 51: 919–925
30 Folkman J (1995) Angiogenesis in cancer, vascular, rheumatoid, and other disease. *Nat Med* 1: 27–31
31 Tamura T, Udagawa N, Takahashi N, Miyaura C, Tanaka S, Yamada Y, Koishihara Y, Ohsugi Y, Kumaki K, Taga T et al (1993) Soluble interleukin-6 receptor triggers osteoclast formation by interleukin 6. *Proc Natl Acad Sci USA* 90: 11924–11928
32 Houssiau FA, Devogelaer JP, Van Damme J, De Deuxchaisies CN, Van Snick J (1988) Interleukin-6 in synovial fluid and serum of patients with rheumatoid arthritis and other inflammatory arthritis. *Arthritis Rheum* 31: 784–788
33 Sack U, Kinne R, Marx T, Heppt P, Bender S, Emmrich F (1993) Interleukin-6 in synovial fluid is closely associated with chronic synovitis in rheumatoid arthritis. *Rheumatol Int* 13: 45–51
34 Madhok R, Crilly A, Watson J, Capell HA. Serum interleukin 6 levels in rheumatoid arthritis: correlation with clinical and laboratory indices of disease activity. *Ann Rheum Dis* (1993) 52: 232–234
35 Yamasaki K, Taga T, Hirata Y, Yawata H, Kawanishi Y, Seed B, Taniguchi T, Hirano T, Kishimoto T (1988) Cloning and expression of the human interleukin-6 (BSF-2/INF β2) receptor. *Science* 241: 825–828
36 Taga T, Hibi M, Hirata Y, Yamasaki, K, Yasukawa K, Matsuda T, Hirano T, Kishimoto T (1989) Interleukin-6 triggers the association of its receptor with a possible signal transducer, gp130. *Cell* 58: 573–581
37 Hibi M, Murakami M, Saito M, Hirano T, Taga T, Kishimoto T (1990) Molecular cloning and expression of an IL-6 signal transducer, gp130. *Cell* 63: 1149–1157
38 Uson J, Balsa A, Pascual-Salcedo D, Cabezas JA, Gonzalez-Tarrio JM, Martin-Mola E, Fontan G (1997) Soluble interleukin 6 (IL-6) receptor and IL-6 levels in serum and synovial fluid of patients with different arthropathies. *J Reumatol* 24: 2069–2075
39 Desgeorges A, Gabay C, Silacci P, Novick D, Roux-Lombard P, Grau G, Dayer JM, Vis-

cher T, Guerne PA (1997) Concentration and origins of soluble interleukin 6 receptor-a in serum and synovial fluid. *J Reumatol* 24: 1510–1516
40 Wendling D, Racadot E, Wijenes J (1993) Treatment of severe rheumatoid arthritis by anti-interleukin 6 monoclonal antibody. *J Rheumatol* 20: 259–262
41 Sato K, Tsuchiya M, Saldanha J, Koishihara Y, Ohsugi Y, Kishimoto T, Bendig MM (1993) Reshaping a human antibody to inhibit the interleukin 6-dependent tumor cell growth. *Cancer Res* 53: 851–856
42 Takai Y, Seki N, Senoh H, Yokota T, Lee F, Hamaoka T, Fujiwara H (1989) Enhanced production of interleukin-6 in mice with type II collagen-induced arthritis. *Arthritis Rheum* 32: 594–600
43 Sugita T, Furukawa O, Ueno M, Murakami T, Takata I, Tosa T (1993) Enhanced expression of interleukin 6 in rat and murine arthritis models. *Int J Immunopharmacol* 15: 469–476
44 Alonzi T, Fattori E, Lazzaro D, Costa P, Probert L, Kollias G, De Benedetti F, Poli V, Ciliberto G (1998) Interleukin-6 is required for the development of collagen-induced arthritis. *J Exp Med* 187: 461–468
45 Sasai M, Saeki Y, Ohshima S, Nishioka K, Mima T, Tanaka T, Katada Y, Yoshizaki K, Suemura M, Kishimoto T (1999) Delayed onset and reduced severity of collagen-induced arthritis in interleukin-6-deficient mice. *Arthritis Rheum* 42: 1635–1643
46 Takagi N, Mihara M, Moriya Y, Nishimoto N, Yoshizaki K, Kishimoto T, Takeda Y, Ohsugi Y (1998) Blockage of Interleukin-6 receptor ameliorates joint disease in murine collagen-induced arthritis. *Arthritis Rheum* 41: 2117–2121
47 Mihara M, Kotoh M, Nishimoto N, Oda Y, Kumagai E, Takagi N, Tsunemi K, Ohsugi Y, Kishimoto T, Yoshizaki K, Takeda Y (2001) Humanized antibody to human interleukin-6 receptor inhibits the development of collagen arthritis in cynomolgus monkeys. *Clin Immunol* 98: 319–326
48 Williams RO, Feldmann M, Maini RN (1992) Anti-tumor necrosis factor ameliorates joint disease in murine collagen-induced arthritis. *Proc Natl Acad Sci USA* 89: 9784–9788
49 van den Berg WB, Joosten LA, Helsen M, van de Loo FA (1994) Amelioration of established murine collagen-induced arthritis with anti-IL-1 treatment. Clin Exp Immunol 95: 237–243
50 Matsuno H, Sawai T, Nezuka T, Uzuki M, Tsuji H, Nishimoto N, Yoshizaki K (1998) Treatment of rheumatoid synovitis with anti-reshaping human interleukin-6 receptor monoclonal antibody. *Arthritis Rheum* 41: 2014–2021
51 Atreya R, Mudter J, Finotto S, Mullberg J, Jostock T, Wirtz S, Schutz M, Bartsch B, Holtmann M, Becker C et al (2000) Blockade of interleukin 6 trans signaling suppresses T-cell resistance against apoptosis in chronic intestinal inflammation: evidence in Crohn disease and experimental colitis *in vivo*. *Nat Med* 6: 583–588
52 Nishimoto N, Yoshizaki K, Maeda K, Kuritani T, Deguchi H, Sato B, Imai N, Suemura M, Kakehi T, Takagi N et al (2003) Toxicity, pharmacokinetics, and dose finding study

of repetitive treatment with humanized anti-interleukin 6 receptor antibody, MRA, in rheumatoid arthritis -Phase I/II clinical study. *J Rheumatol* 30: 1426–1435
53 Nishimoto N, Kishimoto T, Yoshizaki K. Anti-interleukin 6 antibody treatment in rheumatic disease. *Ann Rheum Dis* (2000) 59: i21–i27
54 Choy EH, Isenberg DA, Garrood T, Farrow S, Ioannou Y, Bird H, Cheung N, Williams B, Hazleman B, Price R et al (2002) Therapeutic benefit after blocking interleukin-6 activity in rheumatoid arthritis with an anti-interleukin-6 receptor monoclonal antibody. *Arthritis Rheum* 46: 3143–3150
55 Nishimoto N, Yoshizaki K, Miyasaka N, Kazuhiko Yamamoto, Kawai S, Takeuchi T, Hashimoto J, Azuma J, Kishimoto T (2002) A multi-center, randomized, double-blind, placebo-controlled trial of humanized anti-interleukin 6 (IL-6) receptor monoclonal antibody (MRA) in rheumatoid arthritis (RA). *Arthritis Rheum* 46 (Suppl) S559: 1499
56 Maini RN, Taylor PC, Pavelka K, Emery P, Szechinski J, Balint G, Broll J, CHARISMA study group (2003) Efficacy of IL-6 receptor antagonist MRA in rheumatoid arthritis patients with an incomplete response to methotrexate (CHARISMA). *Arthritis Rheum* 48 (Suppl) S652: 1704

The clinical use of anti-cytokine therapy in rheumatic diseases

Ernest Choy

Department of Rheumatology, GKT School of Medicine, 2nd Floor, Weston Education Centre King's College Hospital, Cutcombe Road, London SE5 9PJ, UK

Introduction

Chronic inflammatory arthropathies such as rheumatoid arthritis (RA) and psoriatic arthritis are common. They are a major cause of disability globally. Although the precise aetiopathogenesis of these inflammatory arthropathies have not been elucidated, inflammation and damage are common features. Active joint inflammation results in pain, stiffness and dysfunction. Chronic synovitis leads to cartilage and bone damage through mechanisms that have been discussed in details already in the previous chapters of this book. Cytokines, mediators of inflammation, are found in abundance in the synovial joints of patients with active inflammatory arthropathies. They are produced by activated lymphocytes, monocytes and fibroblasts. Many cytokines such as interleukin (IL)-1, IL-6 and tumor necrosis factor α (TNFα) are also found in the peripheral blood of patients with inflammatory arthropathies and are responsible for some of the systemic features such as fatigue, fever and weight loss. Specific cytokine inhibitors are now available as treatment for chronic inflammatory arthropathies. Although expensive, they are cost effective treatments that have significantly advanced the management of inflammatory arthropathies.

The therapeutic benefit of cytokine blockade was first demonstrated in RA and has been extended to other chronic inflammatory diseases including Crohn's disease, ankylosing spondylitis and psoriatic arthritis. Worldwide, RA is the most common chronic inflammatory arthropathy. Disease incidence is 30–50/100,000/year while prevalence is 5–20/1,000 [1]. It is a systemic disease characterized by a symmetric, destructive arthritis with a predilection for the joints in the hands and feet, although any synovial joint can be affected. Extra-articular and systemic manifestations are common.

Prognosis of RA is poor; 80% of cases are disabled after 20 years despite treatment by disease modifying anti-rheumatic drugs (DMARDs) [2]. In many RA patients, disability occurs early. Ten percent of patients require home adaptation and/or wheelchairs, while 17% undergo orthopedic surgery after five years of dis-

Cytokines and Joint Injury, edited by Wim B. van den Berg and Pierre Miossec

ease [3]. In the UK, 22% of RA patients lost employment within 12 month of diagnosis [3]. This increases to a third of patients after three years [4]. Moreover, overall premature mortality is increased. Life expectancy is reduced by 8–13 years with a Standardized Mortality Ratio (SMR) of over 1.5. Mortality is highest in those with severe disease where it is comparable to triple vessel coronary artery disease [5]. Mortality also correlates with severity, disability and persistently active disease [6]. Effective treatment by DMARDs reduces mortality [7]. In methotrexate responders, SMR is 1.64 (95% CI 1.11–2.17), compared with 4.11 (95% CI 2.56–5.66) in non-responders and 5.56 (95% CI 3.29–7.83) in those who had methotrexate stopped within the first year.

RA imposes a substantial economic burden on patients, their families and society, which is quantified in terms of direct and indirect costs [8]. Direct costs are those for which actual payments are made, such as treatment and hospitalization costs, while indirect costs are those resulting from loss of resources, mainly productivity. In the UK, direct medical costs of RA averaged approximately £2,600/case/year in 1992 [9]. While, in the United States, the direct medical costs averaged $5,919/case/year in 1995–96 with a further $2,582 in medical costs for co-morbidities [10]. The most consistent and strongest determinant of the costs of RA is the functional status of the patients. Those with poor function experience much higher costs than those with mild impairment [10]. The direct medical and total hospital costs of those RA patients in the worst quartile of function, as measured by the health assessment questionnaire (HAQ), were 2.55 and 6.97 times higher than those in the best quartile. In the UK, indirect cost, in particular work disability in patients under 65 costs £650 M/year [9]. RA disability costs community and personal social services over £160 M/year. Treatment can reduce the number of days off work due to sickness, therefore is cost effectiveness.

Current treatment strategy

In RA, the treatment objectives are to reduce mortality, lessen joint damage, prevent disability and improve symptoms. Historically non-steroidal anti-inflammatory drugs (NSAIDs) were the first-line treatment for RA, while DMARDs had been second-line treatment. The main reason was that DMARDs were perceived to be more toxic than NSAIDs. However, this treatment strategy failed to improve the outcome of RA [2]. Most patients were severely disabled after 20 years of disease despite treatment by DMARDs. Three main factors account for the failure of this "pyramid" approach to improve the outcome of RA:

- DMARDs are introduced too late.
- DMARDs are insufficiently powerful.
- The side effects of DMARDs are too frequent.

Treating RA early with DMARDs

In longitudinal cohorts and cross-sectional studies, functional disability correlates closely with radiologic joint damage [11]. Inception cohort studies in early RA showed that 30% and 75% of patients have joint erosions developed within the 1 and 2 years of symptom onset [12, 13]. This can be reduced if DMARDs are given early. Evidence that DMARDs, when well supervised, have less toxicity that NSAIDs, strengthens the case for their early use [14]. In early RA, most of these trials showed that DMARDs are superior to NSAIDs in improving symptom and suppressing inflammation but the evidence for reducing radiological damage by early DMARD treatment is limited. However, a recent clinical effectiveness trial in early RA showed that early treatment by sulfasalazine reduce radiological joint damage and preserve function [15]. Furthermore, recent analysis of primary data from 14 trials in early RA suggested that treatment response is dependent on the time of treatment from the onset of symptom. Moreover, an inception cohort study of 622 patients with early RA found that patients treated early with DMARDs showed no increase in mortality within the first 10 years when compared with the normal population [16]. Overall, current evidence support the use of DMARD once disease is diagnosed.

DMARDs have limited efficacy

Although DMARDs improves symptoms and signs in RA, disease remission is uncommon. Wolfe el al. studied 458 RA patients followed-up prospectively for 1,131 patient years; only 18% of patients had one remission or more [17]. In these patients, remission only represents a third of their follow-up period. The median length of remission was less than one year. Furthermore, not all DMARDs reduce radiological joint damage [18]. Even among those that reduce joint damage, none halts radiological damage completely. Indeed, many patients have partial response to DMARDs with insufficient suppression of synovitis that may explain why DMARDs does not abort joint damage in RA.

Toxicity of DMARDs

Current DMARDs are associated with frequent side effects. Toxicity is a common reason for stopping DMARD treatment. In a retrospective study by Galindo-Rodriguez et al., toxicity was the reason for discontinuing DMARDs in 20% of patients, while 25% of patient discontinued because of treatment was ineffective [19]. Two percent of patients discontinued treatment for both reasons. The median time to discontinuation was 16 months. Felson et al. studied the toxicity of

Table 1 - The American College Of Rheumatology 1987 Criteria For The Diagnosis Of Rheumatoid Arthritis

Patients must have at least four of the following seven criteria.

1. Morning stiffness lasting more than 1 hour
2. Arthritis in at least 3 joint areas (out of 14 possible areas: right or left proximal interphalangeal, metacarpo-phalangeal, wrist, elbow, knee, ankle, metatarso-phalangeal joints)
3. Arthritis of hand joints (wrist, metacarpo-phalangeal, or proximal interphalangeal joints)
4. Symmetric arthritis (simultaneous, symmetric involvement of the joint areas defined in (2); bilateral involvement of proximal interphalangeal, metacarpo-phalangeal, or metatarso-phalangeal joints without absolute symmetry is acceptable)
5. Presence of rheumatoid nodules
6. Positive rheumatoid factor
7. Radiographic changes (changes on hand and wrist radiographs, which must include erosions or periarticular osteopenia)

DMARDs in randomized control trials and found that injectable gold has the highest toxicity index [20]. The safest was anti-malarial. However, anti-malarial monotherapy does not reduce radiological joint damage [18]. This was reflected in a longitudinal study of 2,479 RA patients in the United States [21]. Other studies have confirmed that toxicity related discontinuation rate was high for injectable gold: 48% after one year [22]. Similar discontinuation rate was observed with for D-penicillamine and sulfasalazine [23]. Methotrexate fares better that other DMARDs with 71% of patients remaining on treatment after one year [23]. The median drug survival for methotrexate is 4.25 years.

Revised treatment standard in RA

In light of these problems, national and international guidelines have been drawn up recently to improve the management of RA [24]. The initial steps in the management of RA are to establish the diagnosis, perform a baseline evaluation, and estimate the prognosis. The currently accepted diagnostic criteria for RA are those developed in 1987 by the American College of Rheumatology (Tab. 1). However, these criteria were not designed for the diagnosis of early RA and are fairly insensitive in patients with disease duration of less than 12 months. Once the diagnosis is established; a baseline evaluation of the patient should be performed to document the disease activity, extra-articular features, the functional status of the patient, serum inflammatory markers such as erythrocyte sedimentation rate or C reactive

Table 2 - Baseline evaluation of disease activity and damage in patients with rheumatoid arthritis

Clinical
1. Degree of joint pain
2. Duration of morning stiffness
3. Duration of fatigue
4. Limitation of function
5. Physical examination
6. Actively inflamed joints (tender and swollen joint counts)
7. Mechanical joint problems: loss of motion, crepitus, instability,
8. malalignment, and/or deformity
9. Extraarticular manifestations
Laboratory
1. Erythrocyte sedimentation rate/C-reactive protein level
2. Rheumatoid factor
3. Complete blood cell count
4. Electrolyte levels
5. Creatinine level
6. Hepatic enzyme levels (AST, ALT, and albumin)
7. Urinalysis
8. Synovial fluid analysis
9. Stool guaiac
Other
1. Functional status or quality of life assessments using standardized questionnaires
2. Physician's global assessment of disease activity
3. Patient's global assessment of disease activity
Radiography
Radiographs of selected involved joints

protein (CRP) and evidence of radiologic damage (Tab. 2). Patients should be treated with DMARDs as soon as the diagnosis is established. Furthermore, sustained control of disease activity is vital. Fries et al., advocated the sawtooth approach in management of RA in which patients are switched from one DMARD to another if disease is insufficiently controlled [25]. However, many patients only achieved partial response to a single DMARD and monotherapy seldom induce complete disease remission. Therefore more powerful treatments are needed in RA.

Table 3 - Differences between TNF inhibitors

	Etanercept	Infliximab	Adalimumab
Class	TNFα IgG1 receptor construct	Chimeric monoclonal antibody	Human monoclonal antibody
	Binding TNFα + TNFβ	TNFα	TNFα
Half-life (days)	4.8 (median)	8–9	10–14
Administration	subcutaneous	intravenous	subcutaneous
Dose	25 mg twice per week	3 mg/kg 0, 2, 6 weeks then 8 weekly	40 mg every other fortnight

Anti-cytokine therapy in RA

Currently there are four cytokine antagonists that are licensed for the treatment of RA. Three are TNFα inhibitors and the other is an antagonist of IL-1.

TNFα inhibitors

The three TNFα inhibitors currently licensed for the treatment of RA are etanercept, infliximab and adalimumab. Table 3 summaries the differences between these TNFα inhibitors.

Infliximab

Infliximab (Centocor) is a chimeric anti-TNFα monoclonal antibody with a human immunoglobulin-1 Fc. It was the first anti-TNFα monoclonal antibody tested in patients with RA. In the initial open-label study, 20 patients were treated with a single 20 mg/kg dose of infliximab and followed for eight weeks. Swollen joint count fell from 18 to 5 mg/l and CRP decreased from 40 to 8 mg/l. A subsequent double-blind, randomized placebo-controlled trial of 73 patients with established RA, a single intravenous dose of 10 mg/kg infliximab, confirmed its efficacy in RA. There was a rapid reduction in the number of swollen joints and CRP [26]. Seventy-nine percent of patients achieved significant disease improvement according to the Paulus criteria compared with 8% in the placebo group. Improvement was rapid with statistically significant reduction in symptoms and signs occurred within the first week of treatment. Synovial inflammation decreased in parallel with clinical improvement. Biopsies of the synovia, before and four weeks after treatment, showed significant reduction in inflammatory infiltrates including T cells and decreased expression of adhesion molecules on blood vessels [27].

In another randomized, placebo-controlled trial of 101 patients with RA, infliximab or placebo was given repeatedly with or without methotrexate [28]. Patients were randomized to one of seven groups, receiving intravenous infliximab at 1, 3 or 10 mg/kg with or without 7.5 mg methotrexate weekly, or placebo plus methotrexate only. Methotrexate enhanced the effect of infliximab so that 3 mg/kg of infliximab plus methotrexate seemed as efficacious as 10 mg/kg of infliximab with or without methotrexate. This was associated with a significant reduction in anti-infliximab antibodies production by concomitant treatment with methotrexate. In the infliximab only groups, human anti-infliximab antibody was found in 53%, 21% and 7% of the patients in 1, 3 and 10 mg/kg groups, respectively. This was reduced to 15%, 7% and 0%, respectively, in patients who were taking concomitant methotrexate.

In a Phase III placebo-controlled trial, 428 RA patients who had partial response to methotrexate were randomized to receive placebo or infliximab, 3 or 10 mg/kg every four or eight weeks for 54 weeks [29]. In addition to reduction in the symptoms and signs of RA, radiological joint damage was reduced substantially; mean change in modified Sharp's score in the methotrexate and infliximab groups was 7.0 and 0.6, respectively. In addition, quality of life measured by short form-36 improved significantly in the infliximab treated group.

Infliximab has also been added to patients taking leflunomide. In an open-label trial of 20 patients with active RA [30], patient received leflunomide 100 mg daily for three days followed by 20 mg daily for 32 weeks. At week two all patients were given infliximab 3 mg/kg, and received a further four infusions at weeks 4, 8, 16 and 24. Adverse events were frequent and led to 11 patients withdrawn from the trial. The most common adverse event was pruritis associated with an eczematous rash including a patient with Stevens-Johnson syndrome. In another prospective open-label trial in 40 RA patients who were partial responders to leflunomide, infliximab was added [31]. Adverse events were frequent and treatment had to be withdrawn in many patients. Interestingly, antinuclear antibody was found in 11% of patients at week 0, by week 60 all the patients developed antinuclear antibody. The mean titer of anti-dsDNA antibody was over 100 by week 24, although clinical lupus was not seen.

Etanercept

Etanercept (Enbrel, Amgen) is a recombinant human p75 soluble TNF receptor-human immunologlobulin Fc fusion protein. It contains two TNF-binding domains linked to the Fc portion of human immunoglobulin. The dimeric structure of etanercept makes it approximately 1,000 times more efficient than the monomeric soluble p75 TNF receptor at neutralizing TNF. It is administered subcutaneously (25 mg) twice per week.

In a randomized placebo-controlled trial, 16 mg/m^2 of etanercept was administered subcutaneously thrice weekly for three months to 168 patients with RA. Dose-dependent improvement in symptoms and signs was seen [32]. ACR20 response rate at the end of the trial was 75% in the high dose group compared with 14% in the placebo group. The number of swollen joints decreased by approximately 50% from baseline after treatment for six months. Etanercept treatment was well tolerated, with only minor reactions at the site of injection. Synovial biopsies showed a statistically significant decrease in the number of T cells, plasma cells, vascular cell adhesion molecule 1, and IL-1 expression after one month of treatment [33]. Further study assessed twice-weekly subcutaneous injections of etanercept, 10 or 25 mg, or placebo in 234 RA patients for six months [34]. At the end of the trial, ACR20 response criteria were met by 59% and 11% of patients in the 25 mg etanercept and placebo groups, respectively.

Etanercept can also be added to patients who have partial response to methotrexate. Eighty-nine RA patients with persistently disease despite methotrexate were randomized to placebo or etanercept 25 mg twice per week [35] for 24 weeks. Seventy-one and 27% of patients receiving etanercept and placebo met the ACR20 criteria, respectively.

Etanercept also has disease-modifying property. In a study of 632 patients with early RA, twice-weekly subcutaneous etanercept (10 or 25 mg) or weekly methotrexate were administered for one year [36]. Modified Sharp's score in the etanercept 25 mg twice weekly and placebo groups was 0.57 and 1.06 at six months and 1 and 1.59 at 12 months, respectively.

Etanercept is the only cytokine inhibitor that is licensed for the treatment of juvenile chronic arthritis. In a randomized, placebo-controlled trial of 51 patients with juvenile polyarticular rheumatoid arthritis, etanercept 0.4 mg/kg of body weight or placebo was injected subcutaneously twice weekly for four months or until a flare of the disease occurred [37]. The total number of active joints decreased by 58% from baseline and joint motion improved by 80%.

Adalimumab

Adalimumab (Abbott Laboratory) is a fully human anti-TNFα monoclonal antibody developed by phage display technology. Since it is fully human, it has low antigenicity. It binds to TNF with high affinity and has a plasma half-life of 10–14 days.

Five hundred and forty-four patients with established RA were randomized to receive either placebo or four different doses of adalimumab: 20 mg, 40 mg every other week or 20 or 40 mg weekly given subcutaneously [38]. After 26 weeks, 46% and 53% of patients in the 40 mg fortnightly and 40 mg weekly groups met the ACR20 response criteria compared with 19% in the placebo group. ACR50 response in the 40 mg fortnightly and 40 mg weekly groups were 22% and 35%, respectively.

In the ARMADA trial, fortnightly adalimumab (20 mg, 40 mg or 80 mg subcutaneously) or placebo was administered to 271 RA patients who are methotrexate partial responders [39]. Methotrexate treatment was continued and the dosage unchanged. Treatment was given for 24 weeks. At the end of the trial, ACR20, 50 and 70 response criteria were met in 14.5%, 8.1% and 4.8% of patients in the placebo group. ACR20 response criteria were achieved by 67.2% and 65.8% in the 40 mg and 80 mg adalimumab groups. ACR50 response rates in 40 mg and 80 mg adalimumab groups were 55.2% and 42.5%, respectively. The 40 mg and 80 mg groups achieved ACR70 response of 26.9% and 19.2%, respectively. These differences were statistically significant.

A subsequent trial with similar design assessed the effect of adalimumab on radiologic joint damage. Six hundred and nineteen patients who were partial responders to methotrexate were randomized to either placebo or adalimumab 20 mg weekly or 40 mg every other week for 54 weeks [40]. Median change in Modified Sharp's score was 0 in the active treatment groups compared with 1 in the placebo group. Median erosion scores were 0.4, 0 and 1.7 in the 20 mg weekly, 40 mg every other week and placebo groups respectively.

Side effects of TNF inhibitors

Infection and reactivation of TB

Post-marketing surveillance has revealed that treatment by TNF inhibitors is associated with an increase in the incidence of infections – in particular tuberculosis [41]. Seventy cases were reported initially. Most cases occurred within six months of starting TNFα inhibitor. Hence it is probable that these represent reactivation of latent tuberculosis rather than newly acquired infections. More than half the cases were extra-pulmonary and 33% of patients had disseminated tuberculosis. In some cases, reactivation of tuberculosis had fatal consequences. Hence, screening for tuberculosis prior to starting treatment on TNFα inhibitors is vital. There are national and international guidelines on screening and management for latent tuberculosis [42]. Most guidelines include chest radiograph. The role of tuberculin skin test is controversial especially in countries where populations are vaccinated for tuberculosis.

An international consensus group recommends that patients should be considered as having latent tuberculosis if, on chest radiograph, a Ghon complex is present or pulmonary scarring plus suspicious clinical history or positive intermediate strength tuberculin PPD skin test in the absence of symptoms and signs of tuberculosis [43]. Hence, if the tuberculin skin test is negative, no further treatment is required. If the skin test or chest radiograph is positive, patients should be given single anti-mycobacterial drug therapy, either isoniazid for nine months or rifampin for four months. Multi-antibiotics may be needed if the suspicion of resistant tuberculosis is high. Antibiotics should be started before commencing TNFα inhibitors.

Table 4 - Lymphoma in patients treated by TNFa inhibitors

		Etanercept	Infliximab	Adalimumab
Clinical trial	No. of treated patients	3,389	1,678	2,468
	Patient-year	8,336	3,445	4,870
	Lymphoma cases	9	6	10
	Standardized incidence rate	3.47	6.35	5.52
	(95% confidence interval)	(0.59–6.59)	(1.7–16.3)	(2.6–10.0)
Postmarketing surveillance	No. of treated patients	150,000	365,000	Not available
	Patient-year	>230,000	554,000	Not available
	Lymphoma cases	70	95	Not available

Cardiac failure

Both infliximab and etanercept are contraindicated in patients with significant or unstable cardiac failure. These are based on preliminary data from Phase II trials investigating the therapeutic potential of infliximab and etanercept in patients with congestive heart failure. In the former, seven of the 101 patients treated with infliximab have died compared with none among the 49 patients given placebo. In the case of etanercept, two large clinical trials in heart failure were terminated early due to lack of efficacy and there was a suggestion of worsening of heart failure outcomes in the active treatment arm.

Lymphoma

Since TNFα has a role in the immunosurveillance of tumor, there is a theoretical risk of development of malignancies after long-term treatment. Table 4 summarizes the number of lymphoma observed during clinical trials and post-marketing surveillance. Most cases (> 80%) were non-Hodgkin's lymphoma. However, previous studies have shown that patients with RA have an increased risk of developing lymphoma. Standardized incidence rate was 2.4–8; the majority of these are non-Hodgkin's lymphoma. The risk is higher in patients with more severe disease. Hitherto, patients treated by receiving TNFα inhibitors have a higher incidence of lymphoma compared with the general population (standardized incidence ratio: 2.3–6.3). However, this may be related to the increased risk associated with RA. Long-term registers with control RA patients treated by conventional DMARDs are needed to assess precisely whether treatment by TNFα inhibitors increases the risk of lymphoma.

Demyelination

Post-marketing surveillance found 19 patients have suffered from neurologic side effects following treatment by TNFα inhibitors [44]. Although a causal relationship cannot be established, TNFα inhibitors should be discontinued if a patient develops new neurologic symptoms and signs. For those patients with a history of multiple sclerosis (MS), TNFα inhibitors are best avoided.

Long-term toxicity

The long-term side effects of TNFα inhibitors remain unknown. Post-marketing surveillance raises concern but cannot establish causality. Therefore national and international registries have been established to address these concerns.

TNFα inhibitors in clinical practice

Compared to conventional DMARDs, TNFα inhibitors are expensive treatments. Therefore not surprisingly, there are international and national guidelines that control the use of TNFα inhibitors in RA. These are listed in Table 5. Differences in these guidelines mean that there is significant variation globally on the use of TNFα inhibitors.

International Consensus Statement 2002

The latest International Consensus Statement recommended the use of TNF inhibitors for the treatment of active RA after failure of one DMARD of which the commonest used is methotrexate [45]. It does not recommend the use of TNFα inhibitors as the first DMARD in patients with early RA because of cost and limited data on the long-term toxicity. It highlights the importance of assessing response within 12 weeks of starting therapy and suggested that validated measures such as the disease activity score should be used. Interestingly, it does not recommend the use of the ACR response criteria. If the patient did not response within 12 weeks, treatment should be stopped and alternative DMARD introduced. In patients who have a partial response to treatment, either increasing the dose or more frequent dosing is recommended. It cautions the use of TNF inhibitors during pregnancy and in patients with MS, lymphoproliferative disease, chronic infection and severe congestive cardiac failure. Treatment should be withdrawn if patients develop serious infection, lupus-like symptoms, pancytopenia and demyelinating disorders.

Table 5 - International and National Guidelines for the Treatment of Rheumatoid Arthritis by TNF Inhibitors

International
- International Consensus Statement 2002
- European Consensus Statement 2000

National
- USA
  - American College Rheumatology Subcommittee in RA Guidelines 2002
  - Physician and Insurance-Driven Guidelines
- UK
  - National Institute for Clinical Excellence
  - Scottish Intercollegiate Guidelines Network
- Germany and Austria
  - Pharmacotherapy Committee of the German Rheumatology Society

European Consensus Statement 2000

This Consensus Statement is very similar to the International Consensus Statement except it defines active disease that justify treatment by TNFα inhibitor as a disease activity score ≥ 3.2–5 or more swollen and painful joints and a high CRP [46]. Adequate response to treatment is defined as improvement in the disease activity score of > 1.2, or more than 20% improvement. Efficacy should be apparent within 8–16 weeks.

UK National Institute for Clinical Excellence Guidance

The National Institute for Clinical Excellence is part of the National Health Service in the UK. Its remit is to provide guidance on good clinical practice. The use of etanercept and infliximab was reviewed by NICE in 2000 (47). It recommends the use to TNFα inhibitors to treat patients with RA who have failed at least two standard DMARDs, of which methotrexate must have been one, and have active disease with a disease activity score of > 5.1 on two occasions, one month apart. Patients are excluded if women who are pregnant or breastfeeding; effective contraception must be practiced if fertile, active infection, previous tuberculosis unless they have completed a full course of anti-tuberculosis therapy with modern antibiotics, septic arthritis of a native or prosthetic joint within the last 12 months, persistent or recurrent chest infections, indwelling urinary catheter, multiple sclerosis, malignancy or pre-malignancy states except basal cell carcinoma or malignancies diagnosed and treated more than 10 years previously. Treatment is to be withdrawn if adverse events occur or response to treatment is inadequate after three months. Response to treatment is defined as improvement in the disease activity score by > 1.2, or the achievement of a disease activity score of < 3.2.

Scottish Intercollegiatic Guidelines Network

The Scottish Intercollegiatic Guidelines Network issued recent guidelines on the management of RA [48]. Similar to other guidelines, it does not recommend the use of TNFα in patients with early RA.

Non-responders to TNF inhibitors

Some RA patients do not respond to TNFα inhibitors, they are the so-called primary non-responders. Others have an initial response but disease relapses despite treatment – the so-called secondary non-responders. The causes of primary and secondary non-response to TNFα inhibitors remain unknown. For some non-responders, increasing the dose and reducing the frequency of treatment may improve disease. Based on the Stockholm TNFα antagonist registry, in 254 patients treated by infliximab, 83 patients required increase in dosage at some point during follow-up [49]. Disease activity score values before and after the dosage increase were 4.13 ± 0.23 and 3.62 ± 0.22, respectively ($p < 0.02$). However, only eight patients met ACR20% improvement criteria and 16 patients met response criteria based on disease activity score. Half the patients had more durable responses between infusions. However, nine patients (20%) had no clinical benefit.

An alternative strategy in TNFα non-responder is switching to a different TNFα inhibitor. Data from the Stockholm register suggested that switching may be effective [50]. Eleven patients were treated initially by infliximab and 14 by etanercept. Switching to alternative TNFα inhibitors in these patients found improved efficacy in 67% of patients, some efficacy in 22%. Only 11% of patients were refractory to both TNFα inhibitors.

Anakinra, recombinant human interleukin-1 receptor antagonist (Amgen)

Another cytokine inhibitor that has demonstrated efficacy in clinical trials is the recombinant IL-1 receptor antagonist, anakinra. IL-1 binds to two types of receptors: IL-1 receptor type I (IL-1RI) and IL-1 receptor type II (IL-1RII). However, only IL-1RI transmits signals intracellularly, leading to cell activation while IL-1RII does not. Both IL-1RI and IL-1RII exist as soluble forms. These soluble receptors and cell surface IL1-RII compete with cell bound IL-1RI for IL-1 and inhibit IL-1 mediated cell activation. In addition, there is also a naturally occurring antagonist: interleukin-1 receptor antagonist which binds to IL-1RI but does not deliver any signal. Therefore, it inhibits IL-1 by competing for IL-1RI. However, the binding affinity of interleukin-1 receptor antagonist for IL-1RI is much lower than IL-1. Therefore, interleukin-1 receptor antagonist needs to be given in concentration 500 times in excess that of IL-1 in order to completely abolish the effects of the latter.

In a six month double-blind placebo-controlled study [51], daily subcutaneous injections of either placebo or anakinra at doses of 30, 75 and 150 mg were given to 472 RA patients. ACR20 response criteria were met by 43% of the patients treated with anakinra 150 mg/day compared with 27% of the placebo group. The acute phase response dropped in parallel with clinical improvement. Clinical benefit persisted up to 12 months. In another randomized controlled trial, 46%, 19%, and 5% of patients receiving anakinra 1 mg/kg plus methotrexate achieved the ACR20, 50, and 70 response, respectively, compared with 19%, 4%, and 0% in the control subjects treated with placebo and methotrexate [52]. Evaluation of radiographic damage in four cohorts of RA patients treated with 30, 75, 150 mg/day anakinra or placebo showed that the treatment groups achieved a reduction of 38% in joint erosions and 58% in joint space narrowing [53]. Anakinra is usually well tolerated, the most common effect being local reaction at the site of injection.

International Consensus Statement on Anakinra

Similar to TNFα inhibitors, an international consensus statement was developed by rheumatologists and scientists from many countries. Anakinra is licensed for the treatment of active RA either alone or in combination with methotrexate in America while in Europe anakinra should be used in conjunction with methotrexate. Similar to recommendations for TNFα inhibitors, treatment should be continued if the patients have significant documentable improvement in disease within 16 weeks. It highlights the long-term efficacy and toxicity of anakinra remains unknown. In particular, the safety of anakinra in patients with history of lymphoproliferative disease or malignancies is unclear. It should be avoided if possible in patients who are pregnant or breastfeeding. It also warns that there may be a small increase in the risk of infections, and therefore should not be started or should be discontinued when serious infections occur. Anakinra should not be used in combination to TNFα inhibitors since infection rate appears much higher when both agents are combined (Wyeth and Amgen communications).

Conclusion

RA causes severe disability, but prompt treatment with DMARDs significantly improves the long-term outcome. Early and sustained suppression of disease activity is essential to prevent joint damage. Better knowledge of the immunologic and inflammatory pathways in RA has culminated in the development of the cytokine inhibitors targeting TNFα and IL-1. Trials of anti-cytokine treatments in combination with DMARDs are currently in progress and may improve the prognosis of RA.

Patients who failed initial DMARDs may respond to cytokine inhibitors. Although cytokine inhibitors are not curative, significant clinical amelioration can be achieved in the vast majority of patients. Proper assessment of disease activity is crucial both to identify patients with severe, progressive disease who require therapy, and to monitor response to treatment.

Some patients do not respond to inhibitors of TNFα and IL-1. New cytokine inhibitors including monoclonal antibodies to IL-6 receptor or IL-15 offer hopes of better disease control in these patients.

References

1 MacGregor AJ, Silman AJ (1998) Rheumatoid arthritis: classification and epidemiology. In: Klippel JH, Dieppe PA (eds): *Rheumatology*. Mosby, Philadelphia, 1–6

2 Scott DL, Symmons DP, Coulton BL, Popert AJ (1987) Long-term outcome of treating rheumatoid arthritis: results after 20 years. *Lancet* 1: 1108–1111

3 Young A, Dixey J, Cox N, Davies P, Devlin J, Emery P, Gallivan S, Gough A, James D, Prouse P et al (2000) How does functional disability in early rheumatoid arthritis (RA) affect patients and their lives? Results of 5 years of follow-up in 732 patients from the Early RA Study (ERAS) [see comments]. *Rheumatology (Oxford)* 39(6): 603–611

4 Barrett EM, Scott DG, Wiles NJ, Symmons DP (2000) The impact of rheumatoid arthritis on employment status in the early years of disease: a UK community-based study. *Rheumatology* 39: 1403–1409

5 Wolfe F, Mitchell DM, Sibley JT, Fries JF, Bloch DA, Williams CA, Spitz PW, Haga M, Kleinheskel SM, Cathey MA (1994) The mortality of rheumatoid arthritis. *Arthritis Rheum* 37: 481–494

6 Pincus T, Callahan LF (1986) Taking mortality in rheumatoid arthritis seriously--predictive markers, socioeconomic status and comorbidity. *J Rheumatol* 13: 841–845

7 Krause D, Schleusser B, Herborn G, Rau R (2000) Response to methotrexate treatment is associated with reduced mortality in patients with severe rheumatoid arthritis. [see comments.]. *Arthritis Rheum* 43: 14–21

8 Cooper NJ (2000) Economic burden of rheumatoid arthritis: a systematic review. *Rheumatology (Oxford)* 39: 28–33

9 McIntosh E (1996) The cost of rheumatoid arthritis. *Br J Rheumatol* 35: 781–790

10 Yelin E, Wanke LA (1999) An assessment of the annual and long-term direct costs of rheumatoid arthritis: the impact of poor function and functional decline. *Arthritis Rheum* 42: 1209–1218
11 Pincus T (1995) Long-term outcome in rheumatoid arthritis. *Br J Rheumatol* 34 (Suppl 2): 59–73
12 van der Heijde DM, van Leeuwen MA, van Riel PL, Koster AM, van 't Ho, MA, van Rijswijk MH, van de Putte LB (1992) Biannual radiographic assessments of hands and feet in a three-year prospective followup of patients with early rheumatoid arthritis. *Arthritis Rheum* 35: 26–34
13 Fuchs HA, Kaye JJ, Callahan LF, Nance EP, Pincus T (1989) Evidence of significant radiographic damage in rheumatoid arthritis within the first 2 years of disease [see comments]. *J Rheumatol* 16: 585–591
14 Fries JF, Williams CA, Ramey DR, Bloch DA (1993) The relative toxicity of alternative therapies for rheumatoid arthritis: implications for the therapeutic progression. *Seminars in Arthritis & Rheumatism* 23(2 Suppl 1): 68–73
15 Choy EH, Scott DL, Kingsley GH, Williams P, Wojtulewski J, Papasavvas G, Henderson E, MacFarlane D, Erhardt C, Young A et al (2002) Treating rheumatoid arthritis early with disease modifying drugs reduces joint damage: a randomised double blind trial of sulphasalazine vs diclofenac sodium. *Clin Exp Rheumatol* 20: 351–358
16 Fries JF, Williams CA, Ramey D, Bloch DA (1993) The relative toxicity of disease-modifying antirheumatic drugs. *Arthritis Rheum* 36: 297–306
17 Wolfe F, Hawley DJ (1985) Remission in rheumatoid arthritis. *J Rheumatol* 12: 245–252
18 Iannuzzi L, Dawson N, Zein N, Kushner I (1983) Does drug therapy slow radiographic deterioration in rheumatoid arthritis? *N Engl J Med* 309:1023–1028
19 Galindo-Rodriguez G, Avina-Zubieta JA, Russell AS, Suarez-Almazor ME (1999) Disappointing longterm results with disease modifying antirheumatic drugs. A practice based study.[comment]. *J Rheumatol* 26: 2337–2343
20 Felson DT, Anderson JJ, Meenan RF (1990) The comparative efficacy and toxicity of second-line drugs in rheumatoid arthritis. Results of two metaanalyses [see comments]. *Arthritis Rheum* 33: 1449–1461
21 Singh G, Fries JF, Williams CA, Zatarain E, Spitz P, Bloch DA (1991) Toxicity profiles of disease modifying antirheumatic drugs in rheumatoid arthritis. *J Rheumatol* 18: 188–194
22 Sambrook PN, Browne CD, Champion GD, Day RO, Vallance JB, Warwick N (1982) Terminations of treatment with gold sodium thiomalate in rheumatoid arthritis. *J Rheumatol* 9: 932–934
23 Wolfe F, Hawley DJ, Cathey MA (1990) Termination of slow acting antirheumatic therapy in rheumatoid arthritis: a 14-year prospective evaluation of 1017 consecutive starts. *J Rheumatol* 17: 994–1002
24 American College of Rheumatology Subcommittee on Rheumatoid Arthritis Guidelines

(2002) Guidelines for the management of rheumatoid arthritis: 2002 Update. *Arthritis Rheum* 46: 328–334

25 Fries JF, Williams CA, Morfield D, Singh G, Sibley J (1996) Reduction in long-term disability in patients with rheumatoid arthritis by disease-modifying anti-rheumatic drug-based treatment strategies. *Arthritis Rheum* 39: 616–622

26 Elliott MJ, Maini RN, Feldmann M, Kalden JR, Antoni C, Smolen JS, Leeb B, Breedveld FC, MacFarlane JD, Bijl H et al (1994) Randomised double-blind comparison of chimeric monoclonal antibody to tumour necrosis factor α (cA2) *versus* placebo in rheumatoid arthritis. *Lancet* 344: 1105–1110

27 Tak PP, Taylor PC, Breedveld FC, Smeets TJM, Daha MR, Kluin PM, Meinders AE, Maini RN (1996) Decrease in cellularity and expression of adhesion molecules by anti-tumour necrosis factor α antibody treatment in patients with rheumatoid arthritis. *Arthritis Rheum* 39: 1077–1081

28 Maini RN, Breedveld FC, Kalden JR, Smolen JS, Davis D, Macfarlane JD, Antoni C, Leeb B, Elliott MJ, Woody JN et al (1998) Therapeutic efficacy of multiple intravenous infusions of anti-tumour necrosis factor α monoclonal antibody combined with low-dose weekly methotrexate in rheumatoid arthritis. *Arthritis Rheum* 41: 1552–1563

29 Lipsky PE, van der Heijde DM, St Clair EW, Furst DE, Breedveld FC, Kalden JR, Smolen JS, Weisman M, Emery P, Feldmann M et al (2000) Infliximab and methotrexate in the treatment of rheumatoid arthritis. Anti-Tumor Necrosis Factor Trial in Rheumatoid Arthritis with Concomitant Therapy Study Group. [see comments]. *N Engl J Med* 343: 1594–1602

30 Kiely PD, Johnson DM (2002) Infliximab and leflunomide combination therapy in rheumatoid arthritis: an open-label study. *Rheumatology* 41: 631–637

31 Bingham SJ, Barcelos A, Buch MH, Lindsay S, Emery P (2002) Induction of serological lupus in patients on leflunomide and infliximab. *Arthritis Rheum* 46(Suppl), S168

32 Moreland LW, Baumgartner SW, Schiff MH, Tindall EA, Fleischmann RM, Weaver AL, Ettlinger RE, Cohen S, Koopman WJ, Mohler K et al (1997) Treatment of rheumatoid arthritis with a recombinant human tumor necrosis factor receptor (p75)-Fc fusion protein. *N Engl J Med* 337: 141–147

33 Kempeni J (1999) Preliminary results of early clinical trials with the fully human anti-TNFalpha monoclonal antibody D2E7. [Review] *Ann Rheum Dis* 58 (Suppl) 1: I70–I72

34 Moreland LW, Schiff MH, Baumgartner SW, Tindall EA, Fleischmann RM, Bulpitt KJ, Weaver AL, Keystone EC, Furst DE, Mease PJ et al (1999) Etanercept therapy in rheumatoid arthritis. A randomized, controlled trial. *Ann Int Med* 130: 478–486

35 Weinblatt ME, Kremer JM, Bankhurst AD, Bulpitt KJ, Fleischmann RM, Fox RI, Jackson CG, Lange M, Burge D (1999) A trial of etanercept, a recombinant tumor necrosis factor receptor:Fc fusion protein, in patients with rheumatoid arthritis receiving methotrexate [see comments]. *N Engl J Med* 340: 253–259

36 Bathon JM, Martin RW, Fleischmann RM, Tesser JR, Schiff MH, Keystone EC, Genovese MC, Wasko MC, Moreland LW, Weaver AL et al (2000) A comparison of etaner-

cept and methotrexate in patients with early rheumatoid arthritis. *N Engl J Med* 343: 1586–1593

37 Lovell DJ, Giannini EH, Reiff A, Cawkwell GD, Silverman ED, Nocton JJ, Stein LD, Gedalia A, Ilowite NT, Wallace CA et al (2000) Etanercept in children with polyarticular juvenile rheumatoid arthritis. Pediatric Rheumatology Collaborative Study Group [see comments]. *N Engl J Med* 342: 763–769

38 van de Putte LB, Atkins C, Malaise M, Sany J, Russell AS, van Riel PCL et al (2002) Adalimumab monotherapy in the treatment of patients with severely active rheumatoid arthritis. *Arthritis Rheum* 46 (Suppl): S205

39 Weinblatt ME, Keystone EC, Furst DE, Moreland LW, Weisman MH, Birbara CA, Teoh LA, Fischkoff SA, Chartash EK (2003) Adalimumab, a fully human anti-tumor necrosis factor alpha monoclonal antibody, for the treatment of rheumatoid arthritis in patients taking concomitant methotrexate: the ARMADA trial. *Arthritis Rheum* 48: 35–45

40 Keystone E, Kavanaugh AF, Sharp J, Hua Y, Teoh L, Fischkoff S et al (2002) Adalimumab, a fully human anti-TNF_ monoclonal antibody, inhibits the progression of structural joint damage in patients with active RA despite concomitant methotrexate therapy. *Arthritis Rheum* 46 (Suppl): S205

41 Keane J, Gershon S, Wise RP, Mirabile-Levens E, Kasznica J, Schwieterman WD, Siegel JN, Braun MM (2001) Tuberculosis associated with infliximab, a tumor necrosis factor alpha-neutralizing agent.[comment]. *N Engl J Med* 345: 1098–1104

42 Verhoeven AC, Boers M, van der Linden S (2000) Responsiveness of the core set, response criteria, and utilities in early rheumatoid arthritis. *Ann Rheum Dis* 59: 966–974

43 Furst DE, Cush J, Kaufmann S, Siegel J, Kurth R (2002) Preliminary guidelines for diagnosing and treating tuberculosis in patients with rheumatoid arthritis in immunosuppressive trials or being treated with biological agents. *Ann Rheum Dis* 61 (Suppl 2): ii62–ii63

44 Mohan N, Edwards ET, Cupps TR, Oliverio PJ, Sandberg G, Crayton H, Richert JR, Siegel JN (2001) Demyelination occurring during anti-tumor necrosis factor alpha therapy for inflammatory arthritides. *Arthritis Rheum* 44: 2862–2869

45 Furst DE, Breedveld FC, Kalden JR, Smolen JS, Antoni CE, Bijlsma JW, Burmester GR, Cronstein B, Keystone EC, Kavanaugh A, Klareskog L et al (2002) Updated consensus statement on biological agents for the treatment of rheumatoid arthritis and other rheumatic diseases. *Ann Rheum Dis* 61 (Suppl 2): ii2–ii7

46 Smolen JS, Breedveld FC, Burmester GR, Combe B, Emery P, Kalden JR, Klareskog L, Maini RN, Numo R, van de Putte LB et al (2000) Consensus statement on the initiation and continuation of tumour necrosis factor blocking therapies in rheumatoid arthritis. *Ann Rheum Dis* 59: 504–505

47 National Institute for Clinical Excellence (2002) *Guidance on the use of etanercept and infliximab for the treatment of rheumatoid arthritis*. 36 ed. NICE Technology Appraisal, London

48 Scottish Intercollegiate Guidelines Network (2000) *Management of early rheumatoid arthritis*. 48 ed. Royal College of Physicians of Edinburgh, Edinburgh
49 van Vollenhoven RF, Gullstrom E, Brannemark S, Klareskog L (2002) Dose escalation of infliximab in clinical practice: data from the Stockholm TNF-alpha registry (STURE). *Ann Rheum Dis* 61 (Suppl 1): 177
50 van Vollenhoven RF, Harju A, Brannemark S, Klareskog L (2002) Switching between etanercept and infliximab: data from the Stockholm TNF-alpha antagonist registry (STURE). *Ann Rheum Dis* 61(Suppl 1): 54
51 Bresnihan B, Alvaro-Gracia JM, Cobby M, Doherty M, Domljan Z, Emery P, Nuki G, Pavelka K, Rau R, Rozman B et al (1998) Treatment of rheumatoid arthritis with recombinant human interleukin-1 receptor antagonist. *Arthritis Rheum* 41: 2196–2204
52 Cohen S, Hurd E, Cush J, Schiff M, Weinblatt ME, Moreland LW, Kremer J, Bear MB, Rich WJ, Mccabe D (2002) Treatment of rheumatoid arthritis with anakinra, a recombinant human interleukin-1 receptor antagonist, in combination with methotrexate: results of a twenty-four-week, multicenter, randomized, double-blind, placebo-controlled trial. *Arthritis Rheum* 46: 614–624
53 Jiang Y, Genant HK, Watt I, Cobby M, Bresnihan B, Aitchison R, Mccabe D (2000) A multicenter, double-blind, dose-ranging, randomized, placebo-controlled study of recombinant human interleukin-1 receptor antagonist in patients with rheumatoid arthritis: radiologic progression and correlation of Genant and Larsen scores. *Arthritis Rheum* 43: 1001–1009

Innate response cytokines in inflammatory synovitis: interleukin-15 and interleukin-18

J. Alastair Gracie, Sarah L. Jongbloed, Hilary Wilson and Iain B. McInnes

Centre for Rheumatic Diseases, Division of Immunology, Infection and Inflammation, University of Glasgow, 10 Alexandra Parade, Glasgow G31 2ER, UK

Introduction

In rheumatoid arthritis (RA), synovial membrane exhibits features suggesting that components of innate and acquired immune responses are of functional importance. In particular, the role of both T cell dependent and independent mechanisms in promoting cytokine networks has received much attention [1]. Cytokines mediate a wide variety of immunological actions and are key mediators in the pathogenesis of human autoimmune disease. Their pleiotropic functions and propensity for synergistic interactions and functional redundancy render them intriguing therapeutic targets. Thus far, single cytokine targeting has proven useful in several rheumatic diseases states. Perhaps the clearest evidence for the central role of these molecules in disease is the success of TNFα-blocking therapies. These have elucidated, for the first time, definitive critical checkpoints in proinflammatory cascades. However, non- or partial-responder patients are not infrequent and inflammatory disease usually flares upon discontinuation of treatment [2, 3]. The fact that TNFα independent pathways operate in inflammatory synovitis carries significant pathogenetic implications for existing disease models and exemplifies the clinical necessity for generation of further novel, pathogenesis-lead interventions.

Many cytokines present within synovium are of macrophage and synovial fibroblast derivation [4, 5]. Interleukin-15 (IL-15) and interleukin-18 (IL-18) attracted our attention at an early stage as potential regulators of synovial inflammation, in particular through modulation of T lymphocyte function and TNFα expression (see Table 1 for comparison of key features). Both have been identified in a number of autoimmune diseases including RA, Crohn's disease, multiple sclerosis (MS), insulin-dependent diabetes mellitus (IDDM) [6-12] and psoriatic arthritis (unpublished observations). This short review will summarize work carried out supporting the role for IL-15 and IL-18 in perpetuating chronic inflammatory synovitis in RA.

Cytokines and Joint Injury, edited by Wim B. van den Berg and Pierre Miossec

Table 1 - Comparison of key features of IL-15 and IL-18

	IL-15	IL-18
Cytokine family	IL-2	IL-1
Molecular weight	15 kDa	18 kDa (mature form)
Pro-molecule	no	yes
Intracellular function	probable	probable
Receptor	IL-15Rαβγ	IL-18Rαβ
Principal signalling pathway	STAT3/STAT5	MyD88 /NF-κB
Constitutive mRNA	yes	yes
Native inhibitor	sIL-15Rα	IL-18bp

Interleukin-15

IL-15 is characteristically expressed during the early stages of innate immune responses. We have explored the hypothesis that IL-15 expression in the synovial membrane can promote and sustain T cell responses that in turn contribute directly to TNFα production [6, 13, 14]. Recent data suggest that relevant bioactivities extend beyond this to include neutrophil, NK cell, macrophage and dendritic cell activation [15-17]. We shall address the therapeutic potential of IL-15 in terms of plausible bioactivities, functional expression in target tissues and tractability in rodent models.

IL-15 (14–15 kDa) is a 4-α helix cytokine with structural similarities to IL-2 [18, 19]. IL-15 mRNA is expressed in numerous normal human tissues and cell types, including activated monocytes, dendritic cells and fibroblasts [16, 18]. IL-15 mRNA expression, however, is not synonymous with protein detection in tissues, reflecting tight regulatory control of translation and secretion. IL-15 is subject to significant post-transcriptional regulation *via* 5'UTR AUG triplets, 3' regulatory elements and a further putative C-terminus region regulatory site. Two isoforms of IL-15 subject to altered glycosylation are thus generated – secreted IL-15 (48–aa) from a long signalling peptide and an intracellular IL-15 form localised to non-endoplasmic regions in both cytoplasmic and nuclear compartments from a short signalling peptide (21–aa) [16]. Cell membrane expression may be crucial in mediating extracellular function rather than secretion and in part explains the difficulty in detecting soluble IL-15 in biologic systems. Exogenous agents include human herpes virus 6 and 7, *Mycobacterium leprae*, *Mycobacterium tuberculosis*, *Staphylococcus aureus*, lipopolysaccharide and ultraviolet irradiation [20-25]. IL-15 functions *via* a widely distributed heterotrimeric receptor (IL-15R) which consists of a β-chain (shared with IL-2) and common γ-chain, together with a unique α-chain (IL-15α) that in turn

Table 2 - Key bioactivities of IL-15

Cell	Bioeffect
T cells	T1/T2 maturation, migration, activation, cytokine release, cytotoxicity T cell memory maintenance (CD8+ compartment)
NK cells	NK cell maturation, activation, cytokine release, cytotoxicity
endothelial cell	retard apoptosis
B cell	isotype switching, immunoglobulin production
macrophage	cytokine expression, adhesion molecule expression
neutrophil	activation, cytoskeletal structure, apoptosis regulation, cytokine release
osteoclast	maturation, calcitonin receptor
FLS	retard apoptosis
dendritic cell	activation, co-stimulation, cytokine release

exists in eight isoforms [16]. This complex may form in *cis* or *trans* orientation allowing for receptor component sharing between adjacent cells. Like IL-2, the IL-15Rαβγ complex signals through JAK1/3 and STAT3/5 [26, 27]. Additional signalling through src-related tyrosine kinases and Ras/Raf/MAPK to fos/jun activation is also proposed. High affinity (10^{11} M^{-1}) with slow off-rate render IL-15Rα in soluble form a useful and specific inhibitor in biologic systems. *Via* this receptor complex, IL-15 exhibits broad proinflammatory effects that are summarised in Table 2. The target cells identified are of relevance to current models of synovitis pathogenesis hence IL-15 exhibits plausible bioactivity.

IL-15 has been observed in inflamed synovium by several investigators. Thus, IL-15 mRNA (Taqman PCR, nested RT-PCR) and protein (immunohistochemistry, ELISA, receptor capture assay) are detected in RA synovial membrane at higher levels than in reactive arthritis or psoriatic synovial biopsies [28]. Lining layer macrophages represent the predominant cellular source *in situ*, although expression in fibroblast-like synoviocytes (FLS) and in endothelial cells is also evident [6, 29, 30]. We detected low levels of IL-15 in a majority of RA SF at median concentrations similar to those of TNFα detected in parallel assays [6, 14], although variable levels have been reported in distinct populations [31-34]. Serum IL-15 expression does not correlate with disease subsets thus far recognised. Whereas RA serum TNFα levels correlate with the presence of germinal centres in parallel synovial biopsies, IL-15 levels were elevated with either germinal centres or diffuse lymphocytic infiltrative patterns [35]. IL-15 expression has also recently been detected in synovial membrane derived from juvenile RA patients [36], associated with IL-18, IL-12 and IFNγ expression. Spontaneous production of IL-15 by primary RA synovial membrane cultures and by isolated synovial fibroblasts is reported [32, 37]. In long-term cultures of synovial tissues, outgrowth was found to be dependent upon the presence of T cells that in turn lead to local release of IL-15, FGF-1 and IL-17 [38].

Effector function of IL-15 in synovium is largely predicted from its basic biology (Tab. 2). In particular, IL-15 sustains T cell/macrophage interactions to promote activation and cytokine release by the latter. Following stimulation with non-physiological mitogens, paraformaldehyde-fixed T cells and T cell clones induce proinflammatory cytokine production by macrophages and fibroblasts through cell contact [39]. Similarly, freshly isolated synovial T cells can induce TNFα synthesis by blood- or synovial-derived macrophages through cell-membrane contact, with no requirement for secretory factor synthesis [13]. This activity is maintained *in vitro* by the addition of exogenous IL-15. These data demonstrated for the first time that a fundamental property of synovial T cells is the capacity to drive TNFα production through cell contact, and that IL-15 represents one factor capable of sustaining this activity. IL-15 confers similar properties upon CD45RO$^+$ PB T cells, such that IL-15-activated PB cells from RA patients induce TNFα synthesis in synovial macrophage/synoviocyte co-cultures. Our neutralisation studies [13] implicated at least CD69, recently confirmed by Ortiz [40], LFA-1 and ICAM-1 in these interactions, although other ligand pairs are likely to have involved including at least CD11b and CD40/CD154 [41, 42] with CD154 having been shown to be upregulated by IL-15 [43]. IL-15 apparently operates in synergy with other locally release cytokines, including TNFα, IL-18, IL-12 and IL-6. Moreover, *via* cognate interactions, activated T cells can induce IL-15 expression in macrophages (our unpublished observations). Similar interactions between T cells and fibroblast-like synoviocytes with endogenous positive feedback loops have also been demonstrated. Of interest, secreted IL-15 is usually not detected in these culture systems, indicating that functionally active membrane bound-IL-15 is likely to be functionally important. IL-15 can also promote T cell migration and survival. Ziolkowska and colleagues have shown that IL-15, *via* cyclosporin A sensitive mechanisms may induce the overproduction of IL-17 seen in the joints of RA patients [33]. Thus IL-15 can promote T cell migration, activation, and effector function *via* both cell–cell interactions and soluble cytokine release to promote pathogenesis.

Effects beyond T cell activation include at least synovial neutrophil activation and survival and NK cell activation. Indeed a subset of NK cells has recently been shown to be greatly expanded within inflamed joints [44]. In addition, IL-15 has been shown to protect both synovial fibroblasts and vascular endothelial cells from apoptosis [45]. Effects on synovial B cells are as yet unproven.

Factors that drive synovial IL-15 expression are under investigation. As noted above, T cell/macrophage interactions induce IL-15 expression in macrophages. Moreover, direct exposure of synovial fibroblasts to TNFα or IL-1β induces high levels of IL-15 expression, although soluble IL-15 is rarely detected. In dermal fibroblasts TNFα but not IFNγ induces membrane expression of IL-15 that in turn can sustain T cell growth [46] – a similar pathway therefore likely exists in RA synovium. A further route to IL-15 production has been suggested in studies of synovial embryonic growth factor expression *via* the wingless (*Wnt*)5 and frizzled (*Fz*)5 ligand pair [47, 48].

Considerable information about the implications of cytokine expression in synovitis has been obtained using inflammatory arthritis models. Administration of recombinant IL-15 during priming with type II collagen (in incomplete Freund's adjuvant) induces development of an erosive inflammatory arthritis in DBA/1 mice, which closely resembles that obtained in positive control mice receiving collagen in complete Freund's adjuvant (Leung BP and McInnes IB, unpublished observations). Thus IL-15 can act to promote early necessary arthritogenic immune responses. Soluble murine IL-15 receptor α (smIL-15Rα) administration provides a mechanism to manipulate IL-15 bioactivities *in vivo*. smIL-15Rα administration following antigen challenge suppresses development of collagen-induced arthritis in DBA/1 mice, associated with delayed development of anti-collagen specific antibodies (IgG2a) and with reduced collagen specific IFNγ and TNFα production *in vitro* [49]. Finally, in preliminary studies, we have shown that shIL-15Rα suppresses the development of CIA in a primate model (unpublished). Together these data clearly indicate that IL-15/IL-15R interactions are important in the development of arthritogenic immune responses *in vivo*. Parallel approaches to IL-15 targeting *in vivo* include the use of mutant IL-15:Fc fusion proteins and specific neutralising monoclonal antibodies. The effect of such approaches in rodent arthritis is under investigation.

The foregoing biology has prompted the conduct of a Phase I, proof-of-concept study, in which a fully human monoclonal antibody (HuMax-IL15) was administered to patients with active RA (Abstract OP0008, EULAR, Lisbon 2003). Although not adequately placebo-controlled, this early study manifest ACR response rates approximating to those achieved thus far with TNFα blockade monotherapies. In general, IL-15 blockade was well tolerated in both clinical and laboratory components. Accordingly, randomised, placebo controlled Phase II studies are now ongoing.

Data discussed already clearly allude to the synergistic activities of IL-15 in the context of synovial inflammation. We have reasoned that cytokines expressed with a similar kinetic in the context of early innate responses, for which synergistic activity would carry advantage to the host *via* accelerated generation of host defence mechanisms, might maintain such synergistic activities if co-expressed in chronic lesions. We therefore extended our studies to include other innate cytokines present early in inflammatory responses, particularly IL-18.

Interleukin-18

Interleukin-18 (IL-18), a member of the interleukin-1 cytokine superfamily, is an important regulator of both innate and acquired immune responses. IL-18 is expressed at sites of chronic inflammation, in autoimmune diseases, in a variety of cancers and in the context of numerous infectious diseases. Previously known as IFNγ inducing factor (IGIF), IL-18 was identified as an endotoxin-induced serum factor that stimulated IFNγ production [50]. It was cloned from a murine liver cell cDNA library generated from animals primed with heat-killed *Propionibacterium acnes* and subsequently challenged with LPS [51]. Nucleotide sequencing of murine IL-18 predicted a precursor polypeptide lacking a conventional signal peptide. Subsequent cloning of human IL-18 cDNA revealed 65% homology with murine IL-18 [52] and showed that both contain an unusual leader sequence consisting of 35 amino acids at their N-terminus. Commensurate with a proposed role in a variety of early inflammatory responses, IL-18 has been identified in cells of both haemopoietic and non-haemopoietic lineages. Thus, IL-18 expression has been reported in macrophages, dendritic cells (DC), Kupffer cells, keratinocytes, osteoblasts, adrenal cortex cells, intestinal epithelial cells, microglial cells and synovial fibroblasts [9, 10, 53–58]. However, without the enzymatic machinery necessary for IL-18 processing, expression of IL-18 mRNA, or indeed pro-IL-18 protein, should not necessarily infer the capacity to contribute biologic activity. The nature of native stimuli for IL-18 expression remain under investigation but include at least LPS and FasL [59].

IL-18 is produced as a 24 kD inactive precursor that is cleaved by IL-1β converting enzyme (ICE, caspase-1) to generate a biologically active mature 18 kD moiety [60, 61]. Recent reports indicate that proteinase 3 (PR3) can also generate biological activity from pro-IL-18 [62] whereas we have observed that human neutrophil derived serine proteases, elastase and cathepsin G, may also generate novel IL-18-derived moieties (unpublished data). The biological significance of the latter remains unclear but raises the fascinating possibility that neutrophil activation during early responses may critically regulate the capacity of IL-18 to contribute to the phenotype of subsequent adaptive immune responses. Mature IL-18 in turn binds a heterodimer containing an α (IL-1Rrp) chain responsible for extracellular binding of IL-18 and a non-binding, signal transducing β (AcPL) chain [63-65]. Both chains are required for functional IL-18 signalling [66]]. IL-18R is expressed on a variety of cells including macrophages, neutrophils, NK cells, endothelial and smooth muscle cells [67–70]. The IL-18R complex can be upregulated on naïve T cells, Th1 type cells and B cells by IL-12 [64, 71]. IL-18Rα serves as a marker of mature Th1 cells in human and murine systems [72], whereas T cell receptor (TCR) ligation together with IL-4 downregulates IL-18R [73]. Consistent with this, administration of anti-IL-18Rα antibody *in vivo* results in reduced LPS induced mortality associated with a subsequent shift in balance from a Th1 to a Th2 immune response [72].

IL-18 signals *via* myeloid differentiation 88 (MyD88) and IL-1 receptor-associated kinase (IRAK) to the receptor complex and their activation [74-76]. Thus IL-18 shares downstream effector pathways with critical immune regulatory molecules such as IL-1β and toll-like receptors (TLR), which are in turn implicated in regulating IL-18 expression [77, 78]. Phosphatidyl-inositol 3 kinase (PI 3-kinase) and NF-κB have also been implicated in induction of vascular cell adhesion molecule-1 (VCAM-1) on synovial fibroblasts by IL-18 [79].

Although originally identified as a factor capable of inducing IFNγ production by murine splenocytes, the effector role of IL-18 is rapidly expanding. Consistent effects on lymphoid series, particularly Th1 lineage in combination with IL-12 have emerged [51]. Thus IL-18 enhances T and NK cell maturation, cytokine production and cytotoxicity [51, 71, 80, 81]. IL-18 also increases FasL on NK cells and consequent Fas-FasL-mediated cytotoxicity [82, 83]. IL-18 deficient mice have reduced NK cell cytolytic ability that can be restored by exogenous IL-18 [84]. However, together with IL-2, IL-18 coinduces IL-13 in murine T and NK cells and in the presence of TCR activation induces T cell IL-4, IL-10, IL-13 and IFNγ production [85, 86]. In isolation IL-18 induces high IgE expression by B cells and in combination with IL-2, anti-CD3 and anti-CD28 markedly enhances IL-4 production by $CD4^+$ T cells [87]. When cultured alone or in combination with IL-4, IL-18 is known to induce murine T cell Th2 differentiation. This however is dependent upon genetic influences as spleen cells from BALB/c and C56BL/6 strains of mice stimulated with anti-CD3 and IL-18 exhibit enhanced Th2 and Th1 responses, respectively [88]. Thus IL-18 can promote Th1 or Th2 lineage maturation dependent upon underlying genetic influences and the ambient cytokine milieu.

On non-T cell populations IL-18 in conjunction with IL-3 induces IL-4 and IL-13 production by bone marrow derived basophils [89]. Direct effects on macrophages and DC have also been observed. Stimulation of bone marrow derived macrophages or splenic DC with IL-12 and IL-18 can induce IFNγ production [90, 91]. Studies of knock-out mice also reveal that IL-18 stimulation of peritoneal macrophages induces IL-6 production, independent of the intermediate induction of endogenous cytokines such as TNFα or IL-1β [92]. Similarly, however, other non-haemopoietic cell responses to IL-18 are likely with direct effects on chondrocytes and cartilage matrix degradation having been reported [93]. Keratinocytes and Langerhans cells (LC) also express IL-18 [54, 94, 95].

The discovery, cloning and characterisation of IL-18 binding protein (IL-18BP), a constitutively secreted protein able to bind IL-18 with high affinity, provides a potential mechanism whereby IL-18 activity could be regulated. Indeed, IL-18BP inhibits IL-18 induced IFNγ, IL-8 production and NF-κB activation *in vitro* and LPS induced IFNγ production *in vivo* [96, 97]. *Via* inhibition of IL-18 induced IFNγ production, recombinant IL-18BP has also been shown to augment PBMC prostaglandin production [98]. Furthermore, recent studies have shown that IL-18BP expression itself may be augmented by IFNγ [99]. It is upregulated in sepsis [100] and in endothelial cells and macrophages during active Crohn's disease [101], suggesting the existence of an endogenous IFNγ regulated feedback loop. Local levels of free IL-18 or IL-18 complexed with IL-18BP are therefore likely vital in determining net IL-18 biological activity. Generated as a result of alternative mRNA splicing, four human and two murine IL-18BP isoforms have been identified. Human IL-18BPa and IL-18BPc and murine IL-18BPc and IL-18BPd isoforms are capable of binding with high affinity to and neutralising IL-18 [102]. The recent discovery of IL-1H, a protein with sequence homology to IL-1ra, which is able to bind the IL-18R but not IL-1R [103] suggests the existence of further IL-18R antagonism *in vivo*.

Although originally identified as a factor capable of inducing IFNγ production, the effector role of IL-18 is rapidly expanding [104]. Some of these biologic effects are summarised in Table 3. Many of these activities are of potential functional significance in inflammatory synovitis and prompted us to seek IL-18 expression in RA. IL-18 is expressed in RA and in psoriatic arthritis synovial membrane ([9, 105, 106] and our unpublished observations). 24 kD pro-IL-18 predominates, although mature IL-18 is consistently detected. IL-18 expression is localised in $CD14^+$ and $CD68^+$ macrophages and in fibroblast-like synoviocytes (FLS) *in situ*. IL-18Rα and β chains are detected *ex vivo* on synovial $CD3^+$ lymphocytes and on synovial $CD14^+$ macrophages and *in vitro* on FLS. IL-18BP may also be present in substantial concentrations [107–109]. These data indicate that IL-18 and its receptor system are present in inflammatory synovitis. Its functional activities include promotion of cytokine release (particularly TNFα, GM-CSF and IFNγ). Marked cytokine synergy with IL-12 and IL-15 is observed in this respect. IL-18 acts not only through lymphocyte activation, but also through direct effects on macrophages. It may also amplify TNFα release through enhancing cell–cell interactions between synovial T cells and macrophages. Importantly, dose response studies reveal that only very low concentrations are required to promote significant levels of TNFα production *in vitro*, especially in synergy with IL-15 [110].

IL-18 expression is in turn upregulated in FLS by IL-1β and TNFα, suggesting the existence of positive feedback loops linking monokine predominance in RA with innate cytokine production and Th/c 1 cell activation in synovial immune responses. IL-18 induces NO release by RA SM *in vitro* that, since NO inhibits caspase-1 activity, provides a further potential regulatory loop. IL-18 possesses pro-degradative effects in articular cartilage. IL-18 reduces chondrocyte proliferation, upregulates iNOS, stromelysin and cyclooxygenase 2 expression and increases glycosaminoglycan (GAG) release *in vitro*. Such activities may be IL-1β independent although contradictory data have also emerged [111]. IL-18 promotes neutrophil activation, reactive oxygen intermediate synthesis, cytokine release and degranulation [9, 67]. IL-18 further promotes synovial chemokine synthesis and angiogenesis as well as upregulation of intracellular adhesion molecule-1 (ICAM-1) and VCAM-1 expression on endothelial cells and synovial fibroblasts [79, 112, 113]. Finally, IL-18 effects are not necessarily detrimental. IL-18 inhibits osteoclast maturation through GM-CSF production by T cells, thereby retarding bone erosion [114]]. Suppression of COX expression may also be mediated through IFNγ production with consequent effects upon prostanoid mediated local inflammation. Thus, like IL-15, IL-18 is expressed in synovial membrane and exhibits plausible bioactivities therein.

Table 3 - Key bioactivities of IL-18

Cell	Bioeffect
T cell	Th1/Tc1 maturation, migration, activation, cytokine release, cytotoxicity, Th2/Tc2 activation in absence of IL-12
NK cell	NK cell activation, cytokine release, cytotoxicity
endothelial cell	promotes angiogenesis, adhesion molecule expression
chondrocyte	reduced proliferation, increased COX2, GAG release MMP activation – net effect towards degradation (possible IL-1 intermediate)
neutrophil	activation, cytokine release, migration, ROI release
B cell	antibody secretion enhanced, IgE
macrophage	cytokine expression, adhesion molecule expression
osteoclast	inhibition via GM-CSF
dendritic cell	activation, co-stimulation, cytokine release

IL-18 has been targeted in several arthritis models *in vivo*. Upon challenge with type II collagen (CII) in CFA, IL-18-deficient mice on a DBA/1 background exhibit reduced incidence and severity of arthritis. *Ex vivo* analysis determined that both cellular and humoral responses to CII were suppressed [115, 116]. Moreover, administration of recombinant IL-18 can replace the requirement for CFA in CII induced erosive arthritis in DBA/1 mice [116]. Neutralisation of IL-18 *in vivo* using specific antibodies or IL-18BP effectively reduces developing and established rodent arthritis in both streptococcal cell wall and CIA models [111, 117]. Such effects may operate independent of IFNγ [111]. A feature of both models is suppression not only of inflammation but also of matrix destruction. These data strongly suggest that the net effect of IL-18 expression is proinflammatory, at least in the context of antigen-driven articular inflammation.

IL-18 therefore represents an exciting novel inflammatory mediator that is upregulated in a number of autoimmune rheumatic diseases and that represents a novel therapeutic target. Clinical studies to test this hypothesis in RA are ongoing, although interventions are not yet as far advanced as those targeting IL-15. Several specific biologic agents capable of targeting IL-18 are already in preparation.

Concluding remarks

We have chosen to study IL-18, an IL-1-like cytokine (NF-κB dependent) and IL-15, an IL-2-like cytokine (STAT3/5 dependent) in the context of inflammatory synovitis. Both show enhanced expression beyond RA to include several human inflammatory diseases suggesting that they will likely be of general importance. They exhibit appropriate biology, expression profiles and tractability *in vivo* arthritis models. However fundamental issues remain. Most importantly we lack good models to predict which cytokines will yield the most beneficial effects when targeted *in vivo*. Moreover, it is not clear how best to capitalise on the synergistic activities of cytokine combinations when translating to the clinic. There may be important kinetics, dose responses and inhibitory interactions (with both known and unknown regulatory moieties) that are as yet poorly understood in complex cytokine networks that could confound predictions when moving to disease. Early efforts to combine targeting of TNFα and IL-1 provide ample evidence of such potential problems. We believe that targeting distinct points in the inflammatory cascade that exhibit synergy rather than functional redundancy whenever possible may provide one rational route to appropriate selection of combinations. IL-15 and IL-18 operate at distinct points in the inflammatory cascade from IL-1 or TNFα and as such represent interesting targets. However, such targets may necessarily reside beyond the cytokine cascade to include adhesion molecules, co-stimulatory pathways or alternate tissue regulatory systems e.g., angiogenesis, bone remodelling. The ultimate significance of IL-15 and IL-18 expression in autoimmune disease *in vivo* therefore requires confirmation in human intervention studies before further strong conclusions can be drawn.

Acknowledgement
Work performed in the authors' laboratories is supported by the Arthritis Research Campaign and The Wellcome Trust. Invaluable discussions with Prof Foo Y. Liew are gratefully acknowledged.

References

1 Firestein GS (2003) Evolving concepts of rheumatoid arthritis. *Nature* 423: 356–361
2 Lipsky PE, van der Heijde DM, St Clair EW, Furst DE, Breedveld FC, Kalden JR, Smolen JS, Weisman M, Emery P, Feldmann M et al (2000) Infliximab and methotrexate in the treatment of rheumatoid arthritis. Anti-Tumor Necrosis Factor Trial in Rheumatoid Arthritis with Concomitant Therapy Study Group. *N Engl J Med* 343: 1594–1602
3 Bathon JM, Martin RW, Fleischmann RM, Tesser JR, Schiff MH, Keystone EC, Genovese MC, Wasko MC, Moreland LW, Weaver AL et al (2000) A comparison of etanercept and methotrexate in patients with early rheumatoid arthritis. *N Engl J Med* 343: 1586–1593
4 Feldmann M, Brennan FM, Maini RN (1996) Role of cytokines in Rheumatoid Arthritis. *Annu Rev Immunol* 14: 397–440

5 Arend WP, and Dayer JM (1995) Inhibition of the production and effects of interleukin-1 and tumor necrosis factor alpha in rheumatoid arthritis. *Arthritis Rheum* 38: 151–160
6 McInnes IB, al-Mughales J, Field M, Leung BP, Huang FP, Dixon R, Sturrock RD, Wilkinson PC, Liew FY (1996) The role of interleukin-15 in T-cell migration and activation in rheumatoid arthritis. *Nat Med* 2: 175–182
7 Kirman I, Nielsen OH (1996) Increased numbers of interleukin-15-expressing cells in active ulcerative colitis. *Am J Gastroenterol* 91: 1789–1794
8 Pashenkov M, Mustafa M, Kivisakk P, Link H (1999) Levels of interleukin-15-expressing blood mononuclear cells are elevated in multiple sclerosis. *Scand J Immunol* 50: 302–308
9 Gracie JA, Forsey RJ, Chan WL, Gilmour A, Leung BP, Greer MR, Kennedy K, Carter R, Wei XQ, Xu D et al (1999) A proinflammatory role for IL-18 in rheumatoid arthritis. *J Clin Invest* 104: 1393–1401
10 Pizarro TT, Michie MH, Bentz M, Woraratanadharm J, Smith MF, Jr., Foley E, Moskaluk CA, Bickston SJ, Cominelli F (1999) IL-18, a novel immunoregulatory cytokine, is up-regulated in Crohn's disease: expression and localization in intestinal mucosal cells. *J Immunol* 162: 6829–6835
11 Balashov KE, Rottman JB, Weiner HL, Hancock WW (1999) CCR5(+) and CXCR3(+) T cells are increased in multiple sclerosis and their ligands MIP-1alpha and IP-10 are expressed in demyelinating brain lesions. *Proc Natl Acad Sci USA* 96: 6873–6878
12 Nicoletti F, Di Marco R, Mangano K, Patti F, Reggio E, Nicoletti A, Bendtzen K, Reggio A (2001) Increased serum levels of interleukin-18 in patients with multiple sclerosis. *Neurology* 57: 342–344
13 McInnes IB, Leung BP, Sturrock RD, Field M, Liew FY (1997) Interleukin-15 mediates T cell-dependent regulation of tumor necrosis factor-alpha production in rheumatoid arthritis. *Nat Med* 3: 189–195
14 McInnes IB, Liew FY (1998) Interleukin 15: a proinflammatory role in rheumatoid arthritis synovitis. *Immunol Today* 19: 75–79
15 Cassatella MA, McDonald PP (2000) Interleukin-15 and its impact on neutrophil function. *Curr Opin Hematol* 7: 174–177
16 Waldmann TA, Tagaya Y (1999) The multifaceted regulation of Interleukin-15 expression and the role of this cytokine in NK cell differentiation and host response to intracellular pathogens. *Annu Rev Immunol* 17: 19–49
17 Mohamadzadeh M, Berard F, Essert G, Chalouni C, Pulendran B, Davoust J, Bridges G, Palucka AK, Banchereau J (2001) Interleukin 15 skews monocyte differentiation into dendritic cells with features of Langerhans cells. *J Exp Med* 194: 1013–1020
18 Grabstein KH, Eisenman J, Shanebeck K, Rauch C, Srinivasan S, Fung V, Beers C, Richardson J, Schoenborn MA, Ahdieh M et al (1994) Cloning of a T cell growth factor that interacts with the beta chain of the interleukin-2 receptor. *Science* 264: 965–968
19 Bamford R, Grant A, Burton J, Peters C, Kurys G, Goldman C, Brennan J, Roessler E, Waldman T (1994) The interleukin (IL) 2 receptor {beta} chain is shared by IL-2 and a

cytokine, provisionally designated IL-T, that stimulates T-cell proliferation and the induction of lymphokine-activated killer cells. *PNAS* 91: 4940–4944
20 Flamand L, Stefanescu I, Menezes J (1996) Human herpesvirus-6 enhances natural killer cell cytotoxicity *via* IL-15. *J Clin Invest* 97: 1373–1381
21 Atedzoe BN, Ahmad A, and Menezes J (1997) Enhancement of natural killer cell cytotoxicity by the human herpesvirus-7 *via* IL-15 induction. *J Immunol* 159: 4966–4972
22 Jullien D, Sieling PA, Uyemura K, Mar ND, Rea TH, Modlin RL (1997) IL-15, an immunomodulator of T cell responses in intracellular infection. *J Immunol* 158: 800–806
23 Carson WE, Ross ME, Baiocchi RA, Marien MJ, Boiani N, Grabstein K, Caligiuri MA (1995) Endogenous production of interleukin 15 by activated human monocytes is critical for optimal production of interferon-gamma by natural killer cells *in vitro*. *J Clin Invest* 96: 2578–2582
24 Chehimi J, Marshall JD, Salvucci O, Frank I, Chehimi S, Kawecki S, Bacheller D, Rifat S, Chouaib S (1997) IL-15 enhances immune functions during HIV infection. *J Immunol* 158: 5978–5987
25 Mohamadzadeh M, Takashima A, Dougherty I, Knop J, Bergstresser PR, Cruz PD, Jr. (1995) Ultraviolet B radiation up-regulates the expression of IL-15 in human skin. *J Immunol* 155: 4492–4496
26 Giri JG, Kumaki S, Ahdieh M, Friend DJ, Loomis A, Shanebeck K, DuBose R, Cosman D, Park LS, Anderson DM (1995) Identification and cloning of a novel IL-15 binding protein that is structurally related to the alpha chain of the IL-2 receptor. *Embo J* 14: 3654–3663
27 Waldmann T, Tagaya Y, Bamford R (1998) Interleukin-2, interleukin-15, and their receptors. *Int Rev Immunol* 16: 205–226
28 Kotake S, Schumacher HR, Jr., Yarboro CH, Arayssi TK, Pando JA, Kanik KS, Gourley MF, Klippel JH, Wilder RL (1997) *In vivo* gene expression of type 1 and type 2 cytokines in synovial tissues from patients in early stages of rheumatoid, reactive, and undifferentiated arthritis. *Proc Assoc Am Physicians* 109: 286–301
29 Thurkow EW, van der Heijden IM, Breedveld FC, Smeets TJ, Daha MR, Kluin PM, Meinders AE, Tak PP (1997) Increased expression of IL-15 in the synovium of patients with rheumatoid arthritis compared with patients with *Yersinia*-induced arthritis and osteoarthritis. *J Pathol* 181: 444–450
30 Oppenheimer-Marks N, Brezinschek RI, Mohamadzadeh M, Vita R, Lipsky PE (1998) Interleukin 15 is produced by endothelial cells and increases the transendothelial migration of T cells *in vitro* and in the SCID mouse-human rheumatoid arthritis model *in vivo*. *J Clin Invest* 101: 1261–1272
31 Shah MH, Hackshaw KV, Caligiuri MA (1998) A role for IL-15 in rheumatoid arthritis? *Nat Med* 4: 643
32 Harada S, Yamamura M, Okamoto H, Morita Y, Kawashima M, Aita T, Makino H (1999) Production of interleukin-7 and interleukin-15 by fibroblast-like synoviocytes from patients with rheumatoid arthritis. *Arthritis Rheum* 42: 1508–1516

33 Ziolkowska M, Koc A, Luszczykiewicz G, Ksiezopolska-Pietrzak K, Klimczak E, Chwalinska-Sadowska H, Maslinski W (2000) High levels of IL-17 in rheumatoid arthritis patients: IL-15 triggers *in vitro* IL-17 production *via* cyclosporin A-sensitive mechanism. *J Immunol* 164: 2832–2838
34 Ortiz AM, Garcia-Vicuna R, Sancho D, Laffon A, Sanchez-Madrid F, Gonzalez-Alvaro I (2000) Cyclosporin A inhibits CD69 expression induced on synovial fluid and peripheral blood lymphocytes by interleukin 15. *J Rheumatol* 27: 2329–2338
35 Klimiuk PA, Sierakowski S, Latosiewicz R, Cylwik B, Skowronski J, Chwiecko J (2001) Serum cytokines in different histological variants of rheumatoid arthritis. *J Rheumatol* 28: 1211–1217
36 Scola MP, Thompson SD, Brunner HI, Tsoras MK, Witte D, Van D, Grom AA, Passo MH, Glass DN (2002) Interferon-gamma: interleukin 4 ratios and associated type 1 cytokine expression in juvenile rheumatoid arthritis synovial tissue. *J Rheumatol* 29: 369–378.
37 Cho ML, Kim WU, Min SY, Min DJ, Min JK, Lee SH, Park SH, Cho CS, Kim HY (2002) Cyclosporine differentially regulates interleukin-10, interleukin-15, and tumor necrosis factor a production by rheumatoid synoviocytes. *Arthritis Rheum* 46: 42–51.
38 Wakisaka S, Suzuki N, Nagafuchi H, Takeba Y, Kaneko A, Asai T, Sakane T (2000) Characterization of tissue outgrowth developed *in vitro* in patients with rheumatoid arthritis: involvement of T cells in the development of tissue outgrowth. *Int Arch Allergy Immunol* 121: 68–79
39 Vey E, Burger D, Dayer JM (1996) Expression and cleavage of tumor necrosis factor-alpha and tumor necrosis factor receptors by human monocytic cell lines upon direct contact with stimulated T cells. *Eur J Immunol* 26: 2404–2409
40 Ortiz AM, Laffon A, Gonzalez-Alvaro I (2002) CD69 expression on lymphocytes and interleukin-15 levels in synovial fluids from different inflammatory arthropathies. *Rheumatol Int* 21: 182–188
41 Rezzonico R, Chicheportiche R, Imbert V, Dayer JM (2000) Engagement of CD11b and CD11c beta2 integrin by antibodies or soluble CD23 induces IL-1beta production on primary human monocytes through mitogen-activated protein kinase-dependent pathways. *Blood* 95: 3868–3877
42 Ribbens C, Dayer JM, Chizzolini C (2000) CD40-CD40 ligand (CD154) engagement is required but may not be sufficient for human T helper 1 cell induction of interleukin-2- or interleukin-15-driven, contact-dependent, interleukin-1beta production by monocytes. *Immunology* 99: 279–286
43 Mottonen M, Isomaki P, Luukkainen R, Toivanen P, Punnonen J, Lassila O (2000) Interleukin-15 up-regulates the expression of CD154 on synovial fluid T cells. *Immunology* 100: 238–244
44 Dalbeth N, Callan MF (2002) A subset of natural killer cells is greatly expanded within inflamed joints. *Arthritis Rheum* 46: 1763–1772
45 Yang L, Thornton S, Grom AA (2002) Interleukin-15 inhibits sodium nitroprusside-

induced apoptosis of synovial fibroblasts and vascular endothelial cells. *Arthritis Rheum* 46: 3010–3014

46 Rappl G, Kapsokefalou A, Heuser C, Rossler M, Ugurel S, Tilgen W, Reinhold U, Abken H (2001) Dermal fibroblasts sustain proliferation of activated T cells *via* membrane-bound interleukin-15 upon long-term stimulation with tumor necrosis factor-alpha. *J Invest Dermatol* 116: 102–109

47 Sen M, Lauterbach K, El-Gabalawy H, Firestein GS, Corr M, Carson DA (2000) Expression and function of wingless and frizzled homologs in rheumatoid arthritis. *Proc Natl Acad Sci USA* 97: 2791–2796

48 Sen M, Chamorro M, Reifert J, Corr M, Carson DA (2001) Blockade of Wnt-5A/frizzled 5 signaling inhibits rheumatoid synoviocyte activation. *Arthritis Rheum* 44: 772–781

49 Ruchatz H, Leung BP, Wei XQ, McInnes IB, Liew FY (1998) Soluble IL-15 receptor alpha-chain administration prevents murine collagen-induced arthritis: a role for IL-15 in development of antigen-induced immunopathology. *J Immunol* 160: 5654–5660

50 Nakamura K, Okamura H, Wada M, Nagata K, Tamura T (1989) Endotoxin-induced serum factor that stimulates gamma interferon production. *Infect Immun* 57: 590–595

51 Okamura H, Tsutsi H, Komatsu T, Yutsudo M, Hakura A, Tanimoto T, Torigoe K, Okura T, Nukada Y, Hattori K et al (1995) Cloning of a new cytokine that induces IFN-gamma production by T cells. *Nature* 378: 88–91

52 Ushio S, Namba M, Okura T, Hattori K, Nukada Y, Akita K, Tanabe F, Konishi K, Micallef M, Fujii M et al (1996) Cloning of the cDNA for human IFN-gamma-inducing factor, expression in *Escherichia coli*, and studies on the biologic activities of the protein. *J Immunol* 156: 4274–4279

53 Matsui K, Yoshimoto T, Tsutsui H, Hyodo Y, Hayashi N, Hiroishi K, Kawada N, Okamura H, Nakanishi K, Higashino K (1997) *Propionibacterium acnes* treatment diminishes $CD4^+$ $NK1.1^+$ T cells but induces type I T cells in the liver by induction of IL-12 and IL-18 production from Kupffer cells. *J Immunol* 159: 97–106

54 Stoll S, Jonuleit H, Schmitt E, Muller G, Yamauchi H, Kurimoto M, Knop J, Enk AH (1998) Production of functional IL-18 by different subtypes of murine and human dendritic cells (DC): DC-derived IL-18 enhances IL-12-dependent Th1 development. *Eur J Immunol* 28: 3231–3239

55 Stoll S, Muller G, Kurimoto M, Saloga J, Tanimoto T, Yamauchi H, Okamura H, Knop J, Enk AH (1997) Production of IL-18 (IFN-gamma-inducing factor) messenger RNA and functional protein by murine keratinocytes. *J Immunol* 159: 298–302

56 Udagawa N, Horwood NJ, Elliott J, Mackay A, Owens J, Okamura H, Kurimoto M, Chambers TJ, Martin TJ, Gillespie MT (1997) Interleukin-18 (interferon-gamma-inducing factor) is produced by osteoblasts and acts *via* granulocyte/macrophage colony-stimulating factor and not *via* interferon-gamma to inhibit osteoclast formation. *J Exp Med* 185: 1005–1012

57 Conti B, Jahng JW, Tinti C, Son JH, Joh TH (1997) Induction of interferon-gamma inducing factor in the adrenal cortex. *J Biol Chem* 272: 2035–2037

58 Prinz M, Hanisch UK (1999) Murine microglial cells produce and respond to interleukin-18. *J Neurochem* 72: 2215–2218
59 Nakanishi K, Yoshimoto T, Tsutsui H, Okamura H (2001) Interleukin-18 regulates both Th1 and Th2 responses. *Annu Rev Immunol* 19: 423–474
60 Ghayur T, Banerjee S, Hugunin M, Butler D, Herzog L, Carter A, Quintal L, Sekut L, Talanian R, Paskind M et al (1997) Caspase-1 processes IFN-gamma-inducing factor and regulates LPS-induced IFN-gamma production. *Nature* 386: 619–623
61 Gu Y, Kuida K, Tsutsui H, Ku G, Hsiao K, Fleming MA, Hayashi N, Higashino K, Okamura H, Nakanishi K et al (1997) Activation of interferon-gamma inducing factor mediated by interleukin-1beta converting enzyme. *Science* 275: 206–209
62 Sugawara S, Uehara A, Nochi T, Yamaguchi T, Ueda H, Sugiyama A, Hanzawa K, Kumagai K, Okamura H, Takada H (2001) Neutrophil proteinase 3-mediated induction of bioactive IL-18 secretion by human oral epithelial cells. *J Immunol* 167: 6568–6575
63 Torigoe K, Ushio S, Okura T, Kobayashi S, Taniai M, Kunikata T, Murakami T, Sanou O, Kojima H et al (1997) Purification and characterization of the human interleukin-18 receptor. *J Biol Chem* 272: 25737–25742
64 Hoshino K, Tsutsui H, Kawai T, Takeda K, Nakanishi K, Takeda Y, Akira S (1999) Cutting edge: generation of IL-18 receptor-deficient mice: evidence for IL-1 receptor-related protein as an essential IL-18 binding receptor. *J Immunol* 162: 5041–5044
65 Parnet P, Garka KE, Bonnert TP, Dower SK, Sims JE (1996) IL-1Rrp is a novel receptor-like molecule similar to the type I interleukin-1 receptor and its homologues T1/ST2 and IL-1R AcP. *J Biol Chem* 271: 3967–3970
66 Born TL, Thomassen E, Bird TA, Sims JE (1998) Cloning of a novel receptor subunit, AcPL, required for interleukin-18 signaling. *J Biol Chem* 273: 29445–29450
67 Leung BP, Culshaw S, Gracie JA, Hunter D, Canetti CA, Campbell C, Cunha F, Liew FY, McInnes IB (2001) A role for IL-18 in neutrophil activation. *J Immunol* 167: 2879–2886
68 Hyodo Y, Matsui K, Hayashi N, Tsutsui H, Kashiwamura S, Yamauchi H, Hiroishi K, Takeda K, Tagawa Y, Iwakura Y et al (1999) IL-18 up-regulates perforin-mediated NK activity without increasing perforin messenger RNA expression by binding to constitutively expressed IL-18 receptor. *J Immunol* 162: 1662–1668
69 Gerdes N, Sukhova GK, Libby P, Reynolds RS, Young JL, Schonbeck U (2002) Expression of interleukin (IL)-18 and functional IL-18 receptor on human vascular endothelial cells, smooth muscle cells, and macrophages: implications for atherogenesis. *J Exp Med* 195: 245–257
70 Afkarian M, Sedy JR, Yang J, Jacobson NG, Cereb N, Yang SY, Murphy TL, Murphy KM (2002) T-bet is a STAT1-induced regulator of IL-12R expression in naive CD4(+) T cells. *Nat Immunol* 3: 549–557
71 Yoshimoto T, Takeda K, Tanaka T, Ohkusu K, Kashiwamura S, Okamura H, Akira S, Nakanishi K (1998) IL-12 up-regulates IL-18 receptor expression on T cells, Th1 cells, and B cells: synergism with IL-18 for IFN-gamma production. *J Immunol* 161: 3400–3407

72 Xu D, Chan WL, Leung BP, Hunter D, Schulz K, Carter RW, McInnes IB, Robinson JH, Liew FY (1998) Selective expression and functions of interleukin 18 receptor on T helper (Th) type 1 but not Th2 cells. *J Exp Med* 188: 1485–1492
73 Smeltz RB, Chen J, Hu-Li J, Shevach EM (2001) Regulation of interleukin (IL)-18 receptor alpha chain expression on CD4(+) T cells during T helper (Th)1/Th2 differentiation. Critical downregulatory role of IL-4. *J Exp Med* 194: 143–153
74 Wesche H, Henzel WJ, Shillinglaw W, Li S, Cao Z (1997) MyD88: an adapter that recruits IRAK to the IL-1 receptor complex. *Immunity* 7: 837–847
75 Kanakaraj P, Ngo K, Wu Y, Angulo A, Ghazal P, Harris CA, Siekierka JJ, Peterson PA, Fung-Leung WP (1999) Defective interleukin (IL)-18-mediated natural killer and T helper cell type 1 responses in IL-1 receptor-associated kinase (IRAK)-deficient mice. *J Exp Med* 189: 1129–1138
76 Adachi O, Kawai T, Takeda K, Matsumoto M, Tsutsui H, Sakagami M, Nakanishi K, Akira S (1998) Targeted disruption of the MyD88 gene results in loss of IL-1- and IL-18-mediated function. *Immunity* 9: 143–150
77 Akira S, Takeda K, Kaisho T (2001) Toll-like receptors: critical proteins linking innate and acquired immunity. *Nat Immunol* 2: 675–680
78 Seki E, Tsutsui H, Nakano H, Tsuji N, Hoshino K, Adachi O, Adachi K, Futatsugi S, Kuida K, Takeuchi O et al (2001) Lipopolysaccharide-induced IL-18 secretion from murine Kupffer cells independently of myeloid differentiation factor 88 that is critically involved in induction of production of IL-12 and IL-1beta. *J Immunol* 166: 2651–2657
79 Morel JC, Park CC, Woods JM, Koch AE (2001) A novel role for interleukin-18 in adhesion molecule induction through NF kappa B and phosphatidylinositol (PI) 3-kinase-dependent signal transduction pathways. *J Biol Chem* 276: 37069–37075
80 Micallef MJ, Ohtsuki T, Kohno K, Tanabe F, Ushio S, Namba M, Tanimoto T, Torigoe K, Fujii M, Ikeda M et al (1996) Interferon-gamma-inducing factor enhances T helper 1 cytokine production by stimulated human T cells: synergism with interleukin-12 for interferon-gamma production. *Eur J Immunol* 26: 1647–1651
81 Dao T, Mehal WZ, Crispe IN (1998) IL-18 augments perforin-dependent cytotoxicity of liver NK-T cells. *J Immunol* 161: 2217–2222
82 Dao T, Ohashi K, Kayano T, Kurimoto M, Okamura H (1996) Interferon-gamma-inducing factor, a novel cytokine, enhances Fas ligand-mediated cytotoxicity of murine T helper 1 cells. *Cell Immunol* 173: 230–235
83 Tsutsui H, Nakanishi K, Matsui K, Higashino K, Okamura H, Miyazawa Y, Kaneda K (1996) IFN-gamma-inducing factor up-regulates Fas ligand-mediated cytotoxic activity of murine natural killer cell clones. *J Immunol* 157: 3967–3973
84 Takeda K, Tsutsui H, Yoshimoto T, Adachi O, Yoshida N, Kishimoto T, Okamura H, Nakanishi K, Akira S (1998) Defective NK cell activity and Th1 response in IL-18-deficient mice. *Immunity* 8: 383–390
85 Hoshino T, Wiltrout RH, Young HA (1999) IL-18 is a potent coinducer of IL-13 in NK and T cells: a new potential role for IL-18 in modulating the immune response. *J Immunol* 162: 5070–5077

86 Hoshino T, Yagita H, Ortaldo JR, Wiltrout RH, Young HA (2000) *In vivo* administration of IL-18 can induce IgE production through Th2 cytokine induction and up-regulation of CD40 ligand (CD154) expression on CD4+ T cells. *Eur J Immunol* 30: 1998–2006
87 Yoshimoto T, Mizutani H, Tsutsui H, Noben-Trauth N, Yamanaka K, Tanaka M, Izumi S, Okamura H, Paul WE, Nakanishi K (2000) IL-18 induction of IgE: dependence on CD4+ T cells, IL-4 and STAT6. *Nat Immunol* 1: 132–137
88 Xu D, Trajkovic V, Hunter D, Leung BP, Schulz K, Gracie JA, McInnes IB, Liew FY (2000) IL-18 induces the differentiation of Th1 or Th2 cells depending upon cytokine milieu and genetic background. *Eur J Immunol* 30: 3147–3156
89 Yoshimoto T, Tsutsui H, Tominaga K, Hoshino K, Okamura H, Akira S, Paul WE, Nakanishi K (1999) IL-18, although antiallergic when administered with IL-12, stimulates IL-4 and histamine release by basophils. *Proc Natl Acad Sci USA* 96: 13962–13966
90 Munder M, Mallo M, Eichmann K, Modolell M (1998) Murine macrophages secrete interferon gamma upon combined stimulation with interleukin (IL)-12 and IL-18: A novel pathway of autocrine macrophage activation. *J Exp Med* 187: 2103–2108
91 Fukao T, Matsuda S, Koyasu S (2000) Synergistic effects of IL-4 and IL-18 on IL-12-dependent IFN-gamma production by dendritic cells. *J Immunol* 164: 64–71
92 Netea MG, Kullberg BJ, Verschueren I, Van Der Meer JW (2000) Interleukin-18 induces production of proinflammatory cytokines in mice: no intermediate role for the cytokines of the tumor necrosis factor family and interleukin-1beta. *Eur J Immunol* 30: 3057–3060
93 Olee T, Hashimoto S, Quach J, Lotz M (1999) IL-18 is produced by articular chondrocytes and induces proinflammatory and catabolic responses. *J Immunol* 162: 1096–1100
94 Mee JB, Alam Y, Groves RW (2000) Human keratinocytes constitutively produce but do not process interleukin-18. *Br J Dermatol* 143: 330–336
95 McKenzie RC, Boyce F, Forsey R, Gracie JA, Szepietowski JC, Weller R, Sabin E, Howie SEM (2000) Psoriatic skin expresses high levels of interleukin-18 (IL-18) and IL-18 receptor (IL-18R). *Br J Dermatol* 142: 618
96 Novick D, Kim SH, Fantuzzi G, Reznikov LL, Dinarello CA, and Rubinstein M (1999) Interleukin-18 binding protein: a novel modulator of the Th1 cytokine response. *Immunity* 10: 127–136
97 Aizawa Y, Akita K, Taniai M, Torigoe K, Mori T, Nishida Y, Ushio S, Nukada Y, Tanimoto T, Ikegami H, Ikeda M, Kurimoto M (1999) Cloning and expression of interleukin-18 binding protein. *FEBS Lett* 445: 338–342
98 Reznikov LL, Kim SH, Westcott JY, Frishman J, Fantuzzi G, Novick D, Rubinstein M, Dinarello CA (2000) IL-18 binding protein increases spontaneous and IL-1-induced prostaglandin production *via* inhibition of IFN-gamma. *Proc Natl Acad Sci USA* 97: 2174–2179
99 Paulukat J, Bosmann M, Nold M, Garkisch S, Kampfer H, Frank S, Raedle J, Zeuzem

S, Pfeilschifter J, Muhl H (2001) Expression and release of IL-18 binding protein in response to IFN-gamma. *J Immunol* 167: 7038–7043
100 Novick D, Schwartsburd B, Pinkus R, Suissa D, Belzer I, Sthoeger Z, Keane WF, Chvatchko Y, Kim SH, Fantuzzi G et al (2001) A novel IL-18BP ELISA shows elevated serum IL-18BP in sepsis and extensive decrease of free IL-18. *Cytokine* 14: 334–342
101 Corbaz A, ten Hove T, Herren S, Graber P, Schwartsburd B, Belzer I, Harrison J, Plitz T, Kosco-Vilbois MH, Kim SH et al (2002) IL-18-binding protein expression by endothelial cells and macrophages is up-regulated during active Crohn's disease. *J Immunol* 168: 3608–3616
102 Kim SH, Eisenstein M, Reznikov L, Fantuzzi G, Novick D, Rubinstein M, Dinarello CA (2000) Structural requirements of six naturally occurring isoforms of the IL-18 binding protein to inhibit IL-18. *Proc Natl Acad Sci USA* 97: 1190–1195
103 Pan G, Risser P, Mao W, Baldwin DT, Zhong AW, Filvaroff E, Yansura D, Lewis L, Eigenbrot C, Henzel WJ, Vandlen R (2001) IL-1H, an interleukin 1-related protein that binds IL-18 receptor/IL-1Rrp. *Cytokine* 13: 1–7
104 Dinarello CA (1999) IL-18: A TH1-inducing, proinflammatory cytokine and new member of the IL-1 family. *J Allergy Clin Immunol* 103: 11–24
105 Tanaka M, Harigai M, Kawaguchi Y, Ohta S, Sugiura T, Takagi K, Ohsako-Higami S, Fukasawa C, Hara M, Kamatani N (2001) Mature form of interleukin 18 is expressed in rheumatoid arthritis synovial tissue and contributes to interferon-gamma production by synovial T cells. *J Rheumatol* 28: 1779–1787.
106 Yamamura M, Kawashima M, Taniai M, Yamauchi H, Tanimoto T, Kurimoto M, Morita Y, Ohmoto Y, Makino H (2001) Interferon-gamma-inducing activity of interleukin-18 in the joint with rheumatoid arthritis. *Arthritis Rheum* 44: 275–285.
107 Moller B, Kukoc-Zivojnov N, Kessler U, Rehart S, Kaltwasser JP, Hoelzer D, Kalina U, Ottmann OG (2001) Expression of interleukin-18 and its monokine-directed function in rheumatoid arthritis. *Rheumatology (Oxford)* 40: 302–309
108 Moller B, Kessler U, Rehart S, Kalina U, Ottmann OG, Peter Kaltwasser J, Hoelzer D, Kukoc-Zivojnov N (2002) Expression of interleukin-18 receptor in fibroblast-like synoviocytes. *Arthritis Res* 4: 139–144
109 Kawashima M, Yamamura M, Taniai M, Yamauchi H, Tanimoto T, Kurimoto M, Miyawaki S, Amano T, Takeuchi T, Makino H (2001) Levels of interleukin-18 and its binding inhibitors in the blood circulation of patients with adult-onset Still's disease. *Arthritis Rheum* 44: 550–560
110 Gracie JA, Robertson SE, McInnes IB (2003) Interleukin-18. *J Leukoc Biol* 73: 213–224
111 Lubberts E, Joosten LA, Oppers B, van den Bersselaar L, Coenen-de Roo CJ, Kolls JK, Schwarzenberger P, van de Loo FA, van den Berg WB, Helsen MM et al (2001) IL-1-independent role of IL-17 in synovial inflammation and joint destruction during collagen-induced arthritis. *J Immunol* 167: 1004–1013
112 Morel JC, Park CC, Kumar P, Koch AE (2001) Interleukin-18 induces rheumatoid arthritis synovial fibroblast CXC chemokine production through NFkappaB activation. *Lab Invest* 81: 1371–1383

113 Park CC, Morel JC, Amin MA, Connors MA, Harlow LA, Koch AE (2001) Evidence of IL-18 as a novel angiogenic mediator. *J Immunol* 167: 1644–1653
114 Horwood NJ, Udagawa N, Elliott J, Grail D, Okamura H, Kurimoto M, Dunn AR, Martin T, Gillespie MT (1998) Interleukin 18 inhibits osteoclast formation *via* T cell production of granulocyte macrophage colony-stimulating factor. *J Clin Invest* 101: 595–603
115 Wei X, Leung BP, Arthur HM, McInnes IB, Liew FY (2001) Reduced incidence and severity of collagen-induced arthritis in mice lacking IL-18. *J Immunol* 166: 517–521
116 Leung BP, McInnes IB, Esfandiari E, Wei XQ, Liew FY (2000) Combined effects of IL-12 and IL-18 on the induction of collagen-induced arthritis. *J Immunol* 164: 6495–6502
117 Plater-Zyberk C, Joosten LA, Helsen MM, Sattonnet-Roche P, Siegfried C, Alouani S, van De Loo FA, Graber P, Aloni S, Cirillo R et al (2001) Therapeutic effect of neutralizing endogenous IL-18 activity in the collagen-induced model of arthritis. *J Clin Invest* 108: 1825–1832

IL-17 as a contributor to inflammation and destruction in rheumatoid arthritis

Pierre Miossec

Department of Immunology and Rheumatology, Hôpital Edouard Herriot, 69437 Lyon Cedex 03, France

Introduction

Therapeutic control of TNFα and IL-1 with specific inhibitors has clearly demonstrated the contribution of these cytokines to the destructive and inflammatory patterns associated with rheumatoid arthritis (RA). At the same time, the list of additional cytokines with similar biological effects has been growing extensively. Among these, IL-17 is of interest since it shares properties and interacts with TNFα and IL-1. In addition, being produced by T cells, demonstration of the contribution of IL-17 in arthritis would add some new information on the contribution of T cells to the pathogenesis of RA. These issues are important since they may lead to new therapeutic applications. Since this topic has been the subject of a recent review, we will focus on the latest developments [1].

Discovery and structure

The IL-17 story started recently when two reports describing the molecule later named IL-17 were published in 1995 and 1996 by researchers at Immunex [2] and Schering-Plough [3]. From the beginning the role of IL-17 in inflammation was indicated by its effects on the production of proinflammatory cytokines by mesenchymal cells.

IL-17 is a 17 kDa protein, which is secreted as a homodimer. Mouse and human IL-17 share 62% of their amino acid sequence. Human IL-17 acts on mouse cells whereas the opposite is not the case. Human IL-17 has a 72% homology with HVS13, an open reading frame from the T cell lymphotrophic Herpes virus saimiri, and a 63% homology with murine CTLA8 [2].

IL-17 was first described as a T cell product. Up until now, this specificity remains valid with almost no exception. However, other cells may also be a source of IL-17. Eosinophils, which can produce almost any cytokine, were shown to be a source of IL-17 in bronchial secretions during asthma [4]. More recently, PMN were

Cytokines and Joint Injury, edited by Wim B. van den Berg and Pierre Miossec

found to produce IL-17 in a model of LPS induced airway neutrophilia, even in T cell-deficient Scid mice [5]. The respective contribution of neutrophil-derived IL-17 over that from T cells remains to be clarified.

IL-17 family

More recently, a systematic gene bank analysis found other molecules with sequences with four cysteine residues, which are characteristic of IL-17. These proteins were classified as members, now up to five, of the IL-17 family [1, 6, 7], listed as IL-17A to F [8, 9]. We will focus on the founding father of the IL-17 family, IL-17A unless indicated (Fig. 1).

Regarding the role of other IL-17 family members in inflammation, IL-17B appears the most directly involved in joint physiology. IL-17B is expressed and produced by chondrocytes [7]. This could give it a particular role during cartilage degradation. IL-17C was found to induce PMN-related lung inflammation [9]. IL-17D stimulates the production of proinflammatory cytokines such as IL-6, IL-8, GM-CSF. Conversely, IL-17D was found to inhibit myeloid progenitor cell formation [8]. In sharp contrast, IL-17E appears to contribute to eosinophilia, B cell hyperplasia and altered antibody production with hyper IgE levels [10]. These features were observed in IL-17E transgenic mice which showed an increased Th2 cytokine patterns associated with high levels of IL-4, IL-5 and eotaxin [11]. Similarly IL-17F, also referred to as IL-25, has been classified as a Th2 cytokine, since it induces the production of IL-4, IL-5, IL-13. These effects are associated with eosinophilia, suggesting the contribution of IL-25 to allergy [9]. Mast cells appear to be a potent source of IL-25 leading to increased Th2 response [12].

Mode of action

The mouse IL-17 receptor was first isolated at the time of IL-17 discovery, followed by the description of the human counterpart [13, 14]. Comparison of known sequences and protein structure indicated that the IL-17 receptor is a member of a new family [13]. Contrasting with the strict limitation of IL-17 production to T cells, its receptor is widely expressed on almost any cell type. Affinity measurements showed however that the isolated IL-17 receptor interacts with IL-17 with a rather low affinity, lower than the one observed with other cytokine receptors, which are often composed of more than one chain. This suggests that additional chains may be associated with the IL-17 receptor [7]. We will see below that a soluble form of the mouse low affinity chain is a potent inhibitor of IL-17 function, even with human cells. In addition, four additional receptors, which share sequence homology with the IL-17 receptor, can interact with new IL-17 family members.

The IL-17 family

IL-17 family member	Other name	Major effects
IL-17A	CTLA-8	Inflammation Neutrophil activation Matrix destruction
IL-17B		Cartilage destruction
IL-17C		Neutrophil activation Lung inflammation
IL-17D	IL-27	Inflammation
IL-17E	IL-25	Eosinophil activation IgE secretion Allergic reaction
IL-17F	ML-1	Allergic reaction

Figure 1
The IL-17 family

Intracellular IL-17 signaling pathways have been identified and their pharmacological control is under active research. IL-17 shares transcriptional pathways with IL-1 and TNFα. In particular, as for the later, p38 and NF-κB are the key transcriptional factors involved in IL-17 function. These pathways have been identified in synoviocytes [15], chondrocytes [16] and appear to be the same in osteoblasts and myoblasts [17]. The use of common pathways is of importance when considering the combination of cytokines, which often act in synergy.

Inducers of IL-17

Because of its proinflammatory properties, IL-17 production is under the control of other proinflammatory cytokines (Fig. 2). Among these, IL-15 is a cytokine produced by mesenchymal cells, which contributes to T cell activation and proliferation. IL-15 has been shown to induce IL-17 production by T cells, an effect inhibit-

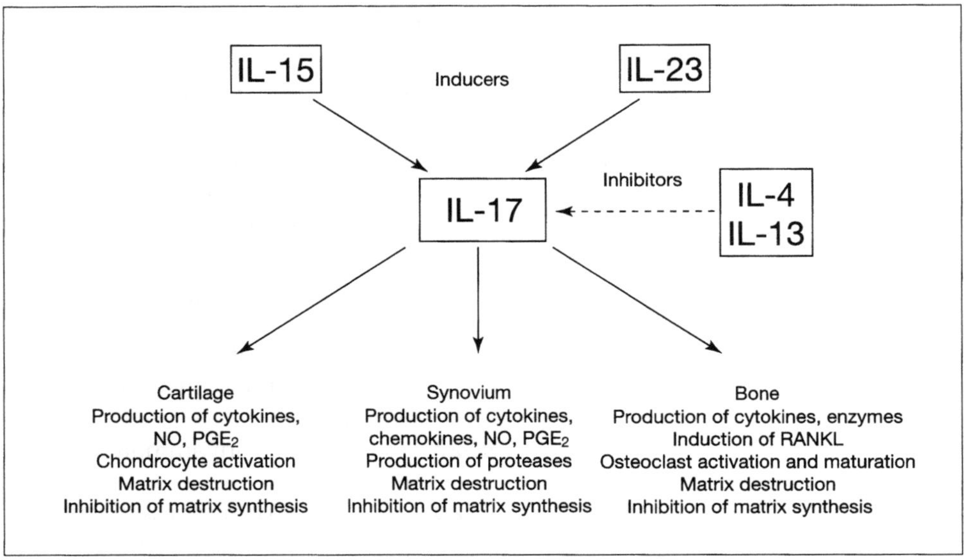

Figure 2
Effects of IL-17 on joint compartments

ed by cyclosporin A [18]. Similar results have been obtained for neutrophil-derived IL-17 [5].

More recently, the contribution of monocyte-derived signals acting on T cells has been clarified with the demonstration of the enhancing effect of the new IL-23. IL-23 is a new member of the IL-12 family, which shares the p40 subunit with IL-12. IL-23 was found to increase IL-17 production by T cells [19]. IL-23 also appears to induce IL-17F production.

IL-17 and the other proinflammatory cytokines

IL-17 was described for its ability to induce the production of IL-6 and IL-8 by fibroblasts [3]. Early studies used synoviocytes suggesting from the beginning a role for IL-17 in joint inflammation. These properties were long ago assigned also to IL-1 and TNFα. When the three cytokines were compared for their effect on IL-6 production, clearly IL-1 was the most potent with a few pg/ml as potent as 10 ng/ml of TNFα or IL-17. The most interesting part of these studies with disease application in mind was the demonstration of an additive and often synergistic effect between IL-1 or TNFα and IL-17 [20]. This observation is important since during synovium inflammation, the combined production of TNFα, IL-1 and IL-17 is to be expected.

As IL-17 is produced by T cells, the synergy will be controlled by the presence of IL-17-producing T cells, whereas the presence of monocytes producing IL-1 and TNFα is far less specific, as also observed in osteoarthritis. In these conditions, IL-17 will have a regulatory role responsible for a fine tuning of the inflammatory response [21]. This synergy was observed first with synoviocytes [20] and later with chondrocytes [22, 23], osteoblasts [24], myoblasts [17].

In addition, these cytokines interact with each other at the level of their production. IL-17 was found to increase the production of IL-1 and TNFα by monocytes [25]. Thus low concentrations of IL-17 will increase the proinflammatory properties of these monocyte-derived cytokines. This feature defines IL-17 as a regulatory cytokine, in accordance to the role assigned to T cells.

The mechanisms responsible for such synergy are not fully understood. The signaling pathways and transcription factors involved with IL-1, TNFα and IL-17 are very similar. However *in vitro* studies have used rather high concentrations of a single cytokine. When lower concentrations, which reflect more directly the *in vivo* situation, were used in combination, a very potent effect on transcription factor activation could be observed. Our observations with various mesenchymal cells such as synoviocytes, osteoblasts or myoblasts indicate that cytokine combination results in a recruitment of additional transcription factors not activated when cytokines are used alone, even at optimal concentrations [17].

It should be noticed that the number of IL-17 producing cells in RA synovium is indeed very small [26]. Although the level of secretion per cell could then be quite high, the biological significance of such secretion appears to imply interactions between IL-17 and other monocyte/synoviocyte derived factors. This could be the way for such a limited cell population to contribute to disease.

IL-17 and host defense

Cytokines have an important effect on systemic response to stress. In addition, T cells are critical for cell-mediated immunity, suggesting the role of IL-17 in host defense. Indeed IL-17 induces the maturation of CD34 hematopoietic lymphoid and myeloid progenitors [27]. Control of acute situation includes the role of IL-17 in neutrophil maturation in combination with stem cell factor and GM-CSF [3]. Circulating neutrophil counts were found to be regulated by a feedback loop involving IL-17 and G-CSF [28]. In the same way, IL-17 induced neutrophil accumulation in infected lungs. In a model of intra-abdominal sepsis, abscess formation was found to be related to the IL-17 production by CD4 T cells as antibody inhibition of IL-17 prevented abscess formation [29]. In addition, IL-17 activated neutrophils to produce chemokines and hematopoietic growth factors [30]. Mice deficient in IL-17 receptor showed an increased sensitivity and mortality to lung bacterial infection, suggesting a mechanism for CD4 T cell defects associated with bacterial pneumonia

[31]. In relation with chronic inflammation, increased IL-17 expression was observed in the inflamed musosa of patients with inflammatory bowel diseases [32]. In Behcet's disease, increased levels of circulating IL-17 were observed in active disease in association with similar findings for IL-18 and IFNα [33].

IL-17 contributes to other features of T cell mediated immunity. Regarding tumor control, IL-17 induced T cell-dependent tumor specific immunity, with an effect on angiogenesis associated with tumor growth in mouse [34, 35]. Regarding graft immunity, inhibition of IL-17 could control experimental allogeneic graft rejection [36]. Accordingly, detection of IL-17 expression may serve as a predictive parameter for subclinical renal allograft rejection [37, 38]. In RA, it is possible that the immune defect associated with TNFα inhibition could be related to a role for IL-17 to control opportunistic infections such as tuberculosis. In this case, inhibition of TNFα modifies the synergy of the cytokines acting on or produced by T cells.

These effects of IL-17 on host defense have to be considered carefully when inhibition of IL-17 is considered in patients where systemic immune defenses are already altered.

IL-17 in arthritis

The first results in arthritis were obtained when an ELISA was made available. Levels of IL-17 were found higher in RA synovial fluids when compared to OA fluids [39]. It was critical to demonstrate that the measured IL-17 in synovial fluid was functional and produced by the synovium itself [26]. Quantification of IL-17 was performed using a specific biological assay looking at the production of IL-6 by synoviocytes stimulated with supernatants from RA synovium first preincubated with a blocking anti-IL-17 antibody. Functional IL-17 was spontaneously produced by many RA samples at high levels, a few OA at very low levels and none of normal synovium explant cultures. IL-17 mRNA expression was demonstrated by RT-PCR in RA but not in OA synovium samples. IL-15 produced by synoviocytes is also a potent inducer of IL-17 production [18]. RA synovium samples with high expression of IL-12 and IL-18, two Th1 inducing cytokines, showed also a high IL-17 expression [40]. It is of interest to note that IL-17 was also found highly expressed in the brain of patients with multiple sclerosis (MS) by micro array analysis [41]. Similar conclusion was observed in experimental autoimmune encephalomyelitis in IL-12 β2 chain receptor deficient mice. These results suggested the contribution of IL-23 in disease expression through the production of proinflammatory cytokines including IL-17 [42]. Recently, the contribution of ICOS, a co-stimulatory receptor expressed by T cells, in the expression of IL-17 in collagen-induced arthritis was shown. ICOS knockout mice were resistant to collagen-induced arthritis [43].

In RA synovium, IL-17 producing cells represent a minor subset of lymphocytes found in lymphocytic infiltrates around vessels. These cells express CD4 but have lost their CD3. On average, they are large cells, with a remote nucleus, giving the appearance of plasma cells [44, 45]. The parallel can be drawn between B cells, which have lost their B cell membrane receptor when they start to produce soluble immunoglobulin as plasma cells. Similarly, this would be in line with a high production of IL-17 per single cell.

The addition of exogenous IL-17 to RA synovium pieces resulted in an increase of IL-6 production, whereas that of a blocking anti-IL-17 antibody reduced it. These first results indicated that IL-17 was present and could contribute to the active proinflammatory and destructive pattern characteristic of RA.

Effects on synovium cell populations

In synovium, T cells are in contact with other cells such as synoviocytes, monocytes and dendritic cells. In bone, the same interactions occur whereas chondrocytes protected by cartilage extracellular matrix are rather the target of soluble factors. These interactions are summarized in Figure 1 and the related functions of IL-17 in Figure 2.

On synoviocytes, IL-1β, TNFα and IL-17 induced IL-6 and LIF production, IL-1 being much more potent than IL-17 or TNFα [20]. Using low concentrations of IL-17 and IL-1β in combination, a synergistic effect was observed for the production of IL-6 with an additive effect for LIF production. Thus, low levels of cytokines produced by monocytes (IL-1) and T cells (IL-17) can act together on synoviocytes. Then some RA synovium T cells producing IL-17 activate mesenchymal cells leading to an increased proinflammatory pattern sensitive in part, to Th2 cytokine regulation.

Increased production of chemokines and other cytokines was observed with IL-17. Synergy was again observed between IL-1, TNFα and IL-17 for the production of MIP-3α now renamed CCL20, pa chemokine involved in the migration of memory T cells and immature dendritic cells [46]. Similar findings were observed with myoblasts [17].

On dendritic cells (DC), IL-17 appears to favor mouse DC progenitor differentiation when cells are stimulated with G-CSF and IL-4 [47]. This favors graft rejection, which is controlled in part by IL-17 inhibition. In RA synovium, a defect in DC maturation was observed with a relative accumulation of immature DC [45]. IL-17 interacts with IL-1 and TNFα to favor DC migration, acting in synergy to induce the production of CCL20/MIP-3α [46, 48]. Such production by synoviocytes was inhibited by the anti-inflammatory cytokines IL-4 and IL-13. CCL20 producing cells were located in the lining layer and perivascular infiltrates in close association with CD1a immature dendritic cells. Addition of exogenous IL-17 to synovium explants increased CCL20 production. Conversely, specific soluble receptors for IL-1, IL-17 and TNFα inhibited CCL20 production to various degrees but maximal inhibition was obtained only when the three receptors were combined. These findings indicate that interactions between monocyte and Th1 cell derived cytokines contribute to the recruitment of T cells and DC by enhancing the production of CCL20 by synoviocytes.

Effects of IL-17 on joint destruction

Proinflammatory cytokines produced by RA synovium have destructive patterns (Fig. 2). As described for IL-1 and TNFα, IL-17 increased the spontaneous production of metalloproteases such as MMP-1 by synoviocytes, IL-1 being again more potent. Addition of IL-4, IL-13 and IL-10 to synoviocytes reduced the spontaneous and IL-17 or/and IL-1β induced production of MMP-1 and induced that of its inhibitor TIMP-1. In the presence of anti-IL-17 blocking antibody, MMP-1 production and collagenase activity by RA synovium was reduced and associated with a reduction of type I collagen C-telopeptide fragments (CTX) released in the supernatants, demonstrating the direct contribution of IL-17 in destruction [49]. From these results, IL-17 and thus its producing T cells appear to contribute to the RA inflammatory destructive process.

On chondrocytes, IL-17 induced PGE_2 and nitric oxide production by cartilage explants in an IL-1 independent but LIF dependent fashion [23, 50–52]. IL-17 increased collagenase-3 production as for IL-1, mainly through AP-1 activation [53]. Here also, a synergistic effect on cartilage degradation was seen when IL-17 was combined with TNF and IL-1 [54]. However differences were observed where IL-17 activates FosB and IL-1β cfos, suggesting differential pathways [53]. Indeed opposite effects have been reported when comparing IL-1 and IL-17 activity on NO effect on cartilage matrix breakdown [55]. IL-17 was found to increase chemokine production with a increasing effect on IL-1β mRNA expression [56]. Other members of the IL-17 family such as IL-17E or IL-25 have also direct cartilage breakdown activity [51].

On osteoblasts, IL-17 induced IL-6 production [57]. As in chondrocytes, but only in combination with TNFα, IL-17 induced NO production in osteoblastic cells and fetal mouse metatarsals through a NF-κB-dependent mechanism [58]. During cell interactions between osteoblasts and osteoclast precursors, the presence of IL-17 induced osteoclastogenesis, a effect found to be PGE_2 mediated [39]. IL-17 and other cytokines stimulating osteoclastogenesis, such as IL-1β and TNFα, increased the expression of RANKL with a decrease of its regulator OPG expression in osteoblasts/stromal cells [59]. IL-17 producing T cells express the membrane form and secrete the soluble form of RANKL [60]. These functions make IL-17 a new cytokine involved in bone resorption [24]. In the context of RA, T cells in juxta-articular bone are the source of IL-17, which then acts locally leading to destruction [44].

The origin and role of IL-17 in cartilage and bone destruction during RA remain to be clarified. In human *ex vivo* models, addition of IL-17 enhanced IL-6 production and collagen destruction and inhibited collagen synthesis by RA synovium explants [44]. In isolated mouse cartilage, IL-17 increased cartilage proteoglycan loss and inhibited its synthesis [44, 61]. On human RA bone explants, IL-17 also increased bone resorption and decreased formation [44]. In these conditions, addition of IL-1 increased IL-17 effect. Blocking of bone derived endogenous IL-17 with specific inhibitors resulted in a protective inhibition of bone destruction. This was observed with the administration of soluble IL-17 receptor to collagen induced arthritis [62]. Conversely, intra-articular administration of IL-17 into a normal mouse joint induced cartilage degradation [63]. These effects were IL-1 independent as the IL-17 effect was still observed in IL-1 deficient mice [64]. This was extended with the continuous administration of IL-17 by gene therapy [65]. In the same model, induction of bone erosion was associated with an unbalance of the RANKL/OPG ratio in favor of an excess of RANKL stimulating osteoclastogenesis and bone erosion [66]. In the same field of bone erosion, IL-17 could contribute to periodontal diseases following local bacterial stimulation [67]. In conclusion, the contribution of IL-17 derived from synovium and bone marrow T cells to joint destruction suggests the control of IL-17 for the treatment of RA.

IL-17 and the Th1/Th2 cytokine balance

As described as a T cell product, it was logical to search for the position of IL-17 in the now classical Th1/Th2 subsets [68]. This classification although oversimplified, is useful to understand cytokine patterns in relation to disease. RA was thus classified as a Th1 disease with high levels of destruction with defective repair whereas conversely, scleroderma is characterized by abnormal repair activity leading to diffuse fibrosis. Accordingly, IL-4 is defective in RA but contributes to matrix formation in scleroderma. Defects have been identified in RA synovium T cells, which appear unable to switch to IL-4 production [69]. Similar conclusion was observed with RA synovium Th1 T cell clones unable to switch to a Th2 phenotype even with Th2 promoting conditions [70].

Results with these T cell lines and clones derived from RA synovium allowed the classification of IL-17 as a Th1 cytokine [71]. Indeed IL-17 was always produced in association with IFNγ and for some clones, in association with a mixed Th0 pattern with IFNγ and low levels of IL-4. This conclusion was extended when looking at function. When fixed T cell clones were incubated with RA synoviocytes, IL-17 producing cells were the most potent inducers of IL-6 production by synoviocytes [72]. Moreover, they inhibited synoviocyte collagen synthesis. Opposite results were obtained with Th2 clones. These results indicate that IL-17 producing Th1 cells contribute to the defective repair activity in joint inflammation.

Pre-incubation of clones in Th1 conditions (IL-12 plus anti-IL-4) increased IL-6 production whereas Th2 conditions (IL-4 plus anti-IL-12) strongly inhibited IL-6 production and restored repair activity [70]. As RA synovium is infiltrated by Th1 cells, local cell interactions result in a proinflammatory pattern with defective repair, which can be reversed at least in part, by a Th2 pattern.

With clones from atopic individuals, it was not possible to associate IL-17 with a particular Th cytokine subset [73]. Such apparent discrepancy may in fact reflect the contribution of the local environment to the cytokine pattern. Indeed induction of a switch from Th1 to Th2 is difficult to obtain with T cells clones from RA synovium. This may be linked to a prolonged *in vivo* exposure to conditions where Th1 cytokines are over represented and Th2 ones rather defective.

IL-17E appears to be in contrast a Th2 promoting cytokine in animal models of allergy [9]. In addition, IL-17E transgenic mice showed both G-CSF dependent neutrophilia and high Th2 cytokine expression [10].

Control of IL-17 and regulation

At present two specific IL-17 inhibitors have been tested *in vitro* and in animal models. They include blocking monoclonal antibodies and a soluble IL-17 receptor (sR). These two tools are the equivalent of infliximab and etanercept for TNFα inhibition respectively.

Although these inhibitors have not been tested in RA or other inflammatory diseases, the rationale for such application has been obtained from studies in animal and human models. Soluble IL-17 receptor reduced severity and joint damage of adjuvant arthritis in the rat [62]. Using RA synovium and bone *ex vivo* models, addition of both types of IL-17 inhibitors reduced inflammation and destruction. As IL-1, TNFα and IL-17 have many additive and/or synergistic effects *in vitro*, the question remains as to whether their combined inhibition would lead to an enhanced effect on *ex vivo* models of synovium inflammation and bone destruction [74]. Indeed anti-TNFα therapy has shown clear efficacy in RA treatment, but some patients do not respond and the treatment is only suspensive.

RA synovium and bone explants were cultured in the presence of etanercept, the TNFα-p75 soluble receptor (sR) alone as in current therapy, and in combination with the type II IL-1 and IL-17 sR. In synovium, all three sR used alone decreased IL-6 production by approximately one third and that of CTX by one half. Combination was more effective, inhibiting the effects up to 70%. In bone, similar results were obtained with the combination on IL-6 production. For CTX release, optimal inhibition was dependent on the presence of sIL-1R. These results support the concept of combination therapy, which may increase the percentage of responding patients as well as the degree of individual patient response.

IL-17 function can also be inhibited by Th2 cytokines, which control at the same time, among other cytokines, IL-1 and TNFα. On T cell clones from RA synovium, addition of Th2-promoting conditions strongly reduced the production of both

IFNγ and IL-17, while Th1-promoting conditions increased the production of these cytokines [70]. These results demonstrate that RA Th2 clones readily switch, while Th1 and Th0 clones are more stable.

This was further extended with the administration of IL-4 to mice with arthritis through gene therapy [75]. Although severe inflammation persisted in treated mice, IL-4 prevented bone erosion and diminished the formation of osteoclast-like cells. Messenger RNA and protein levels of IL-17 in synovium were suppressed. RANKL expression was markedly suppressed leading to reduced osteoclast activation and finally bone destruction. Similarly, bone samples of RA patients showed consistent suppression by IL-4 of type I collagen breakdown. IL-4 also enhanced synthesis of type I procollagen, suggesting an effect on tissue repair.

Regarding potential side effects associated with IL-17 inhibition, blocking of IL-17 in human diseases may have limitations by reducing T cell mediated immunity. This may affect host defense, neutrophil as well as T cell mediated infections, tumor cell control [76].

Conclusion

The demonstration of the activities of IL-17 has added new data in favor of the contribution of T cells to RA, an issue once highly discussed [77]. This proinflammatory cytokine is produced by RA synovial membrane, more particularly by infiltrating Th1 lymphocytes. When in contact with synoviocytes, IL-17 producing T cells have an increased ability to favor destruction and to block repair activity in synoviocytes. Furthermore, IL-17 produced by T lymphocytes from bone marrow contributes locally to juxta-articular bone destruction. Finally, IL-17 acts in synergy or additively with TNFα and IL-1, increasing their production and action. Accordingly, inhibition of IL-17 could be beneficial in human RA as demonstrated in many models of RA. Tools for treatment are already available. Combination therapy with the current cytokine inhibitors may be of interest to increase the rate and duration of response.

References

1 Miossec P (2003) Interleukin-17 in rheumatoid arthritis: if T cells were to contribute to inflammation and destruction through synergy. *Arthritis Rheum* 48: 594–601

2 Yao Z, Painter SL, Fanslow WC, Ulrich D, Macduff BM, Spriggs MK, Armitage RJ (1995) Human IL-17: a novel cytokine derived from T cells. *J Immunol* 155: 5483–5486

3 Fossiez F, Djossou O, Chomarat P, Flores-Romo L, Ait-Yahia S, Maat C, Pin JJ, Garrone P, Garcia E, Saeland S et al (1996) T cell interleukin-17 induces stromal cells to produce proinflammatory and hematopoietic cytokines. *J Exp Med* 183: 2593–2603

4 Molet S, Hamid Q, Davoine F, Nutku E, Taha R, Page N, Olivenstein R, Elias J, Chakir J (2001) IL-17 is increased in asthmatic airways and induces human bronchial fibroblasts to produce cytokines. *J Allergy Clin Immunol* 108: 430–438
5 Ferretti S, Bonneau O, Dubois GR, Jones CE, Trifilieff A (2003) IL-17, produced by lymphocytes and neutrophils, is necessary for lipopolysaccharide-induced airway neutrophilia: IL-15 as a possible trigger. *J Immunol* 170: 2106–2112
6 Aggarwal S, Gurney AL (2002) IL-17: prototype member of an emerging cytokine family. *J Leukoc Biol* 71: 1–8
7 Moseley TA, Haudenschild DR, Rose L, Reddi AH (2003) Interleukin-17 family and IL-17 receptors. *Cytokine Growth Factor Rev* 14: 155–174
8 Starnes T, Broxmeyer HE, Robertson MJ, Hromas R (2002) Cutting edge: IL-17D, a novel member of the IL-17 family, stimulates cytokine production and inhibits hemopoiesis. *J Immunol* 169: 642–646
9 Hurst SD, Muchamuel T, Gorman DM, Gilbert JM, Clifford T, Kwan S, Menon S, Seymour B, Jackson C, Kung TT et al (2002) New IL-17 family members promote Th1 or Th2 responses in the lung: *In vivo* function of the novel cytokine IL-25. *J Immunol* 169: 443–453
10 Pan G, French D, Mao W, Maruoka M, Risser P, Lee J, Foster J, Aggarwal S, Nicholes K, Guillet S et al (2001) Forced expression of murine IL-17E induces growth retardation, jaundice, a Th2-biased response, and multiorgan inflammation in mice. *J Immunol* 167: 6559–6567
11 Kim MR, Manoukian, Yeh R, Silbiger SM, Danilenko DM, Scully S, Sun J, DeRose ML, Stolina M, Chang D et al (2002) Transgenic overexpression of human IL-17E results in eosinophilia, B-lymphocyte hyperplasia, and altered antibody production. *Blood* 100: 2330–2340
12 Ikeda K, Nakajima H, Suzuki K, Kagami SI, Hirose K, Suto A, Saito Y, Iwamoto I (2003) Mast cells produce interleukin-25 upon Fc{epsilon}RI-mediated activation. *Blood* 101: 3594–3596
13 Yao Z, Fanslow WC, Seldin MF, Rousseau AM, Painter SL, Comeau MR, Cohen JI, Spriggs MK (1995) Herpesvirus Saimiri encodes a new cytokine, IL-17, which binds to a novel cytokine receptor. *Immunity* 3: 811–821
14 Yao Z, Spriggs MK, Derry JM, Strockbine L, Park LS, VandenBos T, Zappone JD, Painter SL, Armitage RJ (1997) Molecular characterization of the human interleukin (IL)-17 receptor. *Cytokine* 9: 794–800
15 Kehlen A, Thiele K, Riemann D, Langner J (2002) Expression, modulation and signalling of IL-17 receptor in fibroblast-like synoviocytes of patients with rheumatoid arthritis. *Clin Exp Immunol* 127: 539–546
16 Shalom-Barak T, Quach J, Lotz M (1998) Interleukin-17-induced gene expression in articular chondrocytes is associated with activation of mitogen-activated protein kinases and NF-kappaB. *J Biol Chem* 273: 27467–27473
17 Chevrel G, Page G, Granet C, Streichenberger N, Varennes A, Miossec P (2003) Inter-

leukin-17 increases the effects of IL-1beta on muscle cells: arguments for the role of T cells in the pathogenesis of myositis. *J Neuroimmunol* 137: 125–133

18 Ziolkowska M, Koc A, Luszczykiewicz G, Ksiezopolska-Pietrzak K, Klimczak E, Chwalinska-Sadowska H, Maslinski W (2000) High levels of IL-17 in rheumatoid arthritis patients: IL-15 triggers *in vitro* IL-17 production *via* cyclosporin A-sensitive mechanism. *J Immunol* 164: 2832–2838

19 Aggarwal S, Ghilardi N, Xie MH, de Sauvage FJ, Gurney AL (2003) Interleukin-23 promotes a distinct CD4 T cell activation state characterized by the production of interleukin-17. *J Biol Chem* 278: 1910–1914

20 Chabaud M, Fossiez F, Taupin JL, Miossec P (1998) Enhancing effect of IL-17 on IL-1-induced IL-6 and leukemia inhibitory factor production by rheumatoid arthritis synoviocytes and its regulation by Th2 cytokines. *J Immunol* 161: 409–414

21 Katz Y, Nadiv O, Beer Y (2001) Interleukin-17 enhances tumor necrosis factor alpha-induced synthesis of interleukins 1,6, and 8 in skin and synovial fibroblasts: a possible role as a "fine-tuning cytokine" in inflammation processes. *Arthritis Rheum* 44: 2176–2184

22 Martel-Pelletier J, Mineau F, Jovanovic D, Di Battista JA, Pelletier JP (1999) Mitogen-activated protein kinase and nuclear factor kappaB together regulate interleukin-17-induced nitric oxide production in human osteoarthritic chondrocytes: possible role of transactivating factor mitogen-activated protein kinase-activated proten kinase (MAPKAPK). *Arthritis Rheum* 42: 2399–2409

23 LeGrand A, Fermor B, Fink C, Pisetsky DS, Weinberg JB, Vail TP, Guilak F (2001) Interleukin-1, tumor necrosis factor alpha, and interleukin-17 synergistically upregulate nitric oxide and prostaglandin E2 production in explants of human osteoarthritic knee menisci. *Arthritis Rheum* 44: 2078–2083

24 Van Bezooijen RL, Farih-Sips HC, Papapoulos SE, Lowik CW (1999) Interleukin-17: A new bone acting cytokine *in vitro*. *J Bone Miner Res* 14: 1513–1521

25 Jovanovic DV, Di Battista JA, Martel-Pelletier J, Jolicoeur FC, He Y, Zhang M, Mineau F, Pelletier JP (1998) IL-17 stimulates the production and expression of proinflammatory cytokines, IL-beta and TNF-alpha, by human macrophages. *J Immunol* 160: 3513–3521

26 Chabaud M, Durand JM, Buchs, Fossiez F, Page G, Frappart L, Miossec P (1999) Human interleukin-17: A T cell-derived proinflammatory cytokine produced by the rheumatoid synovium. *Arthritis Rheum* 42: 963–970

27 Schwarzenberger P, La Russa V, Miller A, Ye P, Huang W, Zieske A, Nelson S, Bagby GJ, Stoltz D, Mynatt RL et al (1998) IL-17 stimulates granulopoiesis in mice: use of an alternate, novel gene therapy-derived method for *in vivo* evaluation of cytokines. *J Immunol* 161: 6383–6389

28 Forlow SB, Schurr JR, Kolls JK, Bagby GJ, Schwarzenberger PO, Ley K (2001) Increased granulopoiesis through interleukin-17 and granulocyte colony-stimulating factor in leukocyte adhesion molecule-deficient mice. *Blood* 98: 3309–3314

29 Chung DR, Kasper DL, Panzo RJ, Chtinis T, Grusby MJ, Sayegh MH, Tzianabos AO

(2003) CD4+ T cells mediate abscess formation in intra-abdominal sepsis by an IL-17-dependent mechanism. *J Immunol* 170: 1958–1963
30 Jones CE, Chan K (2002) Interleukin-17 stimulates the expression of interleukin-8, growth-related oncogene-alpha, and granulocyte-colony-stimulating factor by human airway epithelial cells. *Am J Respir Cell Mol Biol* 26: 748–753
31 Ye P, Rodriguez FH, Kanaly S, Stocking KL, Schurr J, Schwarzenberger P, Oliver P, Huang W, Zhang P, Zhang J et al (2001) Requirement of interleukin 17 receptor signaling for lung CXC chemokine and granulocyte colony-stimulating factor expression, neutrophil recruitment, and host defense. *J Exp Med* 194: 519–527
32 Fujino S, Andoh A, Bamba S, Ogawa A, Hata K, Araki Y, Bamba T, Fujiyama Y (2003) Increased expression of interleukin 17 in inflammatory bowel disease. *Gut* 52: 65–70
33 Hamzaoui K, Hamzaoui A, Guemira F, Bessioud M, Hamza M, Ayed K (2002) Cytokine profile in Behcet's disease patients. Relationship with disease activity. *Scand J Rheumatol* 31: 205–210
34 Numasaki M, Fukushi JI, Ono M, Narula SK, Zavodny PJ, Kudo T, Robbins PD, Tahara H, Lotze MT (2003) Interleukin-17 promotes angiogenesis and tumor growth. *Blood* 101: 2620–2627
35 Hirahara N, Nio Y, Sasaki S, Minari Y, Takamura M, Iguchi C, Dong M, Yamasawa K, Tamura K (2001) Inoculation of human interleukin-17 gene-transfected Meth-A fibrosarcoma cells induces T cell-dependent tumor-specific immunity in mice. *Oncology* 61: 79–89
36 Antonysamy MA, Fanslow WC, Fu F, Li W, Qian S, Troutt AB, Thomson AW (1999) Evidence for a role of IL-17 in alloimmunity: a novel IL-17 antagonist promotes heart graft survival. *Transplant Proc* 31: 93
37 Hsieh HG, Loong CC, Lui WY, Chen A, Lin CY (2001) IL-17 expression as a possible predictive parameter for subclinical renal allograft rejection. *Transpl Int* 14: 287–298
38 de Oliveira JG, Xavier PD, Sampaio SM, Tavares IS, Mendes AA (2002) The synthesis by fine-needle aspiration biopsy cultures of IL-7, IL-16 and IL-18 is significantly associated with acute rejection in kidney transplants. *Nephron* 92: 622–628
39 Kotake S, Udagawa N, Takahashi N, Matsuzaki K, Itoh K, Ishiyama S, Saito S, Inoue K, Kamatani N, Gillespie MT et al (1999) IL-17 in synovial fluids from patients with rheumatoid arthritis is a potent stimulator of osteoclastogenesis. *J Clin Invest* 103: 1345–1352
40 Joosten LA, Radstake TR, Lubberts E, van Den Bersselaar LA, van Riel PL, van Lent PL, Barrera P, van den Berg WB (2003) Association of interleukin-18 expression with enhanced levels of both interleukin-1beta and tumor necrosis factor alpha in knee synovial tissue of patients with rheumatoid arthritis. *Arthritis Rheum* 48: 339–347
41 Lock C, Hermans G, Pedotti R, Brendolan A, Schadt E, Garren H, Langer-Gould A, Strober S, Cannella B, Allard J et al (2002) Gene-microarray analysis of multiple sclerosis lesions yields new targets validated in autoimmune encephalomyelitis. *Nat Med* 8: 500–508
42 Zhang GX, Gran B, Yu S, Li J, Siglienti I, Chen X, Kamoun M, Rostami A (2003) Induc-

tion of experimental autoimmune encephalomyelitis in IL-12 receptor-beta2-deficient mice: IL-12 responsiveness is not required in the pathogenesis of inflammatory demyelination in the central nervous system. *J Immunol* 170: 2153–2160

43 Nurieva RI, Treuting P, Duong J, Flavell RA, Dong C (2003) Inducible costimulator is essential for collagen-induced arthritis. *J Clin Invest* 111: 701–706

44 Chabaud M, Lubberts E, Joosten L, van den Berg WB, Miossec P (2001) IL-17 derived from juxta-articular bone and synovium contributes to joint degradation in rheumatoid arthritis. *Arthritis Res* 3: 168–177

45 Page G, Lebecque S, Miossec P (2002) Anatomic localization of immature and mature dendritic cells in an ectopic lymphoid organ: correlation with selective chemokine expression in rheumatoid synovium. *J Immunol* 168: 5333–5341

46 Chabaud M, Page G, Miossec P (2001) Enhancing effect of IL-1, IL-17, and TNF-alpha on macrophage inflammatory protein-3alpha production in rheumatoid arthritis: regulation by soluble receptors and Th2 cytokines. *J Immunol* 167: 6015–6020

47 Antonysamy MA, Fanslow WC, Fu F, Li W, Qian S, Troutt AB, Thomson AW (1999) Evidence for a role of IL-17 in organ allograft rejection: IL-17 promotes the functional differentiation of dendritic cell progenitors. *J Immunol* 162: 577–584

48 Chabaud M, Page G, Miossec P (2001) Enhancing effect of IL-1, IL-17 and TNF-alpha on MIP-3 alpha production in rheumatoid arthritis: Regulation by soluble receptors and by Th2 cytokines. *J Immunol* 167: 6015–6020

49 Chabaud M, Garnero P, Dayer JM, Guerne PA, Fossiez F, Miossec P (2000) Contribution of interleukin 17 to synovium matrix destruction in rheumatoid arthritis. *Cytokine* 12: 1092–1099

50 Attur MG, Patel RN, Abramson SB, Amin AR (1997) Interleukin-17 upregulation of nitric oxide production in human osteoarthritis cartilage. *Arthritis Rheum* 40: 1050–1053

51 Cai L, Yin JP, Starovasnik MA, Hogue DA, Hillan KJ, Mort JS, Filvaroff EH (2001) Pathways by which interleukin 17 induces articular cartilage breakdown *in vitro* and *in vivo*. *Cytokine* 16: 10–21

52 Pacquelet S, Presle N, Boileau C, Dumond H, Netter P, Martel-Pelletier J, Pelletier JP, Terlain B, Jouzeau JY (2002) Interleukin 17, a nitric oxide-producing cytokine with a peroxynitrite-independent inhibitory effect on proteoglycan synthesis. *J Rheumatol* 29: 2602–2610

53 Benderdour M, Tardif G, Pelletier JP, Di Battista JA, Reboul P, Ranger P, Martel-Pelletier J (2002) Interleukin 17 (IL-17) induces collagenase-3 production in human osteoarthritic chondrocytes *via* AP-1 dependent activation: differential activation of AP-1 members by IL-17 and IL-1beta. *J Rheumatol* 29: 1262–1272

54 Koshy PJ, Henderson N, Logan C, Life PF, Cawston TE, Rowan AD (2002) Interleukin 17 induces cartilage collagen breakdown: novel synergistic effects in combination with proinflammatory cytokines. *Ann Rheum Dis* 61: 704–713

55 Cai L, Suboc P, Hogue DA, Fei DT, Filvaroff EH (2002) Interleukin 17 induced nitric

oxide suppresses matrix synthesis and protects cartilage from matrix breakdown. *J Rheumatol* 29: 1725–1736
56 Honorati MC, Bovara M, Cattini L, Piacentini A, Facchini A (2002) Contribution of interleukin 17 to human cartilage degradation and synovial inflammation in osteoarthritis. *Osteoarthritis Cartilage* 10: 799–807
57 Rifas L, Avioli LV (1999) A novel T cell cytokine stimulates interleukin-6 in human osteoblastic cells. *J Bone Miner Res* 14: 1096–1103
58 Van Bezooijen RL, Papapoulos SE, Lowik CW (2001) Effect of interleukin-17 on nitric oxide production and osteoclastic bone resorption: is there dependency on nuclear factor-kappaB and receptor activator of nuclear factor kappaB (RANK)/RANK ligand signaling? *Bone* 28: 378–386
59 Nakashima T, Kobayashi Y, Yamasaki S, Kawakami A, Eguchi K, Sasaki H, Sakai H (2000) Protein expression and functional difference of membrane-bound and soluble receptor activator of NF-kappaB ligand: modulation of the expression by osteotropic factors and cytokines. *Biochem Biophys Res Commun* 275:768–775
60 Romas E, Gillespie MT, Martin TJ (2002) Involvement of receptor activator of NFkappaB ligand and tumor necrosis factor-alpha in bone destruction in rheumatoid arthritis. *Bone* 30: 340–346
61 Lubberts E, Joosten LA, van de Loo FA, van den Bersselaar LA, van den Berg WB (2000) Reduction of interleukin-17-induced inhibition of chondrocyte proteoglycan synthesis in intact murine articular cartilage by interleukin-4. *Arthritis Rheum* 43: 1300–1306
62 Bush KA, Farmer KM, Walker JS, Kirkham BW (2002) Reduction of joint inflammation and bone erosion in rat adjuvant arthritis by treatment with interleukin-17 receptor IgG1 Fc fusion protein. *Arthritis Rheum* 46: 802–805
63 Dudler J, Renggli-Zulliger N, Busso N, Lotz M, So A (2000) Effect of interleukin 17 on proteoglycan degradation in murine knee joints. *Ann Rheum Dis* 59: 529–532
64 Lubberts E, Joosten LA, Oppers B, van den Bersselaar L, Coenen-de Roo CJ, Kolls JK, Schwarzenberger P, van de Loo FA, van den Berg WB (2001) IL-1-independent role of IL-17 in synovial inflammation and joint destruction during collagen-induced arthritis. *J Immunol* 167: 1004–1013
65 Lubberts E, Joosten LA, van de Loo FA, Schwarzenberger P, Kolls J, van den Berg WB (2002) Overexpression of IL-17 in the knee joint of collagen type II immunized mice promotes collagen arthritis and aggravates joint destruction. *Inflamm Res* 51: 102–104
66 Lubberts E, van den Bersselaar L, Oppers-Walgreen B, Schwarzenberger P, Coenen-de Roo C, Kolls JK, Joosten LA, van den Berg WB (2003) IL-17 promotes bone erosion in murine collagen-induced arthritis through loss of the receptor activator of NF-kappaB ligand/osteoprotegerin balance. *J Immunol* 170: 2655–2662
67 Oda T, Yoshie H, Yamazaki K (2003) Porphyromonas gingivalis antigen preferentially stimulates T cells to express IL-17 but not receptor activator of NF-kappaB ligand *in vitro*. *Oral Microbiol Immunol* 18: 30–36
68 Miossec P, van den Berg WB (1997) Th1/Th2 cytokine balance in arthritis. *Arthritis Rheum* 40: 2105–2115

69 Davis LS, Cush JJ, Schulze-Koops H, Lipsky PE (2001) Rheumatoid synovial CD4+ T cells exhibit a reduced capacity to differentiate into IL-4-producing T-helper-2 effector cells. *Arthritis Res* 3: 54–64
70 Aarvak T, Chabaud M, Kallberg E, Miossec P, Natvig JB (1999) Change in the Th1/Th2 phenotype of memory T cell clones from rheumatoid arthritis synovium. *Scand J Immunol* 50: 1–9
71 Aarvak T, Chabaud M, Miossec P, Natvig JB (1999) IL-17 is produced by some proinflammatory Th1/Th0 cells but not by Th2 cells. *J Immunol* 162: 1246–1251
72 Chabaud M, Aarvak T, Garnero P, Natvig JB, Miossec P (2001) Potential contribution of IL-17-producing Th1 cells to defective repair activity in joint inflammation: partial correction with Th2 promoting conditions. *Cytokine* 13: 113–118
73 Albanesi C, Scarponi C, Cavani A, Federici M, Nasorri F, Girolomoni G (2000) Interleukin-17 is produced by both Th1 and Th2 lymphocytes, and modulates interferon-gamma- and interleukin-4-induced activation of human keratinocytes. *J Invest Dermatol* 115: 81–87
74 Chabaud M, Miossec P (2001) The combination of tumor necrosis factor alpha blockade with interleukin-1 and interleukin-17 blockade is more effective for controlling synovial inflammation and bone resorption in an *ex vivo* model. *Arthritis Rheum* 44: 1293–1303
75 Lubberts E, Joosten LA, Chabaud M, van den Bersselaar L, Oppers B, Coenen-de Roo CJ, Richards CD, Miossec P, van den Berg WB (2000) IL-4 gene therapy for collagen arthritis suppresses synovial IL-17 and osteoprotegerin ligand and prevents bone erosion. *J Clin Invest* 105: 1697–1710
76 Benchetrit F, Ciree A, Vives V, Warnier G, Gey A, Sautes-Fridman C, Fossiez F, Haicheur N, Fridman WH, Tartour E (2002) Interleukin-17 inhibits tumor cell growth by means of a T cell-dependent mechanism. *Blood* 99: 2114–2121
77 Firestein GS, Zvaifler NJ (2002) How important are T cells in chronic rheumatoid synovitis? II. T cell-independent mechanisms from beginning to end. *Arthritis Rheum* 46: 298–308

Cell–cell contact in chronic inflammation: the importance to cytokine regulation in tissue destruction and repair

Danielle Burger, Pascale Roux-Lombard, Carlo Chizzolini and Jean-Michel Dayer

Division of Immunology and Allergy, Clinical Immunology Unit (Hans Wilsdorf Laboratory), Department of Internal Medicine, University Hospital, 24 rue Micheli-du-Crest, CH-1211 Geneva 14, Switzerland

Background

In many chronic diseases, inflammation is characterized by the influx into the target tissue of migratory cells such as T and B lymphocytes, dendritic cells, neutrophils and mononuclear phagocytes. This influx of inflammatory cells in the target tissue is associated with the proliferation of invading and resident cells and frequently with destruction and remodeling of the extracellular matrix. In the joint structures, the destruction of the organic phase is due to proteases (mainly MMP whose activity is controlled by specific inhibitors) and the resorption of the inorganic phase of bone, mainly due to the action of the receptor activator of NF-κB and its ligand (RANK–RANKL) and prostanoids. The expression of these proteases and their inhibitors is regulated by different types of stimuli, including soluble factors (i.e., cytokines, hormones), contact with extracellular matrix components and direct cellular interactions [1–3]. In pathologic conditions, the production of cytokines and MMP by infiltrating and resident tissue cells escapes regulatory mechanisms. The activity of proinflammatory cytokines is counterbalanced by numerous mechanisms of which cytokine inhibitors – IL-1 receptor antagonist (IL-1Ra), TNF soluble receptors (TNFsR) – and tissue inhibitor of MMP (TIMP) are examples. It is generally acknowledged that the imbalance between cytokines and their respective inhibitors is responsible for the persistence of chronic inflammatory conditions and maybe even necessary for their initiation. There is now considerable evidence that cytokines such as TNFα and IL-1 are involved in many diseases resulting in tissue destruction. This has been demonstrated by human clinical trials in rheumatoid arthritis (RA) (for review see [4–8]). Cytokine blockade in the clinic has also proved useful in many other diseases including juvenile idiopathic arthritis, Crohn's disease, spondyloarthropathy, vasculitis and psoriasis [9–16]. The above cytokines are also involved in the pathogenesis of several other diseases including multiple sclerosis (MS) [17, 18], systemic lupus erythematosus (SLE) [19, 20], and atherosclerosis [21, 22].

Cytokines and Joint Injury, edited by Wim B. van den Berg and Pierre Miossec

IL-1 and TNF in RA

Considering that IL-1 and TNF are described in detail in other chapters in this book, we will only point out some of their particularities in order to clarify the mechanisms described below.

IL-1 and TNFα are both synthesized transiently after cell activation and can be present in soluble and cell membrane-associated forms [23–25]. They share many effects (redundancy) and can act in synergy. Two types of receptors have been described for each IL-1 and TNFα that can have disparate functions. For example, both IL-1α and IL-1β bind with equal affinity to types I and II IL-1 receptors (IL-1R) [26]. The type I IL-1R is solely responsible for signal transduction while the so-called "decoy" type II IL-1R sequestrates both IL-1α and IL-1β on the cell surface or, when shed in soluble form, can inhibit IL-1 activities [27]. In addition to type II soluble receptor there is the IL-1Ra that competitively binds the IL-1 receptor without inducing signaling and thus inhibits IL-1 activity [28]. Interestingly, although IL-1Ra might bind the IL-1 "decoy" receptor, the two do not interfere with each other [29]. TNFα and lymphotoxin-α (LTα) – although only 28% homologous in amino acid sequence – share the two receptors, each of which activates distinct cellular activities. Binding of TNFα or LTα to the p55 receptor results mainly in cytotoxic events, while binding to the p75 receptor favors lymphoproliferative responses [30]. Both TNF receptors can be shed and thus act as specific TNFα inhibitors. Multiple studies have been devoted to the determination of cytokine levels in biological fluids in various diseases, but have often yielded contradictory results, particularly concerning the relationship with disease activity. Overall, TNFα tends to be elevated at the systemic level during acute infections, but to a lesser extent in chronic diseases, while IL-1 is hardly found at all in plasma or serum. However, IL-1 and TNFα are detectable locally, for instance in synovial fluids and synovium of RA patients as assessed by immunochemistry and by *in situ* hybridization for mRNA expression [31–33]. In contrast, IL-1Ra [34–36] and the two TNFsRs [34, 37–41] are readily detectable in patients' serum or plasma. In some patients anti-IL-1α antibodies have been detected, although their function remains elusive [42, 43]. In animal models the balance between IL-1 and IL-1Ra appears to be crucial to the development and remission of arthritis [44].

From *in vitro* experiments with cultured cells and from animal models the consensus has been reached that IL-1 and TNFα are the main proinflammatory and destructive cytokines, primarily due to their stimulation of MMP production by fibroblast-like synoviocytes in the inflamed synovium. Patients with severe RA, or patients with a high synovial fluid leukocyte count, have elevated levels of TNFα. Macrophages in the synovial lining layer in RA and in the subsynovium contain high levels of TNFα and IL-1 (reviewed in [45]). Although it has been claimed that TNFα is of primary importance in RA because it is the major dominant regulator of IL-1 [4], the principal impact of TNFα on tissue destruction is due to the synergism between TNFα and IL-1. Contrary to TNFα – predominantly detected in the early stages of disease – both IL-1α and IL-1β are detected long after the onset of RA [33, 46, 47]. If TNFα can induce significant amounts of IL-1, IL-1 can also induce TNFα to some extent. In addition, many mechanisms independent of TNFα (such as direct contact with other cells, denatured proteins, hormones, immune complexes and complement components) are able to induce IL-1β production. A logical therapeutic approach would therefore be to aim at inhibiting at a proximal level the production of both IL-1 and TNFα.

Triggering of cytokine production in RA

Although infiltration of T lymphocytes (T_L) into the target tissue precedes tissue damage – suggesting their pathogenic effect – Mϕ, the main producers of IL-1 and TNFα in RA, are also present at an early stage in the lesion, and interactions are likely to occur between T_L and Mϕ [48]. In RA, T_L displaying a mature helper phenotype (i.e., $CD3^+CD4^+CD45RO^+$) are the main infiltrating cells in the pannus, at percentages ranging from 16% of total cells in "transitional areas" and 75% in "lymphocyte-rich perivascular areas" [49, 50]. T_L extravasation occurs at the level of vessels presenting the characteristics of high endothelial venules [51]. Angiogenesis and proliferation of resident cells accompany this infiltration. Like other chronic destructive inflammatory diseases, RA is thought to be a Th1-mediated disease [52–55]. However, the mechanisms by which T cells exert their pathogenic role in RA remains elusive. Our previous seminal studies, and subsequent works by others, strongly argue that direct cellular contact with stimulated T cells is a major pathway for the production of IL-1 and TNFα in Mϕ [56–61]. Indeed, contact-mediated activation of Mϕ by stimulated T_L is as potent as optimal doses of lipopolysaccharide (LPS) and phorbol-myristate-acetate (PMA) in inducing IL-1β and TNFα production in monocytes and cells of the monocytic lineage such as THP-1 cells [2]. We therefore postulate that this mechanism is highly relevant to the pathogenesis and maintenance of chronic destructive, inflammatory reactions in RA.

T cell-signaling of monocyte-macrophages by direct cell-cell contact

The activation of effector cells mediated by stimulated T_L has been abundantly substantiated in the T–B lymphocyte model by the induction of cell proliferation and antibody secretion; this activation requires both direct cell–cell contact and soluble signals. Indeed, B cells can be activated in the absence of antigen by direct contact with activated T cells. A considerable amount of signaling "crosstalk" was observed between T and B cells, which triggered various synchronized signals resulting in the activation of effector functions in both cell types [62]. Similarly, ample data exist regarding the crosstalk between dendritic cells and T_L. Below we will review what is known about the crosstalk between T_L and Mϕ regardless of antigen presentation.

In the mid-1980s it was observed that the expression of membrane-associated IL-1α in mouse Mϕ was mediated by both soluble factors and direct contact with T cells [63]. The importance of cellular contact was confirmed by experiments showing that IL-1 was induced upon T cell–Mϕ contact with both Th1 and Th2 mouse cells in the absence of lymphokine release [64]. We and others further established that direct contact with stimulated T cells is also a potent stimulus of cytokine production by Mϕ in humans [65–67].

It has been established that when stimulated by phytohemagglutinin (PHA) and phorbol 12-myristate 13-acetate (PMA) most T cell types including T cell clones, freshly isolated T_L and T cell lines such as HUT-78 cells induce IL-1 and TNFα in monocytes [65, 66, 68]. Various stimuli other than PHA/PMA induce T_L to activate monocytes by direct cellular contact:

1. Cross-linking of CD3 by immobilized anti-CD3 mAb with or without cross-linking of the co-stimulatory molecule CD28 [57, 69–71].
2. Antigen-recognition on antigen-specific T cell clones [70].
3. Cytokine cocktails (TNFα, IL-6, IL-15 and IL-2) [56, 58, 72].

Although all T cells may express the ability to stimulate Mϕ by direct cell–cell contact, different types of products are induced in Mϕ depending on T cell type and T cell stimulus. Peripheral blood T_L, T cell lines, and T cell clones stimulated with PHA and PMA induce the production of TNFα, IL-1β, IL-1α, IL-6, IL-8, IL-1Ra, TNF soluble receptors, MMP-1, MMP-9, and TIMP-1 in Mϕ and/or monocytic cell lines [65, 66, 68, 69, 73–75]. Peripheral T_L stimulated by phorbol 12,13-dibutyrate and ionomycine induce the production of TNFα and IL-10 in Mϕ [57]. Antibodies to CD3 stimulate T_L to induce TNFα, IL-10, IL-1β, IL-12 and MMP-1 in Mϕ and monocytic cell lines [57, 61, 69, 71, 76]. Cytokine cocktails (IL-6, TNFα, IL-15) induce the production of TNFα and IL-1β in Mϕ [56, 58, 60, 71, 72]. Of interest, an imbalance in production between proinflammatory and anti-inflammatory cytokines has been observed where Th1 cell clones preferentially induce IL-1β rather than IL-1Ra production [70]. Cytokine-stimulated T_L induces TNFα production while failing to produce IL-10 [58, 70]. This suggests that multiple ligands and counter-ligands are involved in the contact-mediated activation of Mϕ, which is differentially induced in T cells depending on the stimulus. It was proposed to classify T_L into two groups, in terms of their different capacities depending on the type of stimulus and their effect on Mϕ: T_{ck} (cytokine-activated T_L) and T_{TCR} (T cell receptor-activated T_L), T_{ck} being likely to be present in RA synovium [71].

In addition to T cell type and stimuli, contact-mediated cytokine production in Mϕ might be regulated by intrinsic factors. Contact-activated THP-1 cells express membrane-associated protease(s) neutralizing TNFα activity both by degrading the

Table 1 - Some factors controlling the balance between IL-1β and IL-1Ra in monocytes

I	Soluble products
	(a) Increase in IL-1Ra levels and decrease in IL-1β levels
	- IL-4, IL-13, IL-10, TGFβ, IFNβ
	(b) Increase in IL-1Ra levels and unchanged IL-1β levels
	- adherent IgG
II	Intracellular mechanisms
	(a) Increase in IL-1Ra levels and decrease in IL-1β levels
	- PI3K activation
III	Contact with stimulated T lymphocytes
	- T lymphocyte subsets: Th1/Th2 → IL-1β/IL-1Ra
	- Types of T lymphocyte stimuli
	- Types of ligands-counter-ligands between T_L and Mϕ
	- CD40–CD40L in co-culture
	- Blockade: ↓ IL-1 levels, unchanged IL-1Ra
IV	Therapeutic agents
	- Leflunomide: increase in IL-1Ra/IL-1 ratio
	- IFNβ: increase in IL-1Ra and decrease in IL-1β

latter cytokine and by cleaving its receptors at the cell surface [74]. Finally, upon contact with stimulated T cells the balance between IL-1β and IL-1Ra production in monocytes is ruled by Ser/Thr phosphatase(s) [75]. Thus the triggering of these intra- and extra-cellular processes by direct contact with stimulated T_L may regulate the proinflammatory cytokines and their inhibitors, and the balance of their production in monocytes might dictate in part the outcome of the inflammatory process. Some factors controlling the balance between IL-1β and IL-1Ra and discussed in this chapter are summarized in Table 1.

Cell surface molecules involved in contact-mediated monocyte activation

A crucial aspect arising from these observations is the missing or incomplete characterization of molecules on the T cell surface, which are involved in contact-mediated signaling of Mϕ activation as well as their counter-ligands. It has been postulated that T cell membrane-associated TNFα was involved in Mϕ activation [77–79]. However, fixed, stimulated Th2 cells from a T cell line that failed to express membrane-associated TNF induced both TNF and IL-1 production in Mϕ [80], demonstrating that TNFα might only play a partial role. We have shown that neither soluble TNFα receptors nor IL-1Ra block T cell-signaling of the monocytic cell line THP-1. Besides, neutralizing antibodies to TNFα, IL-1, IL-2, IFNγ and GM-CSF all failed to affect monocyte activation by membranes from stimulated T cells [65, 66, 73]. Similarly, although LT-β receptor is expressed in Mϕ [81, 82], membrane-associated LT is unlikely to be involved in Mϕ signaling upon contact with stimulated T cells, since Th2 clones do not express the LTαβ2 at their surface nor secrete LTα3 [83, 84].

Surface molecules other than membrane-associated cytokines have also been assessed as to their ability to activate Mϕ upon contact with stimulated T cells, e.g., LFA-1/ICAM-1, CD2/LFA3, CD40/CD40L and lymphocyte activation-antigen-3 (LAG-3). Thus CD40/CD40L interaction was shown to be involved in the contact activation of both human and mouse Mϕ by T_L [76, 85]. However, T_L isolated from CD40L knockout mice triggered Mϕ activation although to a lower extent than their wild type counterpart [86]. Furthermore, we did not observe inhibition of contact-induced cytokine production, by using blocking antibodies to CD40L or soluble CD40 when Mϕ was activated by membranes isolated from stimulated human T_L. Furthermore, HUT-78 cells, which efficiently induce cytokine production in Mϕ, do not express CD40L mRNA in resting or activated conditions [87], and THP-1 cells that respond to contact-mediated activation by membranes of stimulated T cells do not express CD40. We recently confirmed this since CD40L-trimer (provided by Dr. E. Williamson, Immunex Corp., Seattle, USA) did not induce the production of cytokines in peripheral blood monocytes and THP-1 cells, although it induced the production of TNFα in IFNγ-primed monocytes, i.e., macrophages (Fig. 1, open bars). However, CD40L does play a role in co-cultures of living Mϕ and T cells stimulated with IL-15 or IL-2 [72]. Indeed, when living T cell clones and monocytes or THP-1 cells were cultured together, IL-1β production was induced almost exclusively by Th1 cells and was dependent on the presence and dose of IL-2 or IL-15, as well as on cell–cell contact, as demonstrated by double-chamber cultures [72]. In contrast, low levels of IL-1Ra were induced by Th1 and higher levels by Th2 cells. Blockade of the CD40–CD40L interaction resulted in the inhibition of IL-1β-inducing capacity, IL-1Ra induction remaining unaffected. Although this differential effect indicates the selective relevance of CD40–CD40L engagement in inflammatory monocyte responses upon activation by T cells, the premise that CD40L expression levels did not differ in Th1 and Th2 cell clones suggests that additional, unidentified molecule(s) preferentially expressed by Th1 cells are involved in their IL-1β induction capacity [72]. Therefore, CD40/CD40L might be a co-factor in contact-mediated activation of Mϕ by stimulated T_L. LAG-3 might also be one of the latter factors since it is able to synergize with low amounts of CD40L in inducing TNFα and IL-12 on monocyte-derived dendritic cells [59]. In our system LAG-3 alone did not induce the production of IL-1β and TNFα (unpublished results). Recombinant soluble CD23 induced cytokine production on monocytes [88–91]. In monocytes, the counter-ligands of CD23 are CD11b/CD18 and CD11c/CD18 [92] and indeed the engagement of CD11b and CD11c in human

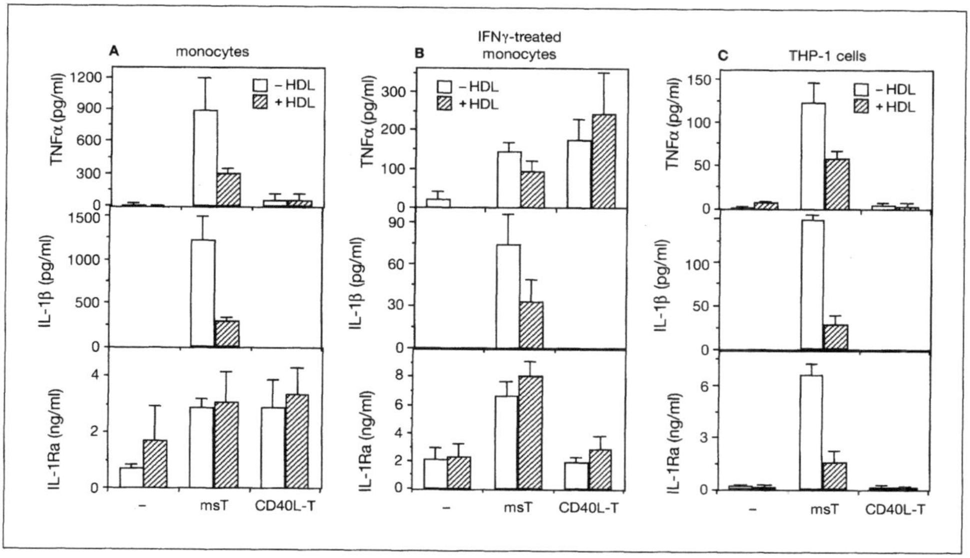

Figure 1
Cytokine production by monocytes, IFNγ-treated monocytes and THP-1 cells
Cells (50 × 10³ cells/200 µl/well) were cultured with the indicated stimulus in the presence (dashed bars) or absence (open bars) of 0.2 mg/ml HDL for 48 hours. A) Freshly isolated human blood monocytes were activated by 6 µg/ml proteins of isolated plasma membranes from stimulated HUT-78 cells (msT) or 3 µg/ml CD40L-trimer (CD40L-T). B) Isolated peripheral blood monocytes were treated for seven days with 500 U/ml IFNγ and then activated with 60 µg/ml proteins of msT or 3 µg/ml CD40L-T. C) THP-1 cells were activated with 60 µg/ml proteins of msT or 3 µg/ml CD40L-T. Cytokines in supernatants were measured by ELISA.

monocytes induced the production of IL-1β and macrophage inflammatory protein 1α (MIP-1α) and MIP-1β [90, 91]. Furthermore, LFA-1 (CD11a/CD18) and CD69 contribute to the activation of human monocytic cells by stimulated T cells [65, 93]. This was substantiated by a study showing that IL-15 induced synovial T cells from RA patients to activate the production of TNFα by Mϕ. This effect was partially inhibited by antibodies to CD69, LFA-1 and ICAM-1 [56].

Together these studies suggest that some known surface molecules are involved in T cell signaling of Mϕ. However, inhibitors (e.g., antibodies) of these molecules fail to abolish monocyte activation altogether, emphasizing the necessity for shedding more light on the factor(s) involved in T cell signaling of human monocytes by direct contact, which process leads to the production of IL-1 and TNF. Furthermore, hierarchy and sequence of events during this crosstalk still need to be established.

Identification of an inhibitor of T cell contact-mediated activation of monocyte-macrophages

The inhibition of T cell-signaling of monocytes might be important in that it would maintain a low level of monocyte activation within the blood stream. This hypothesis was sustained by an initial observation showing that contact-mediated activation of IL-1β and TNFα production on either monocytes or THP-1 cells was inhibited in the presence of human adult serum but not of fetal calf or human cord serum [94]. After human serum fractionation, we identified apolipoprotein A-I (apo A-I) – the principal protein of high-density lipoprotein (HDL) – as being an inhibitor of contact-mediated activation of monocytes [94]. Due to the configuration of the apo A-I molecule, when reconstituted with the lipid component of HDL its inhibitory activity is more potent and reproducible (unpublished). The serum concentrations of apo A-I, a "negative acute-phase protein", undergo marked changes in several inflammatory diseases [94]. In RA, the levels of circulating apo A-I and HDL-cholesterol in untreated patients were lower than in normal controls [95–97]. In contrast, apo A-I was increased in the synovial fluid of RA patients [98], although its concentrations remained 10-fold lower in synovial fluid than in plasma. The elevation of apo A-I levels in synovial fluid of RA patients was accompanied by an enhancement in cholesterol, suggesting an infiltration of HDL particles in the inflamed joint. Indeed, immunohistochemistry revealed an accumulation of apo A-I in the perivascular regions of the RA synovium [99]. The observation that apo A-I can infiltrate and be retained at the inflammatory site suggests that apo A-I might inhibit the local triggering of IL-1β and TNFα release by Mϕ that are in direct contact with stimulated T cells in these areas [49]. It was also shown that the inflammation in juvenile RA was associated with hypo-high density lipoproteinemia and a significant decrease in apo A-I concentration in patient plasma [100]. In SLE, apo A-I plasma concentrations are also diminished [101]. This decrease is associated with the presence of anti-apo A-I antibodies in 32% of patients [102]. In multiple sclerosis patients undergoing IFNβ therapy, levels of apo A-I proved to be lower in a subgroup of patients experiencing relapses and/or progression [103]. Despite the paucity of studies reporting levels of HDL in chronic inflammatory diseases, it seems that chronic inflammation is associated with low levels of HDL-associated apo A-I. We thus hypothesize that in these diseases, a vicious circle sets in that is responsible for the persistence of inflammation [104].

The identification of HDL-associated apo A-I ligand(s) on stimulated T cells might lead to the elucidation of the mechanisms and molecules involved in T cell-signaling of monocytes. HDL-associated apo A-I has been shown to bind specifically to a number of cell-surface molecules (for review see [105]). Among these HDL-binding proteins (HBP) are scavenger receptor B1 (SR-B1), HB2, cubillin [106], ATP-binding cassette (ABC) A1 transporter [107], and a 95 kDa protein at the surface of human fetal hepatocytes [108]. They all display a high molecular weight

(≥ 80 kDa) and have not been identified on T cells to date. Thus, the identity of the apo A-I ligand at the surface of stimulated T cells is still unknown, although a specific HDL binding site on human T_L has been described for which molecular identity has not been characterized [109, 110].

Effects of anti-inflammatory cytokines and drugs on contact-mediated induction of cytokines in monocytes by stimulated T lymphocytes

Since the contact-mediated activation of Mϕ is a major pathway of cytokine production, the modulation of this mechanism, i.e., the blockade of IL-1 and TNFα production at the triggering level of contact-mediated activation, would be of therapeutic interest. We established that therapeutic agents administered to RA and multiple sclerosis patients, i.e., leflunomide, an inhibitor of dihydroorotate dehydrogenase [111–113] and IFNβ [114–116], respectively, affected the contact-mediated activation of monocytes. Leflunomide inhibits the ability of stimulated T_L to trigger IL-1β production in monocytes, resulting in an enhancement of the IL-1Ra/IL-1β molar ratio [117]. Similar results were obtained with IFNβ. Indeed, upon contact-mediated activation of monocytes, IFNβ not only inhibited IL-1β and TNFα but it also stimulated IL-1Ra [118, 119], due to the fact that IFNβ interfered with the activation of both T_L and monocytes. Noticeably, surface molecules of T_L that were found to be involved in contact-signaling of monocytes (i.e., TNFα, CD25, CD69, CD18, CD11a, CD11b, CD11c, CD40L and LAG-3) were not modulated by IFNβ or leflunomide, suggesting that other cell-surface molecules on T_L (i.e., apo A-I ligand(s)) are involved in the contact-mediated activation of monocytes by stimulated T_L.

Dissociation of IL-1β and IL-1Ra production in monocytes in contact with stimulated T lymphocytes

Although HDL inhibited the production of TNFα and IL-1β in both peripheral blood monocytes and THP-1 cells, it did not affect IL-1Ra. Indeed, in isolated peripheral blood monocytes, HDL only inhibited IL-1β and TNFα production, IL-1Ra production being only slightly inhibited or not at all, depending on the experiment/donor. This was not the case in THP-1 cells where all three IL-1β, TNFα and IL-1Ra were inhibited (Fig. 1). This recent observation suggests that molecules involved in contact-mediated activation of blood monocytes and THP-1 cells may differ at least in terms of IL-1Ra induction and deserve further investigation. The differential inhibition of IL-1β and IL-1Ra also occurred in cell co-cultures. Indeed, in PBMC stimulated by either PHA or an antigen (tetanus toxoid, TT) IL-1Ra production was not significantly inhibited, contrasting with the obvious inhibition of

IL-1β and TNFα production [120]. This further indicates that apo A-I inhibit contact-mediated monocyte activation when T_L is stimulated by either non-specific stimuli or antigens.

A dissociation in the contact-mediated induction of IL-1β and IL-1Ra has been observed not only in the presence of IFNβ and HDL-associated apo A-I [120, 121] but also with Th1 and Th2 clones [70, 72] suggesting that T cell surface molecules might be differentially involved in the induction of the proinflammatory cytokine and its inhibitor. Recent results confirm this hypothesis. Indeed, using a specific phosphoinositide-3 kinase (PI3K) inhibitor, we demonstrated that PI3K is a checkpoint in the production of IL-1β and IL-1Ra, decreasing IL-1β while inducing IL-1Ra production [122]. This further confirms that in addition to the involvement of different molecules at the surface of T cells, the production of IL-1β and IL-1Ra is ruled by different signaling pathways.

T cell-signaling of fibroblasts, neutrophils and endothelial cells by direct cell–cell contact

Over the past 50 years, a tremendous amount of information was generated identifying a variety of cell types involved in RA. Endothelial cells are present with an activated phenotype resembling high-endothelial venule cells [123]. Infiltrating cells include T lymphocytes mainly in the perivascular region, monocyte-macrophages and fibroblast-like synoviocytes mainly in the pannus lining layer and also B lymphocytes, mast cells and dendritic cells [122–128]. In contrast, neutrophils are mainly found in synovial fluid. Cytokines and contact with neighboring cells may affect the stage of activation of these various cell types. At some stages of the inflammation, stimulated T_L can potentially contact various cells other than mononuclear phagocytes that are involved in the pathogenic mechanisms. Such target cell types include interstitial fibroblast-like cells (synoviocytes), endothelial cells and maybe migrating neutrophils. We observed that contact with stimulated T lymphocytes potently affected the latter cells [68, 69, 129–131].

Signaling of stromal cells by contact with mononuclear cells has been scarcely studied. It has been shown that co-culture of living and resting mononuclear cells with autologous dermal fibroblasts resulted in a dose-dependent increase of collagen synthesis, unlike that observed with fibroblasts cultured alone [132, 133]. In addition, T cells cultured in the presence of mycobacterial antigen and IL-2 induce *in vitro* outgrowth of synovial fibroblasts [134] and co-culture of T cells and fibroblasts induces the production of IL-6 and IL-8 by fibroblasts [135]. These studies, however, did not distinguish the effects of direct cellular contact from those of soluble factors released by mononuclear cells. In addition, contrasting results on the stimulatory or inhibitory effects of resting or mitogen-activated mononuclear cells are reported in the literature [132, 136, 137]. Contact between fibroblasts and T lymphocytes *in vitro* mediates survival of the latter cells [138], and it is possible that T cells within the synovium may be anergized by contact with HLA-DR$^+$CD80$^-$ fibroblast-like synoviocytes [139].

Cellular contact induces an imbalance between MMP-1 and TIMP-1 production by dermal fibroblasts and monocytic cells [69]. This was due to the length of the stimulation time of T lymphocytes. Indeed, T lymphocytes stimulated with PHA/PMA for 2–4 hours enhanced both TIMP-1 and MMP-1 production by fibroblasts and synoviocytes, whereas T lymphocytes stimulated for 12–48 hours increased only the production of MMP-1 without affecting the expression of TIMP-1. Since direct cell–cell contact with stimulated T lymphocytes induced an imbalance between the production of MMPs and TIMP-1 by both monocytes and fibroblasts *in vitro*, it may, in analogy, favor tissue destruction *in vivo*. In addition to MMP-1 and TIMP-1, direct cell–cell contact with stimulated T lymphocytes induces PGE_2 production on human dermal fibroblasts and synoviocytes [131] that also contribute to tissue destruction by favoring bone resorption. The T cell surface molecules involved in MMP induction in fibroblasts are mainly membrane-associated IL-1 and TNFα. Similarly, membrane-associated cytokines are involved in the inhibition of deposition of the major extracellular matrix components such as types I and III collagen. Indeed, direct contact with either plasma membranes or fixed, stimulated T cells markedly inhibited the synthesis of types I and III collagen in dermal fibroblasts and synoviocytes, whether untreated or treated with transforming growth factor-β (TGFβ). This inhibition was associated with a marked decrease in steady-state levels of pro-α I and pro-α III collagen mRNAs, which was due to a diminished transcription rate but not to a significant alteration of the alpha 1(I) and alpha 1(III) transcript stability. This inhibition of extracellular matrix production mediated by T cell contact was partially due to additive effects of T cell membrane-associated IFNγ, TNFα and IL-1α [130]. Thus, direct contact with stimulated T cells favors extracellular matrix catabolism by enhancing MMP production while diminishing collagen synthesis in fibroblasts and synoviocytes. Interestingly, similar membrane-associated cytokines were involved in both these processes.

It appears that the expression of intercellular adhesion molecule-1 (ICAM-1), vascular cell adhesion molecule-1 (VCAM-1) and E-selectin is increased in the synovium of RA patients [140–142]. A similar phenomenon was observed in brain microvessels of MS patients [143]. In the latter context, we demonstrated that membranes of stimulated T lymphocytes induce the expression of ICAM-1, VCAM-1, and E-selectin on microvascular endothelial cells from human brain (HB-MVEC) [129]. Beside cell adhesion molecules, contact-mediated activation of HB-MVEC induced the production of IL-6 and IL-8 (Fig. 2). The cell contact-induced expression of cell-adhesion molecules and the production of IL-6 and IL-8 were inhibited by TNFα inhibitors, demonstrating that membrane-associated TNFα was largely responsible for the activation of endothelial cells [129]. This suggested that both soluble and membrane-associated TNF can activate microvascular endothelium and

Direct cellular contact with stimulated T cells

Target cells:	Monocyte-macrophages		Synoviocytes Fibroblasts		PMN	Endothelial cells
Production of:	IL-1 ↑ TNFα ↑	IL-1Ra ↑	MMPs ↑ PGE_2 ↑	Collagen ↓	ROS ↑	CAMs ↑ IL-6 and IL-8 ↑
Inhibited or reversed by:						
IL-1Ra	no	no	yes	yes	no	–
TNFsR	no	no	yes	yes	no	yes
Anti-IFNγ	no	no	no	yes	–	–
HDL-associated apo A-I	yes	no*	no	–	yes	–

Figure 2
Effect of direct cellular contact with stimulated T cells on various cell types involved in RA
The production of IL-1 and TNFα in monocyte-macrophages was inhibited by HDL whereas the production of IL-1Ra was inhibited only in THP-1 cells, the production of MMPs and PGE_2 in fibroblast-like synoviocytes was inhibited by IL-1 and TNFα inhibitors but not by HDL whereas the inhibition of collagen synthesis was reversed by inhibitors of IL-1, TNFα, and IFNγ. The induction of reactive oxygen species (ROS) in neutrophils (PMN) was inhibited by HDL. The induction of cell adhesion molecules (CAMs) and IL-6 and IL-8 in endothelial cells was inhibited by TNFα inhibitors. (*) indicate that IL-1Ra production in THP-1 cells was inhibited by HDL (see Fig. 1).

that these two types of activation might be relevant to RA [144], although other cell surface molecules are likely to play a part in the activation of endothelial cells upon direct cell–cell contact with stimulated T lymphocytes. These studies demonstrated that stimulated T lymphocytes could activate endothelial cells by direct contact, implying that this mechanism may be critical for the extravasation of stimulated T lymphocytes into the target tissue. Furthermore, the induction of cytokine and chemokine production by endothelial cells upon contact with stimulated T lymphocytes might facilitate the invasion of the target tissue by other mononuclear cells [129]. Thus, TNFα inhibitors decrease endothelial cell activation and MMP-1 and PGE_2 production and reverse the inhibition of collagen synthesis in synoviocytes upon direct contact with stimulated T lymphocytes [129–131] as summarized in Figure 2. This might explain in part the beneficial effects of TNF blockade in RA [145, 146].

At an early stage of inflammation T_L could be also present simultaneously with PMN in the tissue. PMN produce reactive oxygen radicals, a large variety of proteolytic enzymes and various cytokines, especially IL-8 and IL-1Ra. PMN oxidative metabolism leads to the production of highly reactive oxygen species and contributes to the elimination of pathogenic micro-organisms, but it could be harmful to host tissue at the site of inflammation, by causing membrane lipidic peroxidation leading to cell and tissue damage [147]. In previous studies, we showed that cellular contact with activated T cells stimulated and primed neutrophil oxygen-dependent respiratory burst [148] and that this activity on PMN correlated with that on THP-1 cells, suggesting that similar factors at the surface of stimulated T cells were involved in the activation of neutrophils and monocytes [68]. We recently confirmed this and demonstrated that purified HDL inhibited T cell contact-induced respiratory burst in neutrophils [149]. This strongly suggests that analogous molecules at the surface of T cells are involved in the activation of both monocytes and neutrophils. However, lipid-free apo A-I, but not HDL, proved to inhibit IgG-induced superoxide production by PMN [150], suggesting a specific inhibitory activity of HDL directed to contact-mediated activation of PMN. Thus, the contact with stimulated T cells may activate several cell types involved in RA. While cytokine blockade inhibits most of the activation of synoviocytes and endothelial cells, HDL inhibits the activation of neutrophils and monocytes (Fig. 2). This suggests that contact-mediated activation of neighboring cells is a process largely used by T cells to influence the activation status of other cells, even though the surface molecules involved in these processes might be different.

Conclusions

To date, direct cell–cell contact with stimulated T_L is considered the main pathway triggering activation of Mϕ in the absence of infectious agents. The potency of this mechanism suggests that it is a major pathway by which T lymphocytes exert their pathogenic effect in chronic inflammatory diseases of autoimmune etiology. Additional investigations are needed to identify the surface molecules – ligands and counter-ligands – involved in this process. However, the control of contact-mediated signaling of monocytes and neutrophils by HDL-associated apo A-I might represent an interesting step towards developing a novel approach for interfering with the inflammatory response induced by cell–cell contact and leading to tissue destruction in chronic inflammatory diseases. It also extends the concept of inflammation in connection with lipid metabolism and other diseases such as arteriosclerosis. Stimulated T lymphocytes in contact with fibroblasts or synoviocytes are also responsible for the decrease in collagen synthesis and may therefore play a part in the lack of repair process.

Acknowledgments
The authors are indebted to Mrs Rachel Chicheportiche, Marie-Thérèse Kaufmann, Carmelina de Luca and Lyssia Gruaz for their outstanding technical skills and help and to Mrs Roswitha Rehm for skilful reading of the manuscript. Credit for apo A-I sequencing is due to Carl K. Edwards, III and Tadahiko Kohno (Amgen, Inc., Thousand Oaks). Unpublished results reported above were part of projects supported by grant #3200.068286.02 and #3100.058558.99/1 from the Swiss National Science Foundation as well as grants from the Swiss Society for Multiple Sclerosis.

References

1 Birkedal-Hansen H (1995) Proteolytic remodeling of extracellular matrix. *Curr Opin Cell Biol* 7: 728–735
2 Burger D (2000) Cell contact-mediated signaling of monocytes by stimulated T cells: a major pathway for cytokine induction. *Eur Cytokine Netw* 11: 346–353
3 Dayer JM, Burger D (1999) Cytokines and direct cell contact in synovitis: relevance to therapeutic intervention. *Arthritis Res* 1: 17–20
4 Feldmann M, Brennan FM, Maini RN (1996) Role of cytokines in rheumatoid arthritis. *Annu Rev Immunol* 14: 397–440
5 Harris ED (1997) Cytokines, lymphokines, growth factors, and chemokines. In: ED Harris (ed): *Rheumatoid arthritis*. W.B. Saunders, Philadelphia, 105–125
6 Dayer JM, Arend WP (1997) Cytokines and growth factors. In: WN Kelley, ED Jr Harris, S Ruddy, CS Sledge (eds): *Textbook of rheumatology*. W.B. Saunders, Philadelphia, 267–286
7 Dayer JM, Feige U, Edwards III CK, Burger D (2001) Anti-interleukin-1 therapy in rheumatic diseases. *Curr Opin Rheumatol* 13: 170–176
8 Maini RN, St Clair EW, Breedveld F, Furst D, Kalden J, Weisman M, Smolen J, Emery P, Harriman G, Feldmann M et al (1999) Infliximab (chimeric anti-tumour necrosis factor alpha monoclonal antibody) *versus* placebo in rheumatoid arthritis patients receiving concomitant methotrexate: a randomised phase III trial. ATTRACT Study Group. *Lancet* 354: 1932–1939
9 Bartolucci P, Ramanoelina J, Cohen P, Mahr A, Godmer P, Le Hello C, Guillevin L (2002) Efficacy of the anti-TNFα antibody infliximab against refractory systemic vasculitides: an open pilot study on 10 patients. *Rheumatology (Oxford)* 41: 1126–1132
10 Lorenz HM, Kalden JR (2002) Perspectives for TNFα-targeting therapies. *Arthritis Res* 4 (Suppl 3): S17–S24
11 Kalden JR (2002) Emerging role of anti-tumor necrosis factor therapy in rheumatic diseases. *Arthritis Res* 4 (Suppl 2): S34–S40
12 Brandt J, Haibel H, Reddig J, Sieper J, Braun J (2002) Successful short term treatment of severe undifferentiated spondyloarthropathy with the anti-tumor necrosis factor-alpha monoclonal antibody infliximab. *J Rheumatol* 29: 118–122

13 Wollina U, Konrad H (2002) Treatment of recalcitrant psoriatic arthritis with anti-tumor necrosis factor-alpha antibody. *J Eur Acad Dermatol Venereol* 16: 127–129

14 Sfikakis PP (2002) Behçet's disease: a new target for anti-tumour necrosis factor treatment. *Ann Rheum Dis* 61 (Suppl 2): ii51–ii53

15 Lee RZ, Veale DJ (2002) Management of spondyloarthropathy: new pharmacological treatment options. *Drugs* 62: 2349–2359

16 Murray KJ, Lovell DJ (2002) Advanced therapy for juvenile arthritis. *Best Pract Res Clin Rheumatol* 16: 361–378

17 Robinson WH, Genovese MC, Moreland LW (2001) Demyelinating and neurologic events reported in association with tumor necrosis factor alpha antagonism: by what mechanisms could tumor necrosis factor alpha antagonists improve rheumatoid arthritis but exacerbate multiple sclerosis? *Arthritis Rheum* 44: 1977–1983

18 Skurkovich S, Boiko A, Beliaeva I, Buglak A, Alekseeva T, Smirnova N, Kulakova O, Tchechonin V, Gurova O, Deomina T et al (2001) Randomized study of antibodies to IFNγ and TNFα in secondary progressive multiple sclerosis. *Mult Scler* 7: 277–284

19 Dean GS, Tyrrell-Price J, Crawley E, Isenberg DA (2000) Cytokines and systemic lupus erythematosus. *Ann Rheum Dis* 59: 243–251

20 Segal R, Dayan M, Zinger H, Mozes E (2001) Suppression of experimental systemic lupus erythematosus (SLE) in mice *via* TNF inhibition by an anti-TNFα monoclonal antibody and by pentoxiphylline. *Lupus* 10: 23–31

21 Kaul D (2001) Molecular link between cholesterol, cytokines and atherosclerosis. *Mol Cell Biochem* 219: 65–71

22 Teplyakov AI, Pryschepova EV, Kruchinsky NG, Chegerova TI (2000) Cytokines and soluble cell adhesion molecules. Possible markers of inflammatory response in atherosclerosis. *Ann NY Acad Sci* 902: 320–322

23 Dinarello CA (2000) IL-1α. In: JJ Oppenheim, M Feldmann (eds): *Cytokine feference.* Academic Press, London, 307–318

24 Dinarello CA (2000) IL-1β. In: JJ Oppenheim, M Feldmann (eds): *Cytokine reference.* Academic Press, London, 351–374

25 Aggarwal BB, Samanta A, Feldmann M (2000) TNFα. In: JJ Oppenheim, M Feldmann (eds): *Cytokine reference.* Academic Press, London, 413–434

26 Sims JE (2002) IL-1 and IL-18 receptors, and their extended family. *Curr Opin Immunol* 14: 117–122

27 Mantovani A, Locati M, Vecchi A, Sozzani S, Allavena P (2001) Decoy receptors: a strategy to regulate inflammatory cytokines and chemokines. *Trends Immunol* 22: 328–336

28 Burger D, Dayer JM (2000) IL-1Ra. In: JJ Oppenheim, M Feldmann (eds): *Cytokine reference.* Academic Press, London, 319–336

29 Burger D, Chicheportiche R, Giri JG, Dayer JM (1995) The inhibitory activity of human interleukin-1 receptor antagonist is enhanced by type II interleukin-1 soluble receptor and hindered by type I interleukin-1 soluble receptor. *J Clin Invest* 96: 38–41

30 Aggarwal BB, Samanta A, Feldmann M (2000) TNF receptors. In: JJ Oppenheim, M Feldmann (eds): *Cytokine reference*. Academic Press, London, 1619–1632
31 Kahle P, Saal JG, Schaudt K, Zacher J, Fritz P, Pawelec G (1992) Determination of cytokines in synovial fluids: correlation with diagnosis and histomorphological characteristics of synovial tissue. *Ann Rheum Dis* 51: 731–734
32 Nouri AM, Panayi GS, Goodman SM (1984) Cytokines and the chronic inflammation of rheumatic disease. *Clin Exp Immunol* 55: 295–302
33 Firestein GS, Alvaro-Gracia JM, Maki R (1990) Quantitative analysis of cytokine gene expression in rheumatoid arthritis. *J Immunol* 144: 3347–3353
34 Seitz M, Loetscher P, Dewald B, Towbin H, Rordorf C, Gallati H, Gerber NJ (1996) Interleukin-1 (IL-1) receptor antagonist, soluble tumor necrosis factor receptors, IL-1β, and IL-8: Markers of remission in rheumatoid arthritis during treatment with methotrexate. *J Rheumatol* 23: 1512–1516
35 Gabay C, Gay Crosier F, Roux-Lombard P, Meyer O, Mainetti C, Guerne PA, Vischer T, Dayer JM (1994) Elevated serum levels of interleukin-1 receptor antagonist in polymyositis/dermatomyositis. A biologic marker of disease activity with a possible role in the lack of acute-phase protein response. *Arthritis Rheum* 37: 1744–1751
36 Suzuki H, Takemura H, Kashiwagi H (1995) Interleukin-1 receptor antagonist in patients with active systemic lupus erythematosus. Enhanced production by monocytes and correlation with disease activity. *Arthritis Rheum* 38: 1055–1059
37 Heilig B, Wermann M, Gallati H, Brockhaus M, Berke B, Egen O, Pezutto A, Hunstein W (1992) Elevated TNF receptor plasma concentrations in patients with rheumatoid arthritis. *Clin Invest* 70: 22–27
38 Aderka D, Wysenbeek A, Engelmann H, Cope AP, Brennan FM, Molad Y, Hornik V, Levo Y, Maini RN, Feldmann M et al (1993) Correlation between serum levels of soluble tumor necrosis factor receptor and disease activity in systemic lupus erythematosus. *Arthritis Rheum* 36: 1111–1120
39 van der Poll T, van Deventer SJ, Hack CE, Wolbink GJ, Aarden LA, Buller HR, ten Cate JW (1992) Effects on leukocytes after injection of tumor necrosis factor into healthy humans. *Blood* 79: 693–698
40 Steiner G, Studnicka Benke A, Witzmann G, Hofler E, Smolen J (1995) Soluble receptors for tumor necrosis factor and interleukin-2 in serum and synovial fluid of patients with rheumatoid arthritis, reactive arthritis and osteoarthritis. *J Rheumatol* 22: 406–412
41 Roux-Lombard P, Punzi L, Hasler F, Bas S, Todesco S, Gallati H, Guerne PA, Dayer JM (1993) Soluble tumor necrosis factor receptors in human inflammatory synovial fluids. *Arthritis Rheum* 36: 485–489
42 Hansen MB, Andersen V, Rohde K, Florescu A, Ross C, Svenson M, Bendtzen K (1995) Cytokine autoantibodies in rheumatoid arthritis. *Scand J Rheumatol* 24: 197–203
43 Jouvenne P, Fossiez F, Garrone P, Djossou O, Banchereau J, Miossec P (1996) Increased incidence of neutralizing autoantibodies against interleukin-1 alpha (IL-1α) in nondestructive chronic polyarthritis. *J Clin Immunol* 16: 283–290

44 Gabay C, Marinova-Mutafchieva L, Williams RO, Gigley JP, Butler DM, Feldmann M, Arend WP (2001) Increased production of intracellular interleukin-1 receptor antagonist type I in the synovium of mice with collagen-induced arthritis: a possible role in the resolution of arthritis. *Arthritis Rheum* 44: 451–462
45 Arend WP, Dayer JM (1995) Inhibition of the production and effects of interleukin-1 and tumor necrosis factor alpha in rheumatoid arthritis. *Arthritis Rheum* 38: 151–160
46 Ulfgren AK, Grondal L, Lindblad S, Khademi M, Johnell O, Klareskog L, Andersson U (2000) Interindividual and intra-articular variation of proinflammatory cytokines in patients with rheumatoid arthritis: potential implications for treatment. *Ann Rheum Dis* 59: 439–447
47 Kimura N, Du RP, Mak TW (1987) Rearrangement and organization of T cell receptor gamma chain genes in human leukemic T cell lines. *Eur J Immunol* 17: 1653–1656
48 Harris ED (ed) (1997) *Rheumatoid arthritis*. W.B. Saunders, Philadelphia
49 Tak PP, Smeets TJM, Daha MR, Kluin PM, Meijers KAE, Brand R, Meinders AE, Breedveld FC (1997) Analysis of the synovial cell infiltrate in early rheumatoid synovial tissue in relation to local disease activity. *Arthritis Rheum* 40: 217–225
50 Smeets TJ, Kraan MC, Galjaard S, Youssef PP, Smith MD, Tak PP (2001) Analysis of the cell infiltrate and expression of matrix metalloproteinases and granzyme B in paired synovial biopsy specimens from the cartilage-pannus junction in patients with RA. *Ann Rheum Dis* 60: 561–565
51 Davis LS, Geppert TD, Meek K, Oppenheimer-Marks N, Lipsky PE (1997) Immune and inflammatory responses. In: WN Kelley, EDJr Harris, S Ruddy, CS Sledge (eds): *Textbook of rheumatology*. W.B. Saunders, Philadelphia, 95–127
52 Schulze-Koops H, Kalden JR (2001) The balance of Th1/Th2 cytokines in rheumatoid arthritis. *Best Pract Res Clin Rheumatol* 15: 677–691
53 Elliott CL, El Touny SY, Filipi ML, Healey KM, Leuschen MP (2001) Interferon β1a treatment modulates TH1 expression in gammadelta + T cells from relapsing-remitting multiple sclerosis patients. *J Clin Immunol* 21: 200–209
54 Laman JD, Thompson EJ, Kappos L (1998) Balancing the Th1/Th2 concept in multiple sclerosis. *Immunol Today* 19: 489–490
55 Wong WM, Vakis SA, Ayre KR, Ellwood CN, Howell WM, Tutt AL, Cawley MI, Smith JL (2000) Rheumatoid arthritis T cells produce Th1 cytokines in response to stimulation with a novel trispecific antibody directed against CD2, CD3, and CD28. *Scand J Rheumatol* 29: 282–287
56 McInnes IB, Leung BP, Sturrock RD, Field M, Liew FY (1997) Interleukin-15 mediates T cell-dependent regulation of tumor necrosis factor-alpha production in rheumatoid arthritis. *Nat Med* 3: 189–195
57 Parry SL, Sebbag M, Feldmann M, Brennan FM (1997) Contact with T cells modulates monocyte IL-10 production. Role of T cell membrane TNFα. *J Immunol* 158: 3673–3681
58 Sebbag M, Parry SL, Brennan FM, Feldmann M (1997) Cytokine stimulation of T lymphocytes regulates their capacity to induce monocyte production of tumor necrosis fac-

tor-alpha, but not interleukin-10: Possible relevance to pathophysiology of rheumatoid arthritis. *Eur J Immunol* 27: 624–632

59 Avice MN, Sarfati M, Triebel F, Delespesse G, Demeure CE (1999) Lymphocyte activation gene-3, a MHC class II ligand expressed on activated T cells, stimulates TNFα and ILα production by monocytes and dendritic cells. *J Immunol* 162: 2748–2753

60 Avice MN, Demeure CE, Delespesse G, Rubio M, Armant M, Sarfati M (1998) IL-15 promotes IL-12 production by human monocytes *via* T cell-dependent contact and may contribute to IL-12-mediated IFNγ secretion by CD4+ T cells in the absence of TCR ligation. *J Immunol* 161: 3408–3415

61 Chabot S, Charlet D, Wilson TL, Yong VW (2001) Cytokine production consequent to T cell-microglia interaction: the PMA/IFNγ-treated U937 cells display similarities to human microglia. *J Neurosci Methods* 105: 111–120

62 Takemura S, Klimiuk PA, Braun A, Goronzy JJ, Weyand CM (2001) T cell activation in rheumatoid synovium is B cell dependent. *J Immunol* 167: 4710–4718

63 Weaver CT, Unanue ER (1986) T cell induction of membrane IL-1 on macrophages. *J Immunol* 137: 3868–3873

64 Weaver CT, Duncan LM, Unanue ER (1989) T cell induction of macrophage IL-1 during antigen presentation: characterization of a lymphokine mediator and comparison of Th1 and Th2 subsets. *J Immunol* 142: 3469–3476

65 Vey E, Zhang JH, Dayer J-M (1992) IFNγ and 1,25(OH)2D3 induce on THP-1 cells distinct patterns of cell surface antigen expression, cytokine production, and responsiveness to contact with activated T cells. *J Immunol* 149: 2040–2046

66 Isler P, Vey E, Zhang JH, Dayer JM (1993) Cell surface glycoproteins expressed on activated human T cells induce production of interleukin-1 beta by monocytic cells: a possible role of CD69. *Eur Cytokine Netw* 4: 15–23

67 Landis CB, Friedman ML, Fisher RI, Ellis TM (1991) Induction of human monocyte IL-1 mRNA and secretion during anti-CD3 mitogenesis requires two distinct T cell-derived signals. *J Immunol* 146: 128–135

68 Li JM, Isler P, Dayer JM, Burger D (1995) Contact-dependent stimulation of monocytic cells and neutrophils by stimulated human T cell clones. *Immunology* 84: 571–576

69 Miltenburg AMM, Lacraz S, Welgus HG, Dayer JM (1995) Immobilized anti-CD3 antibody activates T cell clones to induce the production of interstitial collagenase, but not tissue inhibitor of metalloproteinases, in monocytic THP-1 cells and dermal fibroblasts. *J Immunol* 154: 2655–2667

70 Chizzolini C, Chicheportiche R, Burger D, Dayer JM (1997) Human Th1 cells preferentially induce interleukin (IL)-1 beta while Th2 cells induce IL-1 receptor antagonist production upon cell/cell contact with monocytes. *Eur J Immunol* 27: 171–177

71 Brennan FM, Hayes AL, Ciesielski CJ, Green P, Foxwell BM, Feldmann M (2002) Evidence that rheumatoid arthritis synovial T cells are similar to cytokine-activated T cells: involvement of phosphatidylinositol 3-kinase and nuclear factor κB pathways in tumor necrosis factor alpha production in rheumatoid arthritis. *Arthritis Rheum* 46: 31–41

72 Ribbens C, Dayer JM, Chizzolini C (2000) CD40-CD40 ligand (CD154) engagement is

required but may not be sufficient for human T helper 1 cell induction of interleukin-2- or interleukin-15-driven, contact-dependent, interleukin-1β production by monocytes. *Immunology* 99: 279–286
73 Lacraz S, Isler P, Vey E, Welgus HG, Dayer JM (1994) Direct contact between T lymphocytes and monocytes is a major pathway for induction of metalloproteinase expression. *J Biol Chem* 269: 22027–22033
74 Vey E, Burger D, Dayer JM (1996) Expression and cleavage of tumor necrosis factor-alpha and tumor necrosis factor receptors by human monocytic cell lines upon direct contact with stimulated T cells. *Eur J Immunol* 26: 2404–2409
75 Vey E, Dayer JM, Burger D (1997) Direct contact with stimulated T cells induces the expression of IL-1β and IL-1 receptor antagonist in human monocytes. Involvement of serine/threonine phosphatases in differential regulation. *Cytokine* 9: 480–487
76 Wagner DH, Stout RD, Suttles J (1994) Role of the CD40-CD40 ligand interaction in CD4(+) T cell contact-dependent activation of monocyte interleukin-1 synthesis. *Eur J Immunol* 24: 3148–3154
77 Sypek JP, Matzilevich MM, Wyler DJ (1991) Th2 lymphocyte clone can activate macrophage antileishmanial defense by a lymphokine-independent mechanism *in vitro* and can augment parasite attribution *in vivo*. *Cell Immunol* 133: 178–186
78 Sypek JP, Wyler DJ (1991) Antileishmanial defense in macrophages triggered by tumor necrosis factor expressed on CD4+ T lymphocyte plasma membrane. *J Exp Med* 174: 755–759
79 Grell M (1996) Tumor necrosis factor (TNF) receptors in cellular signaling of soluble and membrane-expressed TNF. *J Inflamm* 47: 8–17
80 Suttles J, Miller RW, Tao X, Stout RD (1994) T cells which do not express membrane tumor necrosis factor-alpha activate macrophage effector function by cell contact-dependent signaling of macrophage tumor necrosis factor-alpha production. *Eur J Immunol* 24: 1736–1742
81 Ware CF, Vanarsdale TL, Crowe PD, Browning JL (1995) The ligands and receptors of the lymphotoxin system. *Curr Top Microbiol Immunol* 198: 175–218
82 Murphy M, Walter BN, Pike-Nobile L, Fanger NA, Guyre PM, Browning JL, Ware CF, Epstein LB (1998) Expression of the lymphotoxin beta receptor on follicular stromal cells in human lymphoid tissues. *Cell Death Differ* 5: 497–505
83 Mosmann TR, Coffman RL (1989) TH1 and TH2 cells: different patterns of lymphokine secretion lead to different functional properties. *Annu Rev Immunol* 7: 145–173
84 Gramaglia I, Mauri DN, Miner KT, Ware CF, Croft M (1999) Lymphotoxin αβ is expressed on recently activated naive and Th1- like CD4 cells but is downregulated by IL-4 during Th2 differentiation. *J Immunol* 162: 1333–1338
85 Malik N, Greenfield BW, Wahl AF, Kiener PA (1996) Activation of human monocytes through CD40 induces matrix metalloproteinases. *J Immunol* 156: 3952–3960
86 Stout RD, Suttles J, Xu J, Grewal IS, Flavell RA (1996) Impaired T cell-mediated macrophage activation in CD40 ligand-deficient mice. *J Immunol* 156: 8–11

87 Gauchat J-F, Aubry J-P, Mazzei G, Life P, Jomotte T, Elson G, Bonnefoy J-Y (1993) Human CD40-ligand: molecular cloning, cellular distribution and regulation of expression by factors controlling IgE production. *FEBS Lett* 315: 259–266
88 Armant M, Rubio M, Delespesse G, Sarfati M (1995) Soluble CD23 directly activates monocytes to contribute to the antigen-independent stimulation of resting T cells. *J Immunol* 155: 4868–4875
89 Armant M, Ishihara H, Rubio M, Delespesse G, Sarfati M (1994) Regulation of cytokine production by soluble CD23: co-stimulation of interferon gamma secretion and triggering of tumor necrosis factor alpha release. *J Exp Med* 180: 1005–1011
90 Rezzonico R, Chicheportiche R, Imbert V, Dayer JM (2000) Engagement of CD11b and CD11c β2 integrin by antibodies or soluble CD23 induces IL-1β production on primary human monocytes through mitogen-activated protein kinase-dependent pathways. *Blood* 95: 3868–3877
91 Rezzonico R, Imbert V, Chicheportiche R, Dayer JM (2001) Ligation of CD11b and CD11c β(2) integrins by antibodies or soluble CD23 induces macrophage inflammatory protein 1α (MIP-1α) and MIP-1β production in primary human monocytes through a pathway dependent on nuclear factor-κB. *Blood* 97: 2932–2940
92 Lecoanet-Henchoz S, Gauchat JF, Aubry JP, Graber P, Life P, Paul-Eugene N, Ferrua B, Corbi AL, Dugas B, Plater-Zyberk C et al (1995) CD23 regulates monocyte activation through a novel interaction with the adhesion molecules CD11b-CD18 and CD11c-CD18. *Immunity* 3: 119–125
93 Manié S, Kubar J, Limouse M, Ferrua B, Ticchioni M, Breittmayer JP, Peyron JF, Schaffar L, Rossi B (1993) CD3-stimulated Jurkat T cells mediate IL-1β production in monocytic THP-1 cells: role of LFA-1 molecule and participation of CD69 T cell antigen. *Eur Cytokine Netw* 4: 7–13
94 Hyka N, Dayer JM, Modoux C, Kohno T, Edwards CK, III, Roux-Lombard P, Burger D (2001) Apolipoprotein A-I inhibits the production of interleukin-1β and tumor necrosis factor-alpha by blocking contact-mediated activation of monocytes by T lymphocytes. *Blood* 97: 2381–2389
95 Park YB, Lee SK, Lee WK, Suh CH, Lee CW, Lee CH, Song CH, Lee J (1999) Lipid profiles in untreated patients with rheumatoid arthritis. *J Rheumatol* 26: 1701–1704
96 Doherty NS, Littman BH, Reilly K, Swindell AC, Buss JM, Anderson NL (1998) Analysis of changes in acute-phase plasma proteins in an acute inflammatory response and in rheumatoid arthritis using two-dimensional gel electrophoresis. *Electrophoresis* 19: 355–363
97 Lakatos J, Harsagyi A (1988) Serum total, HDL, LDL cholesterol, and triglyceride levels in patients with rheumatoid arthritis. *Clin Biochem* 21: 93–96
98 Ananth L, Prete PE, Kashyap ML (1993) Apolipoproteins A-I and B and cholesterol in synovial fluid of patients with rheumatoid arthritis. *Metabolism* 42: 803–806
99 Bresnihan B, Gogarty M, Burger D, Fitzgerald O, Dayer JM (2002) Localization of apolipoprotein A-1 at sites of T lymphocyte and macrophage contact in rheumatoid arthritis tissue. *Rheumatology* 41: 65 (abstract)

100 Tselepis AD, Elisaf M, Besis S, Karabina SA, Chapman MJ, Siamopoulou A (1999) Association of the inflammatory state in active juvenile rheumatoid arthritis with hypo-high-density lipoproteinemia and reduced lipoprotein-associated platelet-activating factor acetylhydrolase activity. *Arthritis Rheum* 42: 373–383
101 Lahita RG, Rivkin E, Cavanagh I, Romano P (1993) Low levels of total cholesterol, high-density lipoprotein, and apolipoprotein AI in association with anticardiolopin antibodies in patients with SLE. *Arthritis Rheum* 36: 1566–1574
102 Dinu AR, Merrill JT, Shen C, Antonov IV, Myones BL, Lahita RG (1998) Frequency of antibodies to the cholesterol transport protein apolipoprotein A1 in patients with SLE. *Lupus* 7: 355–360
103 Sena A, Pedrosa R, Ferret-Sena V, Almeida R, Andrade ML, Morais MG, Couderc R (2000) Interferon β1a therapy changes lipoprotein metabolism in patients with multiple sclerosis. *Clin Chem Lab Med* 38: 209–213
104 Burger D, Dayer JM (2002) High-density lipoprotein-associated apolipoprotein A-I: the missing link between infection and chronic inflammation? *Autoimmunity Rev* 1: 111–117
105 Fidge NH (1999) High-density lipoprotein receptors, binding proteins, and ligands. *J Lipid Res* 40: 187–201
106 Kozyraki R, Fyfe J, Kristiansen M, Gerdes C, Jacobsen C, Cui S, Christensen EI, Aminoff M, de la Chapelle A, Krahe R et al (1999) The intrinsic factor-vitamin B12 receptor, cubilin, is a high-affinity apolipoprotein A-I receptor facilitating endocytosis of high-density lipoprotein. *Nat Med* 5: 656–661
107 Chambenoit O, Hamon Y, Marguet D, Rigneault H, Rosseneu M, Chimini G (2001) Specific docking of apolipoprotein A-I at the cell surface requires a functional ABCA1 transporter. *J Biol Chem* 276: 9955–9960
108 Bocharov AV, Vishnyakova TG, Baranova IN, Patterson AP, Eggerman TL (2001) Characterization of a 95 kDa high affinity human high density lipoprotein-binding protein. *Biochemistry* 40: 4407–4416
109 Jurgens G, Xu QB, Huber LA, Bock G, Howanietz H, Wick G, Traill KN (1989) Promotion of lymphocyte growth by high-density lipoproteins (HDL). Physiological significance of the HDL binding site. *J Biol Chem* 264: 8549–8556
110 Xu Q, Buhler E, Steinmetz A, Schonitzer D, Bock G, Jurgens G, Wick G (1992) A high-density-lipoprotein receptor appears to mediate the transfer of essential fatty acids from high-density lipoprotein to lymphocytes. *Biochem J* 287 (Pt 2): 395–401
111 Williamson RA, Yea CM, Robson PA, Curnock AP, Gadher S, Hambleton AB, Woodward K, Bruneau JM, Hambleton P, Spinella Jaegle S et al (1996) Dihydroorotate dehydrogenase is a target for the biological effects of leflunomide. *Transplant Proc* 28: 3088–3091
112 Tugwell P, Wells G, Strand V, Maetzel A, Bombardier C, Crawford B, Dorrier C, Thompson A (2000) Clinical improvement as reflected in measures of function and health-related quality of life following treatment with leflunomide compared with methotrexate in patients with rheumatoid arthritis: sensitivity and relative efficiency to

detect a treatment effect in a twelve-month, placebo-controlled trial. Leflunomide Rheumatoid Arthritis Investigators Group. *Arthritis Rheum* 43: 506–514
113 Sharp JT, Strand V, Leung H, Hurley F, Loew-Friedrich I (2000) Treatment with leflunomide slows radiographic progression of rheumatoid arthritis: results from three randomized controlled trials of leflunomide in patients with active rheumatoid arthritis. Leflunomide Rheumatoid Arthritis Investigators Group. *Arthritis Rheum* 43: 495–505
114 Smeets TJ, Dayer JM, Kraan MC, Versendaal J, Chicheportiche R, Breedveld FC, Tak PP (2000) The effects of interferon-beta treatment of synovial inflammation and expression of metalloproteinases in patients with rheumatoid arthritis. *Arthritis Rheum* 43: 270–274
115 Chofflon M (2000) Recombinant human interferon beta in relapsing-remitting multiple sclerosis: a review of the major clinical trials. *Eur J Neurol* 7: 369–380
116 Arnason BG (1999) Treatment of multiple sclerosis with interferon beta. *Biomed Pharmacother* 53: 344–350
117 Déage V, Burger D, Dayer JM (1998) Exposure of T lymphocytes to leflunomide but not to dexamethasone favors the production by monocytic cells of interleukin-1 receptor antagonist and the tissue-inhibitor of metalloproteinases-1 over that of interleukin-1β and metalloproteinases. *Eur Cytokine Netw* 9: 663–668
118 Coclet-Ninin J, Dayer JM, Burger D (1997) Interferon-beta not only inhibits interleukin-1β and tumor necrosis factor-alpha but stimulates interleukin-1 receptor antagonist production in human peripheral blood mononuclear cells. *Eur Cytokine Netw* 8: 345–349
119 Franceschini G (1996) Apolipoprotein function in health and disease: insights from natural mutations. *Eur J Clin Invest* 26: 733–746
120 Burger D, Dayer JM (2002) The role of human T lymphocyte-monocyte contact in inflammation and tissue destruction. *Arthritis Res* 4 (Suppl 3): S169–S176
121 Jungo F, Dayer JM, Modoux C, Hyka N, Burger D (2001) IFNβ inhibits the ability of T lymphocytes to induce TNFα and IL-1β production in monocytes upon direct cell-cell contact. Cytokine 14: 272–282
122 Burger D, Molnarfi N, Gruaz L, Dayer JM (2002) PI3-K is a checkpoint in the production of interleukin-1β (IL-1β) and IL-1 receptor antagonist. *Eur J Biochem* 269 (Suppl 1): 106 (abstract)
123 Dinther-Janssen AC, Pals ST, Scheper R, Breedveld F, Meijer CJ (1990) Dendritic cells and high endothelial venules in the rheumatoid synovial membrane. *J Rheumatol* 17: 11–17
124 Kang YM, Zhang X, Wagner UG, Yang H, Beckenbaugh RD, Kurtin PJ, Goronzy JJ, Weyand CM (2002) CD8 T cells are required for the formation of ectopic germinal centers in rheumatoid synovitis. *J Exp Med* 195: 1325–1336
125 De Vita S, Zaja F, Sacco S, De Candia A, Fanin R, Ferraccioli G (2002) Efficacy of selective B cell blockade in the treatment of rheumatoid arthritis: evidence for a pathogenetic role of B cells. *Arthritis Rheum* 46: 2029–2033
126 Kim HJ, Berek C (2000) B cells in rheumatoid arthritis. *Arthritis Res* 2: 126–131

127 Tetlow LC, Woolley DE (1995) Mast cells, cytokines, and metalloproteinases at the rheumatoid lesion: dual immunolocalisation studies. *Ann Rheum Dis* 54: 896–903
128 Tetlow LC, Woolley DE (1995) Distribution, activation and tryptase/chymase phenotype of mast cells in the rheumatoid lesion. *Ann Rheum Dis* 54: 549–555
129 Lou J, Dayer JM, Grau GE, Burger D (1996) Direct cell/cell contact with stimulated T lymphocytes induces the expression of cell adhesion molecules and cytokines by human brain microvascular endothelial cells. *Eur J Immunol* 26: 3107–3113
130 Rezzonico R, Burger D, Dayer JM (1998) Direct contact between T lymphocytes and human dermal fibroblasts or synoviocytes downregulates types I and III collagen production *via* cell-associated cytokines. *J Biol Chem* 273: 18720–18728
131 Burger D, Rezzonico R, Li JM, Modoux C, Pierce RA, Welgus HG, Dayer JM (1998) Imbalance between interstitial collagenase and tissue inhibitor of metalloproteinases 1 in synoviocytes and fibroblasts upon direct contact with stimulated T lymphocytes: involvement of membrane-associated cytokines. *Arthritis Rheum* 41: 1748–1759
132 Gonzalez-Amaro R, Alarcon-Segovia D, Alcocer-Varela J, Diaz de Leon L, Rosenstein Y (1988) Mononuclear cell-fibroblast interactions in scleroderma. *Clin Immunol Immunopathol* 46: 412–420
133 Hibbs MS, Postlethwaite AE, Mainardi CL, Seyer JM, Kang AH (1983) Alterations in collagen production in mixed mononuclear leukocyte-fibroblast cultures. *J Exp Med* 157: 47–59
134 Holoshitz J, Kosek J, Sibley R, Brown DA, Strober S (1991) T lymphocyte-synovial fibroblast interactions induced by mycobacterial proteins in rheumatoid arthritis. *Arthritis Rheum* 34: 679–686
135 Spörri B, Bickel M, Limat A, Waelti ER, Hunziker T, Wiesmann UN (1996) Juxtacrine stimulation of cytokine production in co-cultures of human dermal fibroblasts and T cells. *Cytokine* 8: 631–635
136 Korn JH, Halushka PV, LeRoy EC (1980) Mononuclear cell modulation of connective tissue function: suppression of fibroblast growth by stimulation of endogenous prostaglandin production. *J Clin Invest* 65: 543–554
137 Postlethwaite AE, Smith GN, Mainardi CL, Seyer JM, Kang AH (1984) Lymphocyte modulation of fibroblast function *in vitro*: stimulation and inhibition of collagen production by different effector molecules. *J Immunol* 132: 2470–2477
138 Scott S, Pandolfi F, Kurnick JT (1990) Fibroblasts mediate T cell survival: a proposed mechanism for retention of primed T cells. *J Exp Med* 172: 1873–1876
139 Corrigall VM, Solau-Gervais E, Panayi GS (2000) Lack of CD80 expression by fibroblast-like synoviocytes leading to anergy in T lymphocytes. *Arthritis Rheum* 43: 1606–1615
140 Dinther-Janssen AC, Horst E, Koopman G, Newmann W, Scheper RJ, Meijer CJ, Pals ST (1991) The VLA-4/VCAM-1 pathway is involved in lymphocyte adhesion to endothelium in rheumatoid synovium. *J Immunol* 147: 4207–4210
141 Matsuyama T, Kitani A (1996) The role of VCAM-1 molecule in the pathogenesis of rheumatoid synovitis. *Hum Cell* 9: 187–192

142 Postigo AA, Garcia-Vicuna R, Diaz-Gonzalez F, Arroyo AG, de Landazuri MO, Chi-Rosso G, Lobb RR, Laffon A, Sanchez-Madrid F (1992) Increased binding of synovial T lymphocytes from rheumatoid arthritis to endothelial-leukocyte adhesion molecule-1 (ELAM-1) and vascular cell adhesion molecule-1 (VCAM-1). *J Clin Invest* 89: 1445–1452
143 Washington R, Burton J, Todd RF, Newman W, Dragovic L, Dore-Duffy P (1994) Expression of immunologically relevant endothelial cell activation antigens on isolated central nervous system microvessels from patients with multiple sclerosis. *Ann Neurol* 35: 89–97
144 Burger D, Lou J, Dayer JM, Grau GE (1997) Both soluble and membrane-associated TNF activate brain microvascular endothelium: relevance to multiple sclerosis. *Mol Psychiatry* 2: 113–116
145 Feldmann M, Maini RN (2001) Anti TNFα therapy of rheumatoid arthritis: What have we learned? *Annu Rev Immunol* 19: 163–196
146 Maini RN, Feldmann M (2002) How does infliximab work in rheumatoid arthritis? *Arthritis Res* 4 Suppl 2: S22–S28
147 Babior BM (2000) Phagocytes and oxidative stress. *Am J Med* 109: 33–44
148 Zhang JH, Ferrante A, Arrigo AP, Dayer JM (1992) Neutrophil stimulation and priming by direct contact with activated human T lymphocytes. *J Immunol* 148: 177–181
149 Cettour-Rose P, Dayer JM, Burger D, Roux-Lombard P (2003) Activation of respiratory burst in polymorphonuclear leukocytes upon contact with stimulated T cells and inhibition by high-density lipoproteins (HDLs). *Arthritis Res Ther* 5 (Suppl 1): 97 (abstract)
150 Blackburn WDJr, Dohlman JG, Venkatachalapathi YV, Pillion DJ, Koopman WJ, Segrest JP, Anantharamaiah GM (1991) Apolipoprotein A-I decreases neutrophil degranulation and superoxide production. *J Lipid Res* 32: 1911–1918

The role of proteinases in joint tissue destruction

Tim E. Cawston and Andrew D. Rowan

Rheumatology Department, School of Clinical and Medical Sciences, The Medical School, University of Newcastle, Framlington Place, Newcastle upon Tyne NE2 4HH, UK

Introduction

The breakdown of cartilage and bone in the arthritides prevents joints from functioning normally. In severe cases both cartilage and the underlying bone can be destroyed. Cartilage contains different types of collagen [1], rod-shaped molecules that aggregate in a staggered array to form cross-linked fibers giving connective tissues strength and rigidity. Trapped between these collagen fibers are the aggrecan molecules [2] that consist of three globular domains interspersed with heavily glycosylated and sulfated polypeptide. In the presence of hyaluronic acid these form highly charged aggregates, attract water into the tissue and so allow cartilage to resist compression. Chondrocytes in normal adult cartilage maintain a steady state where the rate of matrix synthesis equals the rate of degradation. Any change in this steady state will affect the functional integrity of the cartilage. During growth and development, synthesis of matrix components exceeds the rate of degradation; reduction in the rate of matrix synthesis and an increase in the rate of degradation occur during matrix resorption [3].

The primary cause of cartilage and bone destruction in joint pathology involves the elevated levels of active proteinases, secreted from a variety of cells, which degrade proteoglycans and collagens. While the sources of these proteinases depend on the type of disease, they are regulated by different cytokines and growth factors acting on the cells within the joint. In osteoarthritis (OA) the proteinases produced by chondrocytes play a major role [4]. In contrast, in a highly inflamed rheumatoid joint the proteinases produced by chondrocytes, synovial cells and inflammatory cells all contribute to the loss of tissue matrix [5]. This review will describe cytokines that are produced by a variety of cells that either minimize matrix production or upregulate different classes of proteinases that are implicated in the destruction of joint tissue.

Cytokines and Joint Injury, edited by Wim B. van den Berg and Pierre Miossec

Proteolytic pathways of connective tissue breakdown

Extracellular matrix proteins are broken down by different proteolytic pathways. The five main classes of proteinases are classified according to the chemical group which participates in the hydrolysis of peptide bonds [6], although a more detailed classification system, the MEROPS database [7], is now available. Cysteine, aspartate, and threonine proteinases are predominantly active at acid pH and act intracellularly; the serine and metalloproteinases are active at neutral pH, and act extracellularly. Examples of these enzymes are shown in Figure 1. Some enzymes, such as elastase are released when neutrophils are stimulated, while others may not participate in the cleavage of matrix proteins but activate proenzymes that degrade the matrix. Some proteinases are membrane-bound rather than secreted from the cell. These enzymes are known to be associated with cytokine processing, receptor shedding and the removal of proteins that are associated with cell–cell or cell–matrix interactions [8]. Proteinases can also participate in matrix breakdown by cleaving inhibitors that, once removed, allow destruction of the matrix by other proteinases.

All classes of proteinase play a part in the turnover of connective tissues and one proteinase pathway may precede another. The pathway that predominates will vary with different resorptive situations. The turnover of the extracellular matrix often involves a complex interaction between different cell types. For example the osteoid layer in bone is removed by osteoblast metalloproteinases prior to the attachment of osteoclasts which contain or secrete predominantly cysteine proteinases such as cathepsin B and K. These degrade bone matrix after the removal of mineral. A complex interaction occurs in the rheumatoid joint between T cells, macrophages, synovial fibroblasts and chondrocytes. In septic arthritis neutrophils release both serine and metalloproteinases that exceed the local concentration of inhibitors, resulting in a rapid stripping of the cartilage matrix from the joint cavity.

Extracellular proteolysis – neutral proteinases

The metzincin superfamily

These metalloproteinases are distinguished by a highly conserved motif containing three histidines that bind zinc at the catalytic site and a conserved methionine turn that lies beneath the active-site zinc [9]. This group is further divided into four multigene families:

- Serralysins (large bacterial enzymes)
- Astacins (e.g., bone morphogenetic protein-1 or BMP-1)
- ADAMs (a disintegrin and metalloproteinase)/adamalysins (e.g., TNFα convertase or ADAM 17)
- Matrix metalloproteinases (MMPs) [10].

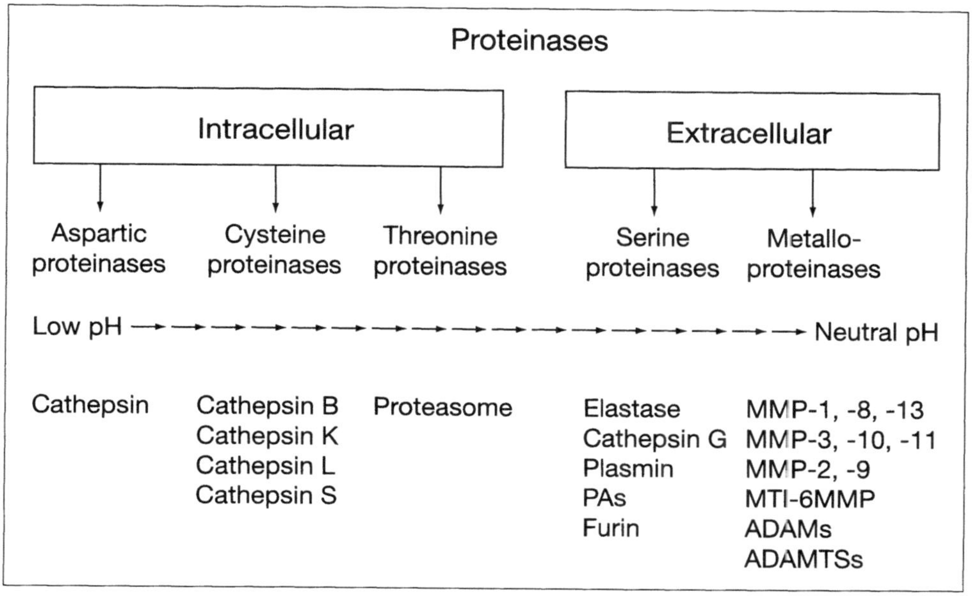

Figure 1
There are five classes of proteinase, three of which act predominantly intracellularly (aspartate, cysteine and threonine) and two predominantly extracellular (metallo and serine). Examples are shown of enzymes from each class.

These families are classified according to the sequence around the three conserved histidines that binds zinc. The third H is followed by either E, D, P or S, respectively in the astacins, reprolysins, serralysins and MMPs. The distance between the third H and the conserved M turn also differs markedly between the families with the MMPs separated by seven residues and the astacins by 44–46 residues. A fifth group, the pappalysins, have been proposed [11] where a V follows the third H and a large gap is found between these residues and the M turn of 63 amino acids. The pappalysins have been shown to cleave IGFBP-4 and -5 [12].

Matrix metalloproteinases

The MMPs constitute a multigene family of over 25 secreted and cell surface enzymes that process or degrade numerous substrates [13]. The MMP family, when activated and acting together, are best known for their ability to cleave components of the extracellular matrix (ECM). They also cleave proteinases, proteinase

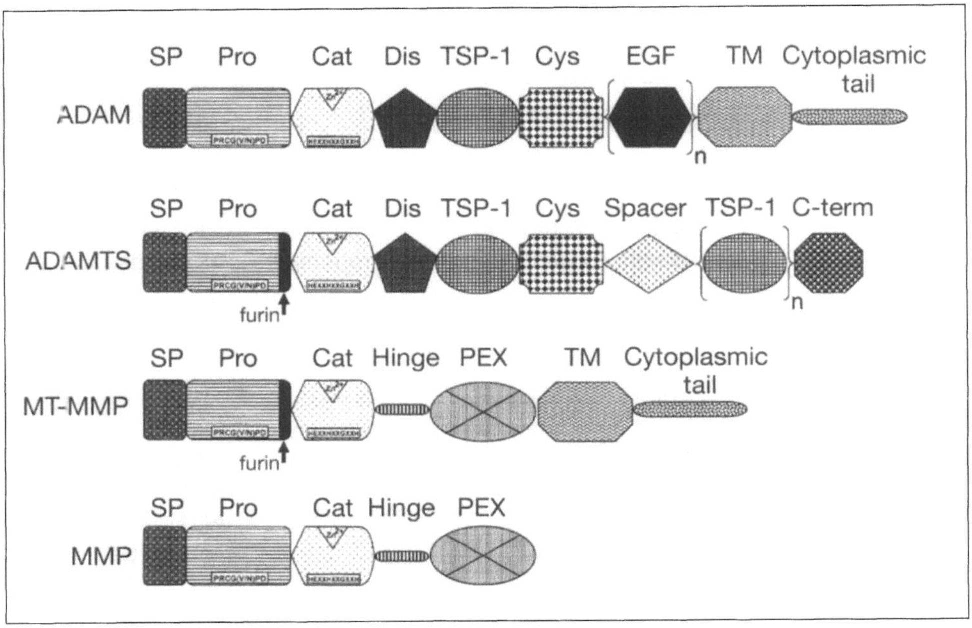

Figure 2
MMPs and ADAMs have a domain structure, with several common domains across the family. All have a zinc binding domain and propeptide that preserves latency. Other domains influence the behaviour of the protein; the transmembrane domain anchors the proteinase on to the membrane surface while other domains bind to different components of the extracellular matrix.

inhibitors, latent growth factors, chemotactic molecules, growth factor binding proteins, cell surface receptors and cell–cell adhesion molecules [14]. Thus MMPs are able to regulate many biological processes and consequently are precisely controlled themselves. MMPs all contain common domains (Fig. 2), zinc is present at the catalytic centre and all are produced in a proenzyme form which is activated by the loss of a propeptide. Some MMPs (e.g., MT-MMPs) have a short peptide insert between the propeptide and the N-terminal domain. This sequence is recognized by furin [15], a serine proteinase located in the Golgi, and such MMPs are therefore secreted in an active form. Two of the MMPs (collagenase-2/MMP-8 and gelatinase B/MMP-9) are found stored within the specific granules of the neutrophil while most other MMPs are produced by different connective tissue cells after stimulation by cytokines [16]. Traditionally, MMPs have been grouped according to their substrates but with the increasing numbers they are often now grouped according to their domain structure [17]. Most MMPs resemble MMP-1 (collagenase-1); MMP-

2 (gelatinase A) and MMP-9 have fibronectin-like inserts, while MMP-21 has a vitronectin-like domain insert. MMP-17 and MMP-25 both have a cytoplasmic GPI anchor, MMP-23 has a C-terminal immunoglobulin-like domain but neither MMP-7 (matrilysin) nor MMP-26 has a haemopexin domain [10]. The domain structure of the MMPs and ADAMs are shown in Figure 2.

MMPs are divided into at least five groups:

- Stromelysins
- Collagenases
- Gelatinases
- Membrane-type MMPs (MT-MMPs)
- GPI anchored enzymes [13, 16, 17]

The stromelysins have broad substrate specificity. The natural substrates of these enzymes are probably proteoglycans, fibronectin and laminin. MMP-3 (stromelysin-1) is not normally widely expressed but can be readily induced by growth factors and cytokines such as interleukin (IL)-1 and tumour necrosis factor (TNF)α. The stromelysins have similar substrate specificity [18] but the expression pattern of these enzymes is often distinct.

There are three mammalian collagenases: MMP-1, MMP-8 and MMP-13 (collagenase-3). These enzymes cleave fibrillar collagens at a single site, producing three-quarter- and one-quarter-sized fragments. MMP-2 and MT-1MMP (MMP-14) can also cleave at this site [19, 20]. The enzymes differ in their specificity for different collagens; collagenase-3 has much broader substrate specificity than the other collagenases [21]. Both MMP-1 and MMP-13 are synthesized by macrophages, fibroblasts and chondrocytes when these cells are stimulated with inflammatory mediators. MMP-8 is predominantly released from neutrophils upon stimulation of the cell but this MMP can also be produced by chondrocytes and all three collagenases are present in diseased cartilage [22]. MMP-13 is often controlled in a different way to collagenase-1; retinoic acid, which downregulates MMP-1, is known to upregulate MMP-13 in some species [23].

The two gelatinases cleave denatured collagens, type IV and V collagen and elastin. MMP-2 is the most widespread of all the MMPs; MMP-9 is expressed in a wide variety of transformed and tumor-derived cells [24]. These gelatinases are involved in the degradation of basement membranes and in angiogenesis.

There is an increase in levels of different MMPs in rheumatoid synovial fluid, in conditioned culture media from rheumatoid synovial tissues and cells, in synovial tissue at the cartilage–pannus junction from rheumatoid joints, in osteoarthritic cartilage and in animal models of arthritis [16, 22, 24, 25]. These proteinases are implicated in the pathological destruction of joint tissue and are involved in the normal turnover of connective tissue matrix that takes place during growth and development. In osteoarthritis, both the rate of matrix synthesis and breakdown are

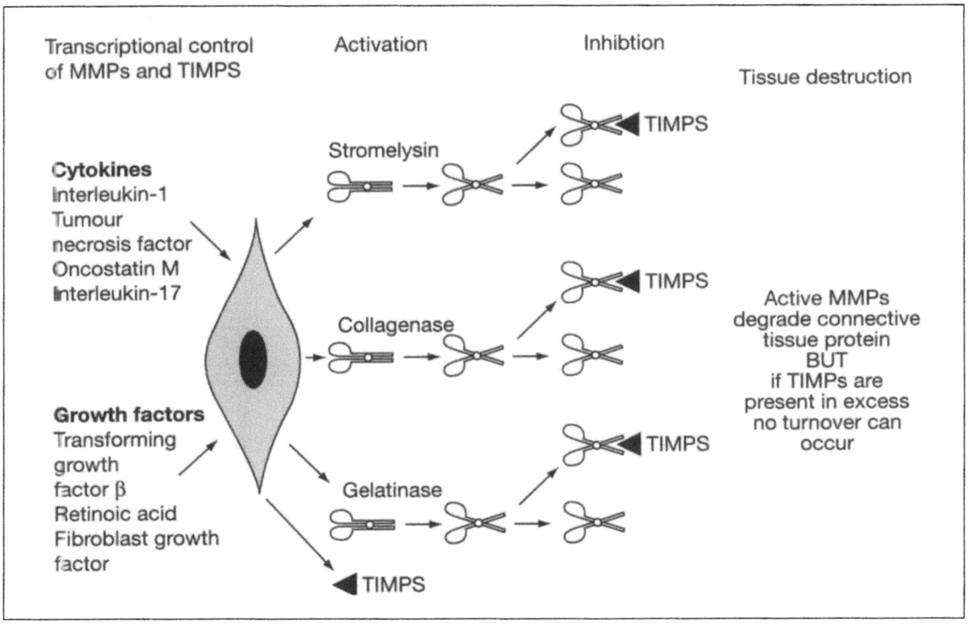

Figure 3
The control of the matrix metalloproteinases occurs at different levels. Different cytokines and growth factors can upregulate or downregulate the transcription of MMPs and/or TIMPs. Protein is sorted through the trans-Golgi network and arrives at the cell surface in a latent form unless it has a furin recognition site in which case it is activated within the secretory vesicle. Activation has to occur before the substrate can be digested and this will only occur if the level of active enzyme exceeds the local levels of TIMPs

increased. This leads to the formation of excess matrix in some regions (osteophytes) with focal loss of matrix in other areas.

The MMPs need careful control and this occurs at a number of critical steps (see Fig. 3) that include: synthesis and secretion, activation of the proenzymes, inhibition of the active enzymes and localisation and clearance of MMPs. In addition, the response of cells is modified by cell–cell and cell–matrix interactions.

Synthesis and secretion

IL-1, TNFα and IL-17 stimulate numerous cell types to produce proteinases with proinflammatory and degradative effects. The synthesis and secretion of MMP-1 and MMP-3 and other MMPs are stimulated by IL-1 and TNFα [26]. Within the arthritic joint there are large numbers of different cell types that produce specific cytokines and growth factors. Synovial fluids from RA patients have been shown to contain a variety of inflammatory mediators [27, 28]. These often differ in their action on individual cell types and many cytokines can synergise to increase the response of the cells. It is therefore difficult to predict the outcome of blocking the action of individual cytokines to prevent cartilage destruction. The blocking of TNFα action in patients with RA successfully reduces inflammation and joint destruction, although some studies suggest that blocking additional cytokines such as IL-1 or IL-17 is likely to confer additional benefit [29]. This combinational anti-cytokine approach is further supported by observations of synergy between cytokine combinations and the profound catabolic effects they have on cartilage [30–36].

Activation of the proenzymes

Activation of MMPs, which are produced in a proenzyme form, is an important and understudied control point in connective tissue breakdown [37, 38]. The propeptide is removed proteolytically; this allows the enzyme to hydrolyze peptide bonds and activation is likely to be achieved in a tightly controlled environment close to the cell surface. Active MMP-3 activates procollagenases and other MMPs [39]. Some MMPs (MT-MMPs and MMP-11, -17, -21, -23, -25 and -28) have a furin recognition sequence between the propeptide and the catalytic domain and these enzymes are often activated as they traverse the Golgi arriving at the cell surface in an active form. Recent data implicate MT1-MMP as an important initiator of MMP activation cascades at the cell surface [40, 41]. Plasmin and possibly other serine proteinases activate some MMPs, and cartilage explant cultures treated with cytokines and an inhibitor of furin have reduced levels of collagenases and low collagen release [41].

Inhibition of the active enzymes

All active MMPs are inhibited by tissue inhibitors of metalloproteinases (TIMPs) [16, 42, 43], stable proteins that bind tightly to active MMPs in a 1:1 ratio. These TIMPs play an important role in controlling connective tissue breakdown by blocking the action of the activated MMPs (see Fig. 3). If TIMP levels exceed those of active enzyme then connective tissue turnover is prevented. TIMP-3 is bound by the extracellular matrix after secretion and also inhibits some members of the ADAM family (see below). TIMP-4 is predominantly localized in heart tissue but can be produced by joint tissues [44]. TIMP-1 and -3 are upregulated by growth factors such as transforming growth factor (TGF)β, insulin-like growth factor (IGF)-1 and oncostatin M (OSM), and these agents also induce matrix synthesis [45]. The mechanisms that control the extracellular activity of the MMPs are illustrated in Figure 3.

Localization and clearance

Much of the proteolysis that occurs takes place in the immediate vicinity of the cell in pericellular pockets close to the cell membrane. MMPs can be secreted to specific areas at the cell surface [46]. This allows a high degree of control and these localisation mechanisms can enhance MMP activity, limit the access of MMP inhibitors, concentrate MMPs to their precise target substrate and limit the extent of proteolysis to a discrete region. Although the MMPs with transmembrane domains are the most important cell surface enzymes, some MMPs can bind to cell surface receptors, to cell surface activating enzymes such as urokinase-type plasminogen activator (uPA) and thrombin, or to pericellular matrix proteins. For example, MMP-7 binds to cell surface and ECM heparan sulphate molecules and this can increase the stability and activity of MMP-7 [47]. These cell surface heparan sulphate molecules can also bind other MMPs and also TIMP-3. MMP-1 can bind to cell surface EMMPRIN [48]. MMPs can also be regulated by means of their own proteolysis which renders them inactive. For example, MMP-1 when cleaved in the hinge region can no longer cleave collagen [49]. After binding to α2-macroglobulin many MMPs are rapidly cleared *via* endocytosis and degradation within the lysosomal system.

Many MMPs and members of the matrixin family are involved in the activation and or release of cytokines and growth factors from the ECM [50, 51]. These released factors have a myriad of effects on cellular proliferation and behaviour. For example, mice deficient in MT-1MMP [52] or ADAM-17 [53] both show gross abnormalities in tissue development and growth demonstrating that these cell surface enzymes have important functions.

ADAM family of proteinases

This family, closely related to MMPs, are also implicated in cartilage biology and in the turnover of the matrix particularly in relation to proteoglycan turnover. ADAMs are usually membrane anchored proteinases with diverse functions conferred by the addition of different protein domains [8, 54] (Fig. 2). Members are associated with sperm-egg fusion and the sheddases that mediate the release of cell surface proteins. ADAM-17 is known for its ability to release TNFα from the cell surface [55]. To date, 29 ADAM genes and 18 ADAMTS (a disintegrin and metalloproteinase with thrombospondin motifs) genes have been described. The disintegrin domain can bind to integrins and prevent cell–cell interactions; cysteine-rich, epidermal growth factor-like, transmembrane and cytoplasmic tail domains are also found (Fig. 2). The ADAMTS family members are distinguished from the ADAMs in that they lack these latter three domains but have additional thrombospondin-1 (TSP-1) domains (which can number up to 13), predominantly at the C-terminus which are thought to mediate interactions with the ECM [56].

Analysis of the major aggrecan fragments from resorbing cartilage revealed that cleavage at a specific glutamyl residue, within the interglobular domain of the aggrecean monomer, had occurred [57]. The proteinase responsible for this was termed 'aggrecanase', and two enzymes that belong to the ADAMTS family, ADAMTS-4 and -5 [58–60], have subsequently been described that are thought to be the pathophysiological mediators of aggrecean catabolism (reviewed in [3, 59]). ADAMTS-4 and -5 [59, 60] can both cleave cartilage proteoglycan, and these enzymes may be associated with the cell surface as some studies suggest that chondrocyte membrane factors cleave aggrecan; it is not clear if an ADAMTS family member is attached tightly *via* thrombospondin domains or a separate enzyme is present [60]. Other ADAMs have also been described in cartilage including ADAM-10, -12 and -15 in addition to ADAM-17 [62]. Interestingly, many of the ADAMs family are inhibited by TIMP-3 [63].

Proteoglycan release from cartilage occurs following stimulation with a variety of mediators such as IL-1, TNFα, IL-17, retinoic acid and fibronectin fragments [64–66]. Some members of the ADAMs family (e.g., ADAMTS-1) are upregulated by IL-1, and the correct arrangement of these enzymes on the cell surface may be important to allow substrate cleavage to occur. Levels of ADAMTS-4 are upregulated in cartilage in response to IL-1 and TNFα, and in synovial fibroblasts in response to TGFβ [59, 67], while ADAMTS-5 appears to be unaffected. In an immortalised chondrocyte line, ADAMTS-1, -4 and -5 were all regulated by a mixture of IL-1 and OSM although the speed of response differed between these enzymes [68]. Aggrecanase activity can be blocked by specific synthetic inhibitors [69] and also by TIMP-3 [63].

Despite the identification of ADAMTS-4 and -5 as aggrecanases, considerable debate still exists as to the proteinases involved in aggrecan catabolism. There are at least five known aggrecanase cleavage sites in aggrecan, and proteolysis at these sites leads to cartilage destruction while MMP cleavages have been proposed to be essentially non-destructive since the products are retained in the tissue [70]. In experimental arthritis, MMP-generated degradation products correlate with severe cartilage damage (i.e., late stage disease) while aggrecanase-mediated products were evident early [71]. These differences in the temporal contribution of MMPs and aggrecanases to aggrecan catabolism have been further supported in a study that suggests that there may also be spatial differences for the site of action of these two enzyme families [72]. The mechanism of aggrecan release from cartilage may vary between different cartilages for a given stimulus, and that, for a given time, different factors may elicit aggrecan release *via* different pathways [73]. For example, it has been proposed that two independent, but concurrent, mechanisms promote IL-1β-induced aggrecan release [74].

Although it is clear that a variety of cytokine stimuli induce aggrecanase activity, it is still less clear as to whether this induction of proteolytic activity is due to gene transcription. Some studies have found evidence of cytokine regulation of

ADAMTS-4 [68, 75] and ADAMTS-5 [76], although most studies suggest that ADAMTS-5 is not inducible [75, 77, 78]. The ADAMTS-4 promoter appears to contain silencer elements that repress transcriptional activity in chondrocytes [79]. Indeed, induction of ADAMTS-4 by IL-1 has been proposed to be through the activation of constitutively active protein [80], and this activation may be MMP-mediated [81]. A role for neprilysin-induced aggrecanase activity *via* the generation of regulatory peptides has also been proposed [82].

There is increasing evidence that several of the ADAM/ADAMTS genes give rise to multiple mRNA transcripts. Many of these mRNA isoforms result from alternative splicing of the primary RNA transcript. In most of the cases analysed, the shorter alternative transcripts encode C-terminally truncated forms of the proteins. In the case of ADAM proteins, this may produce a soluble, secreted form of the enzyme (e.g., ADAM-9 [83–85]) or produce enzymes with variable cytoplasmic domains (e.g., ADAM-22 or MDC2 [86]). For ADAMTS proteins, C-terminal truncation removes some or all of the TSP-1 type repeats (e.g., ADAMTS-7 and -9 [87], ADAMTS-13 [88]). This may alter the ability of the enzymes to bind to the ECM or to potential cofactors. In addition to alternative splicing, some multiple transcripts arise from the use of alternative transcription start sites or alternative polyadenylation signals, leading to transcripts with variable 5' and 3' untranslated regions (e.g., ADAMTS-2 [89], ADAMTS-14 [90]). Changes in the untranslated regions of mRNA may play a role in control of translation, or in the subcellular localisation or stability of the message. It seems likely that the ability to produce multiple messages from ADAM/ADAMTS genes allows the cell to fine tune levels of expression and localisation of these enzymes in response to specific stimuli. In support of this, some studies have demonstrated differential expression of particular splice variants. For example, in ADAM-22, the epsilon variant is expressed in normal human brain tissue, whereas the gamma variant is the major transcript in gliomas [86]. Similarly, both the long and short forms of ADAM-12 are expressed in the human placenta and some tumour cell lines, whereas only the long variant is detectable in normal human adult skeletal, cardiac and smooth muscle.

Serine proteinases

The plasminogen-dependent pathway is implicated in a wide variety of situations where connective tissue matrix is remodeled. Both tissue- and urokinase -type plasminogen activators (tPA and uPA) are found in cartilage and cleave plasminogen to plasmin. Inhibitors can bind and inactivate the released enzymes. Many *in vitro* experiments with tissue or cells point to a role for the plasminogen–plasmin system in the activation of pro-MMPs in some situations of matrix turnover [91]. It is possible that chondrocytes synthesize and secrete other serine proteinases but no enzymes have yet been purified and characterized from cartilage tissue. Plasminogen activators are involved in cartilage degradation since proinflammatory agents that stimulate cartilage resorption such as IL-1, TNFα and retinoic acid reduce plasminogen activator inhibitor-1 (PAI-1) levels while protective factors such as TGFβ and latent derived growth factor (PDGF) increase PAI-1 [92].

Other serine proteinase activities have been implicated in arthritis. IL-1 and TNFα-induced proteoglycan release can be potently blocked with an inactivator of uPA [93], and granzyme B can initiate proteoglycan degradation (but not collagen); granzyme B-positive cells can be detected in synovium and at the invasive front in RA [94]. Inclusion of α1 proteinase inhibitor to cartilage stimulated to resorb effectively blocks the release of collagen implicating serine proteinase(s) in the activation cascades of proinflammatory cytokine-induced proMMPs [38]. Several metalloproteinases have furin recognition motifs in their prodomains, thus enabling the secretion of active proteinases, and recent studies have shown that blocking furin can prevent the release of collagen from resorbing cartilage [41]. Furin and plasmin have both been specifically implicated in the activation of latent TGFβ [95, 96].

RA is considered to be an autoimmune disorder, and complement activation occurs adjacent to the articular surface where abundant co-deposits of immunoglobulins and activated complement components such as C1s are present. This serine proteinase contributes to disease pathogenesis *via* its collagenolytic activity [97] as well as its ability to degrade IGF binding protein (IGFBP)-5 and hence reduce the levels of available IGF-1 [98]. Similar activities have been attributed to various MMPs [51, 99].

The current paradigm on RA pathogenesis involves a T cell-mediated process in the initiation of synovial inflammation, although such a mechanism may not be necessary to sustain chronic inflammation [100]. Activation of the coagulation cascade may represent an important pathway that amplifies and perpetuates synovitis. Tissue factor is the primary initiator of extravascular coagulation [101] which involves the sequential activities of the serine proteinases factor VIIa, factor Xa and thrombin. Increased expression of both tissue factor and thrombin within synovium, combined with reduced levels of coagulation factors and elevated thrombin activity in RA synovium fluids, support ongoing coagulation within the rheumatoid joint [102, 103]. Thrombin converts fibrinogen into fibrin, and abundant fibrin deposits within arthritic joints are a notable pathological feature in RA. Thrombin inhibition significantly reduces synovitis, fibrin deposition and synovial expression of IL-1β and IL-12 in experimental arthritis models [104, 105]. Synovial expression of RANTES has been shown to be thrombin-induced through protease-activated receptor (PAR)-1, implicating thrombin in RANTES-mediated migration of inflammatory cells to the synovium in RA [106]. An essential role for PAR-2 has been demonstrated in mediating chronic inflammation since PAR-2-deficient mice were resistant to adjuvant arthritis with almost no evidence of tissue damage [107].

While the involvement of proinflammatory cytokines in RA is well established their role in OA is less clear. Some studies have suggested that low levels of both IL-

1 and TNFα have a role in driving joint destruction in OA [108]. These cytokines have a marked effect on matrix synthesis at low concentrations, and in OA synovium it is reported that there are low levels of IL-1 receptor antagonist (IL-1RA) and increased expression of IL-1 receptors. Chondrocytes from OA cartilage produce raised levels of the p55 TNFa receptor making OA cartilage more sensitive to TNFα stimulation. Raised levels of ADAM-17 mRNA are also reported in OA cartilage compared to normal. In the experimental dog and rabbit models of OA, intra-articular injections of IL-1RA gene can slow the progression of structural changes [108].

Neutrophil-dependent pathways

Large numbers of polymorphonuclear leukocytes are often found in the rheumatoid joint that contains both serine proteinases (elastase and cathepsin G) and MMPs (MMP-8 and -9). Release from the cell is stimulated by a variety of stimuli. Neutrophil proteinases have been localized in cartilage [109], and enzymes released into the synovial fluid cannot avoid the excess inhibitors present such as α2-macroglobulin (which inhibits all classes of proteinases), α1-antiproteinase inhibitor, TIMP-1 and TIMP-2, and so it is unlikely that damage to the cartilage occurs from the joint space. However, in septic arthritis the normal inhibitory capacity of the synovial fluid is exceeded and neutrophil proteinases rapidly degrade cartilage from the fluid phase [16].

Although it is clear that neutrophils do play an important role in animal models some studies have shown that matrix degradation does not correlate with neutrophil granule release. Animals in which elastase and cathepsin G is absent or depleted can still show severe cartilage loss [110]. However, when neutrophils abut cartilage or adhere to the cartilage surface or where the inhibitory capacity of body fluids is exceeded, then neutrophil proteinases will contribute to matrix degradation.

Intracellular pathways – acid proteinases

Cysteine proteinases

An involvement of lysosomal cysteine proteinases [111] in matrix resorption was first suggested due to their ability to degrade type I collagen at acidic pH [112, 113]. Subsequently, direct evidence of a role was demonstrated when cysteine proteinase inhibitors prevented the resorption of bone explants [114, 115]. Cathepsin D (an aspartate proteinase), and cathepsin B (cysteine proteinase) are raised in OA tissue and raised levels of cathepsins B, L and H (cysteine proteinases) are found in antigen-induced rat arthritis models and within the rheumatoid joint. Aggrecan is susceptible to cathepsins B and L *in vitro* [116].

Incubation of resorbing cartilage with specific cathepsin B inhibitors that penetrate the cell blocked the release of proteoglycan fragments [117] suggesting the involvement of an intracellular route for cartilage proteoglycan breakdown. Everts [118] proposed that collagen resorption was *via* an intracellular route since substantial amounts of fibrillar collagen accumulate intracellularly in tissue culture studies in the presence of cysteine proteinase inhibitors [119]. Cathepsin B has been most strongly associated with an intracellular role, especially in the degradation of soft tissues [120] but also in bone resorption, although this has been suggested to be *via* the activation of other proteinases [121] such as pro-urokinase [122, 123]. The identification of cathepsin K [124, 125] has been a major advancement in our understanding of bone catabolism, although cathepsins B, L, N, S and K are all capable of collagen degradation [126, 127]. A specialised role for cathepsin K in bone resorption (see below) was proposed based on its selective expression in osteoclasts compared to cathepsins B, L or S [128] and anti-sense studies [129]. There is also evidence that it is involved in cartilage degradation [130, 131] where it can be localised to sites of cartilage erosion. Evidence also suggests that cathepsin K has potent aggrecan-degrading activity and that the resulting degradation products specifically potentiate the collagenolytic activity of cathepsin K toward types I and II collagen [132].

Cathepsin S has been linked to major histocompatibility complex II antigen presentation [133], which may have significance in auto-immune diseases such as RA. Moreover, given the potent ECM degrading activity of cathepsin S, and stability at neutral pH, this proteinase may also contribute to cartilage destruction. Both cathepsins K and S are expressed in RA and OA synovia [131].

Both IL-1β and TNFα increase cathepsin K and B expression in synovial fibroblasts, but not cathepsin L or S [132, 134]. Interferon (IFN)α elevates both cathepsin B and L, while PDGF and basic fibroblast growth factor (bFGF) induce cathepsin B and L, respectively [134]. IGF-II, but not IGF-I, appears to compete for the mannose-6-phosphate receptor which targets newly synthesised proteinases to the lysosome. Dysregulation of this system could therefore result in the secretion of procathepsin enzymes that would be stable at neutral pH [135]. Other studies have colocalised IL-1β and TNFα expression in subchondral bone with cathepsins B, K and L [136]. Cathepsin K expression is downregulated by IFNα [137] which is also involved in the synergistic prevention of osteoclastic bone resorption by IL-12/IL-18 (see below) [138]. During fracture repair, both cathepsin K and MMP-9 are predominantly expressed in osteoclasts at the osteochondral junction, while cathepsins B, H, L and S are mainly associated with chondrocyte hypertrophy [139].

A role for the inactivation of TIMPs by cathepsin B has been suggested in neovasularization of tissues [140]. Angiogenesis is a characteristic feature of many joint pathologies, and such observations have marked implications for disease pathogenesis whereby new blood vessel formation and/or increased MMP activity could be a direct consequence of extracellular cathepsin B activity.

Calpain (calcium-dependent neutral cysteine proteinase) was originally considered to be an intracellular proteinase, but has now been shown to also act extracellularly. This externalization may have catalytic relevance since it is unlikely that the specific inhibitor of calpain, calpastatin [141], could penetrate cell membranes. Proteoglycan is a substrate for calpain [142], which could then be replaced by a mineralized matrix [143]. Indeed, the presence of calpain correlates with arthritis and tissue destruction [144] although proinflammatory cytokines have been shown to downregulate both calpain and calpastatin in OA and RA fibroblasts [145]. This may also explain why the presence of calpain did not correlate with IL-1-induced proteoglycan release *in vitro* [146].

Cysteine and serine proteinases have been convincingly linked with the regulation of cytokine production during inflammation. Specifically dipeptidyl peptidase I (DPPI, a lysosomal cysteine proteinase) has been shown to activate the polymorphonuclear neutrophil-derived serine proteinases, cathepsin G and neutrophil elastase. Mice deficient in DPPI were protected against arthritis development, and these defects were also associated with a decrease in the local production of both TNFα and IL-1β [147].

Threonine proteinases

Threonine proteinases represent a relatively new class of proteinases, so termed due to the presence of a threonine residue at the catalytic site [148]. The proteasome is a ubiquitously expressed, essential intracellular protease complex belonging to this new proteinase class. It performs many intracellular roles including the degradation of phosphorylated and ubiquitinylated inhibitor of kappa B (IκB). This phosphorylation is a consequence of proinflammatory cytokines such as IL-1, IL-17 and TNFα which utilize nuclear factor kappa B (NF-κB) as a signal transducing molecule. Proteasomal degradation of IκB allows NF-κB to become transcriptionally active leading to the upregulation of a variety of matrix-degrading proteinases. Inhibition of the proteasome blocks the matrix degrading potential of chondrocytes [149].

Some workers suggest [118] that intracellular pathways predominate in the normal turnover of the extracellular matrix while the extracellular routes are only used in pathologic conditions. A close apposition of intra- and extra-cellular pathways will be found in many conditions where there is connective tissue turnover.

Osteoclastic bone resorption

In RA bone is also destroyed [150], and both the MMPs and cysteine proteinases are involved [151]. Osteoblasts respond to parathyroid hormone (PTH) and other agents that induce bone resorption, such as IL-1 and TNFα, by increasing the secretion of MMPs to remove the osteoid layer on the bone surface. Osteoclast precursors then adhere to the exposed bone surface, differentiate and form a low pH microenvironment beneath their lower surface. This removes mineral, and lysosomal proteinases then resorb the exposed matrix. Cathepsin B and L cleave collagen types II, IX and XI and destroy cross-linked collagen matrix at low pH [151] and were thought to be responsible. However, cathepsin K, a cysteine proteinase produced by osteoclasts, also cleaves type I collagen at the N-terminal end of the triple helix at pH values as high as pH 6.5 [152]. This enzyme plays a key role in the degradation of bone collagen and its expression correlates with bone resorption [153]. Cathepsin K is also produced by synovial fibroblasts and is thought to contribute to synovium-initiated bone destruction in the rheumatoid joint [153] and bone resorption is impaired in situations where cathepsin K is deficient. Interestingly, the defects in bone resorption in tartrate-resistant acid phosphatase (TRAP) mice were not due to reduced cathepsin K levels but rather due to defective intracellular transport [154].

There is clear evidence for a central role for receptor activator of NF-κB ligand (RANKL) in the bone destruction seen in RA. This member of the TNF ligand family of cytokines is abundantly produced by T cells and synovial fibroblasts in RA synovial membrane, and its action is to stimulate osteoclast formation. It is upregulated by a variety of cytokines that include IL-1, TNFα, IL-11, OSM, PTH-related peptide and IL-17. It binds to a specific receptor, RANK, on the surface of osteoclast precursors. The potent activity of IL-17 in osteoclastogenesis is mediated by the upregulation of RANKL and its action is antagonized by osteoprotegerin (OPG). This molecule is effective at blocking bone resorption in rat adjuvant arthritis and arthritis of TNFα transgenic mice [155]. It protects against the development of bone and cartilage destruction. RANKL has been shown to increase cathepsin K expression in osteoclasts [156] and is a potent mediator of osteoclastogenesis [157]. Activated human T cells can directly induce osteoclastogenesis from monocytes [158], suggesting that T cell-derived cytokines may initiate and/or exacerbate this effect. Indeed, the T cell-specific proinflammatory cytokine IL-17 is elevated in RA [158], found in the early stages of the disease [159] can promote osteoclastogenesis [158] and bone and cartilage destruction [35, 160].

Models of joint destruction and cytokine synergies

Animal models have been used to great effect to discover the roles of different cytokines in joint disease. The major effect of TNFα occurs in the inflammatory phase of the disease in causing the early joint swelling but on its own it is not arthritogenic or destructive; these effects are predominantly mediated by IL-1 [161]. Thus, TNFα over-expressing mice develop chronic erosive disease and this effect can be reproduced in mice with local over-expression of this cytokine within the joint

[162, 163]. However this joint inflammation was completely arrested in TNFα transgenic mice when treated with antibodies against IL-1 receptor [164]. TNFα independent IL-1 production is also found in some animal models [165], and IL-1 is more potent than TNFα in inducing cartilage destruction with much lower levels being effective. A total lack of erosive disease is found in IL-1-deficient mice [166]. The incidence and severity of collagen induced arthritis was less in TNF receptor knockout mice, but once the joints were affected then full progression to erosive damage was seen in a TNFα-independent mechanism [167]. Local expression of IL-1RA reduced the incidence of arthritis and severity as well as tissue destruction. This is in contrast to therapy with TNFα soluble receptors which blocked early inflammatory episodes but had no effect in advanced arthritis [168]. Thus, many models point to a crucial role for IL-1 in late-stage erosive joint destruction. IL-1α acts in the acute phase of the disease while IL-1β is the crucial cytokine in advanced disease stages as chronic erosive disease is absent in IL-1β-deficient mice [169]. The importance of IL-1 in arthritis was demonstrated by studies of the IL-1RA knockout mouse where a spontaneous arthritis develops [170]. In TNFα-deficient mice a full-blown arthritis can still develop. IL-1 is crucial to activate chondrocytes and initiate matrix destruction [171, 172]. The relative contributions of IL-1 and TNFα to joint destruction in different animal models have been reviewed [165].

The relationship between the induction of an acute arthritis and the development of the chronic destructive phase was highlighted by studies showing that the local over expression of IL-12 effected this transformation, and that the resulting arthritis was also more susceptible to flares [173].

IL-18 is present in 80% of RA patient synovia and is strongly associated with the expression of IL-1β and TNFα. These data suggest that IL-18 is a primary proinflammatory cytokine that drives the local expression of IL-1 and TNFα [174]. IL-18 (part of the IL-1 family) is a cytokine that with IL-12 can induce IL-1 and TNFα and so promotes the development of joint lesions. When IL-18 was neutralised in mice with collagen induced arthritis the clinical severity was significantly reduced with decreased cartilage breakdown and a reduction in circulating cartilage oligomeric protein (COMP) levels. Treatment was associated with a slowing of disease progression. Neutralisation of IL-18 significantly reduced TNFα, IL-6 and IFNγ secretion by macrophages [175].

IL-17 has joint destruction potential that is independent of IL-1 [176]. Overexpression of IL-17 accelerated the onset and aggravated synovial inflammation and markedly increased joint destruction; blocking of IL-17 in the collagen induced arthritis model resulted in the suppression of arthritis [177]. Osteoclastic bone resorption is stimulated by IL-1 as well as a combination of TNFα and IL-17.

The role of TGFβ is not always clear. Treatment of murine experimental arthritis with soluble TGFβ receptor II almost completely prevents osteophyte formation while enhancing articular cartilage proteoglycan loss with reduced thickness of the articular cartilage surface [178]. These results suggest that TGFβ is crucial in the formation of osteophytes and plays an important part in protecting cartilage from destruction. Osteophyte induction by TGFβ may hamper its use to induce repair within the joint unless it can be specifically targeted to sites within existing cartilage lesions. BMP-2 is only capable of enhancing chondrocyte proteoglycan synthesis in the absence of IL-1 and so its role would appear to be limited [179]. These studies are supported by *in vitro* studies that show cartilage is protected from IL-1 induced resorption by TGFβ [180]. The main cartilage anabolic factor IGF-1 is very effective in promoting matrix synthesis but has little contribution to make in OA diseased tissue since arthritic chondrocytes do not respond to IGF-1 [181]. Anti-inflammatory therapy with IL-4 and IL-10 has been effective. Anti IL-4 treatment provides impressive protection against cartilage and bone erosion. These studies showed that the proliferation of synovial cells was not blocked while joint tissue destruction was inhibited indicating that these two pathways are not directly linked [182]. Treatment with high local doses of IL-10 reduced swelling in the treated joint with some effects in the surrounding joints [183]. The effects of IL-4 *in vivo* mirror those of IL-13 found *in vitro* [184] while IL-10 does not block IL-1-induced cartilage resorption in *in vitro* studies [183].

In vitro, IL-1 and TNFα can initiate cartilage collagen resorption alone, but when these cytokines are combined with OSM a rapid and reproducible release of collagen is found in bovine and porcine cartilage [185]. Synthetic inhibitors of MMPs, as well as TIMP-1 and TIMP-2, are able to prevent the release of collagen fragments from cartilage [69]. This firmly implicates the collagenolytic MMPs in this process and it is known that collagenase-1, -2 and -3 are made by chondrocytes. A novel model for studying the mechanisms of joint destruction *in vivo* was developed by implanting RA synovial fibroblasts with normal articular cartilage under the renal capsule of SCID mice. The aggressive phenotype of RA synovial fibroblasts was maintained; they invaded cartilage expressing degradative enzymes while OA synovium did not invade [186]. This model emphasizes the role that synovial fibroblasts play in destroying both bone and cartilage in RA [187], and methods have also been developed where synovium and cartilage are co-cultured *in vitro*.

Effects on matrix synthesis and repair

Within arthritic tissues there are areas of cartilage where synthesis of new matrix takes place and areas where net loss of the extracellular matrix occurs. Polypeptide growth factors have a direct effect on chondrocytes [188] and it is apparent that these factors play a major role in the regulation of normal matrix synthesis and in the processes that protect cartilage in disease. The effect of any particular growth factor will depend on the state of differentiation of the cell and on the type of cells within the tissues and local factors. These factors can affect cellular proliferation, migration or differentiation and the control of individual genes (Fig. 3).

TGF-β and IGF-1 have profound effects on the synthesis of matrix components by chondrocytes. Details of these growth factors and their role in the joint have been reviewed [179]. TGFβ can downregulate the production of matrix-degrading proteinases and upregulate proteinase inhibitors such as TIMP-1 or PAI-1, suggesting that it may prevent cartilage destruction by both stimulating synthesis and blocking breakdown pathways. This growth factor is locally synthesised by chondrocytes, affects protein synthesis by chondrocytes, and has been found to potentiate the stimulation of DNA synthesis achieved with other growth factors, rather that initiating this itself [187]. IGF-1 mimics many of the actions of TGFβ and has a significant effect on matrix synthesis. Within the cartilage, TGFβ and IGF-1 are stored in considerable quantity bound to different matrix components. Thus the source of these growth factors will vary. They can be produced by cells outside the cartilage and diffuse in; they can be produced locally by the chondrocytes or released from the matrix as it is degraded. The control mechanisms are complex and growth factors also exist in a latent form and other proteins may be present, such as IGFBPs, which sequester the factors and prevent them from binding to cellular receptors.

These growth factors are recognized as protective to cartilage, stimulating matrix synthesis and blocking the effects of the proinflammatory cytokines. Their use in cartilage repair and the problems associated with their use *in vivo* have been reviewed [189].

Acknowledgments
We thank the arthritis research campaign, the Wellcome Trust, FARNE, Dunhill Medical Trust and the Nuffield Foundation (Oliver Bird Fund) for financial support. The contribution of Dr Deborah Bevitt to the section on control of ADAMTS proteins and of Mrs Catherine Smurthwaite for typing the manuscript is gratefully acknowledged.

References

1 Cremer MA, Rosloniec EF, Kang AH (1998) The cartilage collagens: a review of their structure, organization and role in the pathogenesis of experimental arthritis in animals and in human rheumatic disease. *J Mol Med* 76: 275–288
2 Iozzo RV (1998) Matrix proteoglycans: From molecular design to cellular function. *Annu Rev Biochem* 67: 609–652
3 Mort JS, Billington CJ (2001) Articular cartilage and changes in arthritis: matrix degradation. *Arthritis Res* 3: 337–341
4 Van den Berg WB (2000) Pathophysiology of osteoarthritis. *Joint Bone Spine* 67: 555–556
5 Firestein GA (2003) Evolving concepts of rheumatoid arthritis. *Nature* 423: 356–361

6 Barrett AJ, Rawlings ND, Woessner JF (1998) Introduction. In: Barrett AJ, Rawlings ND, Woessner JF (eds): *Handbook of proteolytic enzymes.* Academic Press, London, UK, xxv–xxix

7 Barrett AJ, Rawlings ND, O'Brien EA (2001) The MEROPS database as a protease information system. *J Struct Biol* 134: 95–102

8 Becherer JD, Blobel CP (2003) Biochemical properties and functions of membrane-anchored metalloprotease-disintegrin proteins (ADAMs). *Curr Top Dev Biol* 54: 101–123

9 Stocker W, Grams F, Baumann U, Reinemer P, Gomis-Ruth FX, McKay DB, Bode W (1995) The metzincins – topological and sequential relations between the astacins, adamalysins, serralysins, and matrixins (collagenases) define a superfamily of zinc-peptidases. *Protein Sci* 4: 823–840

10 Egeblad M, Werb Z (2002) New functions for the matrix metalloproteinases in cancer progression. *Nat Rev Cancer* 2: 161–174

11 Boldt HB, Overgaard MT, Laursen LS, Weyer K, Sottrup-Jensen L, Oxvig C (2001) Mutational analysis of the proteolytic domain of pregnancy-associated plasma protein-A (PAPP-A): classification as a metzincin. *Biochem J* 358: 359–367

12 Overgaard MT, Boldt HB, Laursen LS, Sottrup-Jensen L, Conover CA, Oxvig C (2001) Pregnancy-associated plasma protein0A2 (PAPP-A2), a novel insulin-like growth factor-binding protein-5 proteinase. *J Biol Chem* 276: 21849–21853

13 Nagase H, Woessner JF (1999) Matrix metalloproteinases. *J Biol Chem* 274: 21491–21494

14 Sternlicht MD, Werb Z (2001) How matrix metalloproteinases regulate cell behaviour. *Annu Rev Cell Dev Biol* 17: 463–516

15 Pei D, Weiss SJ (1995) Furin-dependent intracellular activation of the human stromelysin-3 zymogen. *Nature* 375: 244–247

16 Cawston TE (1996) Metalloproteinase inhibitors and the prevention of connective tissue breakdown. *Pharm Ther* 3: 163–182

17 Clark IM, Parker AE (2003) Metalloproteinases: their role in arthritis and potential as therapeutic targets. *Expert Opin Ther Targets* 7: 19–34

18 Nagase H (1995) Human stromelysin 1 & 2. *Methods Enzymol* 248: 449–470

19 Aimes RT, Quigley JP (1995) Matrix metalloproteinase-2 is an interstitial collagenases. Inhibitor-free enzyme catalyzes the cleavage of collagen fibrils and soluble native type I collagen generating the specific 3/4- and 1/4-length fragments. *J Biol Chem* 240: 5872–5876

20 Ohuchi E, Imai K, Fujii Y, Sato H, Seiki M, Okada Y (1997) Membrane type 1 matrix metalloproteinase digests interstitial collagens and other extracellular matrix macromolecules. *J Biol Chem* 272: 2446–2451

21 Knauper V, Lopez-Otin C, Smith B, Knight G, Murphy G (1996) Biochemical characterization of human collagenases-3. *J Biol Chem* 271: 1544–1550

22 Tetlow LC, Adlam DJ, Woolley DE (2001) Matrix metalloproteinases and proinflam-

matory cytokine production by chondrocytes of human osteoarthritic cartilage: Association with degenerative changes. *Arthritis Rheum* 44: 585–594
23 Shingleton WD, Ellis AJ, Rowan AD, Cawston TE (2000) Retinoic acid combines with interleukin-1 to promote the degradation of collagen from bovine nasal cartilage: matrix metalloproteinases-1 and -13 are involved in collagen breakdown. *J Cell Biol* 79: 519–531
24 Murphy G, Crabbe T (1995) Gelatinases A and B. *Methods Enzymol* 248: 470–484
25 Konttinen YT, Ainola M, Valleala H, Ma J, Ida H, Mandelin J, Kinne RW, Santavirta S, Sorsa T, Lopez-Otin C, Takagi M (1999) Analysis of 16 different matrix metalloproteinases (MMP-1 to MMP-20) in the synovial membrane: different profiles in trauma and rheumatoid arthritis. *Ann Rheum Dis* 58: 691–697
26 Van den Berg WB (1999) The role of cytokines and growth factors in cartilage destruction: osteoarthritis and rheumatoid arthritis. *Z Rheumatol* 58: 136–141
27 Westacott CI, Whicher JT, Barnes IC, Thompson D, Swan AJ, Dieppe PA (1990) Synovial fluid concentration of five different cytokines in rheumatic diseases. *Ann Rheum Dis* 49: 676–681
28 Manicourt DH, Poilvache P, Van Egeren A, Devogelaer JP, Lenz ME, Thonar EJ (2000) Synovial fluid levels of tumor necrosis factor alpha and oncostatin M correlate with levels of markers of the degradation of cross-linked collagen and cartilage aggrecan in rheumatoid arthritis but not in osteoarthritis. *Arthritis Rheum* 43: 281–288
29 Chabaud M, Miossec P (2001) The combination of tumor necrosis factor alpha blockade with interleukin-1 and interleukin-17 blockade is more effective for controlling synovial inflammation and bone resorption in an *ex vivo* model. *Arthritis Rheum* 44: 1293–1303
30 Chin JE, Hatfield CA, Krzesicki RF, Herblin WF (1991) Interactions between interleukin-1 and basic fibroblast growth factor on articular chondrocytes. Effects on cell growth, prostanoid production, and receptor modulation. *Arthritis Rheum* 34: 314–324
31 Berenbaum F, Jacques C, Thomas G, Corvol MT, Bereziat G, Masliah J (1996) Synergistic effect of interleukin-1 beta and tumor necrosis factor alpha on PGE2 production by articular chondrocytes does not involve PLA2 stimulation. *Exp Cell Res* 222(2): 379–384
32 Cawston TE, Ellis AJ, Humm G, Lean E, Ward D, Curry V (1995) Interleukin-1 and oncostatin M in combination promote the release of collagen fragments from bovine nasal cartilage in culture. *Biochem Biophys Res Commun* 215: 377–385
33 Rowan AD, Koshy PJT, Shingleton WD, Degnan BA, Heath J, Vernallis AB, Spaull JR, Life PF, Hudson K, Cawston TE (2001) Synergistic effects of gp130 binding cytokines in combination with interleukin-1 on cartilage collagen breakdown. *Arthritis Rheum* 44: 1620–1632
34 Rowan AD, Hui W, Cawston TE, Richards CD (2003) Adenoviral gene transfer of interleukin-1 in combination with oncostatin M induces significant joint damage in a murine model. *Am J Pathol* 162: 1975–1984
35 Koshy PJT, Henderson N, Logan C, Life PF, Cawston TE, Rowan AD (2002) Inter-

leukin-17 induces cartilage collagen breakdown: novel synergistic effects in combination with proinflammatory cytokines. *Ann Rheum Dis* 61: 704–713
36 Hui W, Cawston TE, Rowan AD (2003) Transforming growth factor-β1 and insulin-like growth factor-1 block collagen degradation induced by oncostatin M in combination with tumour necrosis factor-α from bovine cartilage. *Ann Rheum Dis* 62: 172–174
37 Kleiner DE Jr, Stetler-Stevenson G (1993) Structural biochemistry and activation of matrix metalloproteinases. *Curr Biol* 5: 891–897
38 Milner JM, Elliott S-F, Cawston TE (2001) Activation of procollagenases is a key control point in cartilage collagen degradation. *Arthritis Rheum* 44: 2084–2096
39 Murphy G, Cockett MI, Stephens PE, Smith BJ, Docherty AJ (1987) Stromelysin is an activator of procollagenase. A study with natural and recombinant enzymes. *Biochem J* 248: 265–268
40 Knauper V, Will H, Lopez-Otin C, Smith B, Atkinson SJ, Stanton L, Hembry RM, Murphy G (1996) Cellular mechanisms for human procollagenase-3 (MMP-13) activation. Evidence that MT1-MMP (MMP-14) and gelatinase (MMP-2) are able to generate active enzyme. *J Biol Chem* 271: 17124–17131
41 Milner JM, Rowan AD, Elliott S-F, Cawston TE (2003) Inhibition of furin-like enzymes blocks interleukin-1/oncostatin M stimulated cartilage degradation. *Arthritis Rheum* 48: 1057–1066
42 Murphy G, Willenbrock F (1995) Tissue inhibitors of matrix metalloendopeptidases. *Methods Enzymol* 248: 496–510
43 Brew K, Dinarkarpandian D, Nagase H (2000) Tissue inhibitors of metalloproteinases: evolution, structure and function. *Biochem Biophys Acta* 1477: 267–283
44 Greene J, Wang M, Liu YE, Raymond LA, Rosen C, Shi YE (1996) Molecular cloning and characterization of human tissue inhibitor of metalloproteinase 4. *J Biol Chem* 271: 30375–30380
45 Varga J, Rosenbloom J, Jimenez SA (1987) Transforming growth factor beta (TGFβ) causes a persistent increase in steady-state amounts of type I and type III collagen and fibronectin mRNAs in normal human dermal fibroblasts. *Biochem J* 247:597–604
46 Zucker S, Pei D, Cao J, Lopez-Otin C (2003) Membrane type-matrix metalloproteinases (MT-MMP). *Curr Top Dev Biol* 54: 1–74
47 Yu WH, Woessner JF Jr (2000) Heparan sulphate proteoglycans as extracellular docking molecules for matrilysin (matrix metalloproteinase 7). *J Biol Chem* 275: 4183–4191
48 Guo H, Li R, Zucker S, Toole BP (2000) EMMPRIN (CD 147), an inducer of matrix metalloproteinase synthesis, also binds interstitial collagenases to the tumor cell surface. *Cancer Res* 60: 888–891
49 Clark IM, Cawston TE (1989) Fragments of human fibroblast collagenases. Purification and characterization. *Biochem J* 263: 201–206
50 Imai K, Hiramatsu A, Fukushima D, Pierschbacher MD, Okada Y (1997) Degradation of decorin by matrix metalloproteinases: identification of the cleavage sites, kinetic analyses and transforming growth factor-β1 release. *Biochem J* 322: 809–814
51 Fowlkes JL, Winkler MK (2002) Exploring the interface between metalloproteinase

activity and growth factor and cytokine bioavailability. *Cytokine Growth Factor Reviews* 13: 277–287
52 Zhou Z, Apte SS, Soininen R, Cao R, Baaklini GY, Rauser RW, Wang J, Cao Y, Tryggvason K. (2000) Impaired endochondral ossification and angiogenesis in mice deficient in membrane-type matrix metalloproteinase I. *Proc Natl Acad Sci USA* 97: 4052–4057
53 Zhao J, Chen H, Peschon JJ, Shi W, Zhang Y, Frank SJ, Warburton D (2001) Pulmonary hypoplasia in mice lacking tumor necrosis necrosis factor-α converting enzyme indicates an indispensable role for cell surface protein shedding during embryonic lung branching morphogenesis. *Dev Biol* 232: 204–218
54 Primakoff P, Myles DG (2000) The ADAM gene family: surface proteins with an adhesion and protease activity packed into a single module. *Trends Genet* 16: 83–87
55 Black R, Rauch CT, Kozlosky CJ, Peschon JJ, Slack JL, Wolfson MF, Castner BJ, Stocking KL, Reddy P, Srinivasan S et al (1997) A metalloprotease disintegrin that releases tumour-necrosis factor-α from cells. *Nature*: 385: 729–733
56 Tang BL (2001) ADAMTS: a novel family of extracellular matrix proteinases. *Int J Biochem Cell Biol* 33: 33–44
57 Sandy JD, Flannery CR, Neame PJ, Lohmander LS (1992) The structure of aggrecan fragments in human synovial fluid: evidence for the involvement in osteoarthritis of a novel proteinase. *J Clin Invest* 89: 1512–1516
58 Tortorella MD, Makfait AM, Deccico C, Arner E (2001) The role of ADAM-TS4 (aggrecanase-1) and ADAM-TS5 (aggrecanase-2) in a model of cartilage degradation. *Osteoarthritis Cartilage* 9: 539–552
59 Caterson B, Flannery CR, Hughes CE, Little CB (2000) Mechanisms involved in cartilage proteoglycan catabolism. *Matrix Biol* 19: 333–344
60 Tortorella MD, Liu RQ, Burn T, Newton RC, Arner E (2002) Characterization of human aggrecanase 2 (ADAM-TS5): substrate specificity studies and comparison with aggrecanase 1 (ADAM-TS4). *Matrix Biol* 21: 499–511
61 Billington CJ, Clark IM, Cawston TE (1998) An aggrecan-degrading activity associated with chondrocyte membranes. *Biochem J* 336: 207–212
62 McKie N, Edwards T, Dallas DJ, Houghton A, Stringer B, Graham R, Russell G, Croucher PI (1997) Expression of members of a novel membrane linked metalloproteinase family (ADAM) in human articular chondrocytes. *Biochem Biophys Res Commun* 230: 335–339
63 Kashiwagi M, Tortorella M, Nagase H, Brew K (2001) TIMP-3 is a potent inhibitor of aggrecanase 1 (ADAM-TS4) and aggrecanase 2 (ADAM-TS5). *J Biol Chem* 276: 12501–12504
64 Arner EC, Hughes CE, Decico CP, Caterson B, Tortorella MD (1998) Cytokine-induced cartilage proteoglycan degradation is mediated by aggrecanse. *Osteoarthritis Cartilage* 63: 214–228
65 Stanton H, Ung L, Fosang AJ (2002) The 45 kDa collagen-binding fragment of fibronectin induces matrix metalloproteinase-13 synthesis by chondrocytes and aggrecan degradation by aggrecanases. *Biochem J* 364: 181–190

66 Dudler J, Renggli-Zulliger N, Busso N, Lotz M, So A (2000) Effect of interleukin-17 on proteoglycan degradation in murine knee joints. *Ann Rheum Dis* 59: 529–532
67 Yamanishi Y, Boyle DL, Clark M, Maki RA, Tortorella MD, Arner EC, Firestein GS (2002) Expression and regulation of aggrecanases in arthritis: the role of TGFβ. *J Immunol* 163: 22201–22208
68 Koshy PJ, Lundy CJ, Rowan AD, Porter S, Edwards DR, Hogan A, Clark IM, Cawston TE. (2002) The modulation of matrix metalloproteinases and ADAM gene expression in human chondrocytes by interleukin-1 and oncostatin M: a time-course study using real time quantitative reverse transcription-polymerase chain reaction. *Arthritis Rheum* 46: 961–967
69 Ellis AJ, Curry VA, Powell EK, Cawston TE (1994) The prevention of collagen breakdown in bovine nasal cartilage by TIMP, TIMP-2 and a low molecular weight synthetic inhibitor. *Biochem Biophys Res Commun* 201: 94–101
70 Sandy JD, Verscharen C (2001) Analysis of aggrecan in human knee cartilage and synovial fluid indicates that aggrecanases (ADAMTS) activity is responsible for the catabolic turnover and loss of whole aggrecan whereas other protease activity is required for C-terminal processing *in vivo*. *Biochem J* 358: 615–626
71 Van Meurs JB, van Lent PL, Holthuysen AE, Singer H, Bayne EK, van den Berg WB (1999) Kinetics of aggrecanases- and metalloproteinase-induced neopitopes in various stages of cartilage destruction in murine arthritis. *Arthritis Rheum* 42(6): 1128–1139
72 Lee ER, Lamplugh L, Davoli MA, Beauchemin A, Chan K, Mort JS, Leblond CP (2001) Enzymes active in the areas undergoing cartilage resorption during the development of the secondary ossification center of the tibiae of rats ages 0–21 days: I. Two groups of proteinase cleave the core protein of aggrecan. *Dev Dyn* 222: 52–70
73 Sztrolovics R, White RJ, Roughley PJ, Mort JS (2002) The mechanism of aggrecan release from cartilage differs with tissue origin and the agent used to stimulate catabolism. *Biochem J* 362: 465–472
74 Sztrolovics R, Recklies AD, Roughley PJ, Mort JS (2002) Hyaluronate degradation as an alternative mechanism for proteoglycan release from cartilage during interleukin-1β stimulated catabolism. *Biochem J* 362: 473–479
75 Yamanishi Y, Boyle DL, Clark M, Maki RA, Tortorella MD, Arner EC, Firestein GS (2002) Expression and regulation of aggrecanases in arthritis: the role of TGF-β. *J Immunol* 168: 1405–1412
76 Flannery CR, Little CB, Hughes CE, Curtis CL, Caterson B, Jones SA (2000) IL-6 and its soluble receptor augment aggrecanase-mediated proteoglycans catabolism in articular cartilage. *Matrix Biol* 19: 549–553
77 Bluteau G, Conrozier T, Mathieu P, Vignon E, Herbage D, Mallein-Gerin F (2001) Matrix metalloproteinase-1, -3, -13 and aggrecanase-1 and -2 are differentially expressed in experimental osteoarthritis. *Biochim Biophys Acta* 1526: 147–158
78 Vankemmelbeke MN, Holen I, Wilson AG, Ilic MZ, Hanley CJ, Kelner GS, Clark M, Liu C, Maki RA, Burnett D, Buttle DJ (2001) Expression and activity of ADAMTS-5 in synovium. *Eur J Biochem* 268: 1259–1268

79 Mizui Y, Yamazaki K, Kuboi Y, Sagane K, Tanaka I (2000) Characterization of 5'-flanking region of human aggrecanase-1 (ADAMTS4) gene. *Mol Biol Rep* 27: 167–173
80 Pratta MA, Scherle PA, Yang G, Liu RQ, Newton RC (2003) Induction of aggrecanase 1 (ADAM-TS4) by interleukin-1 occurs through activation of constitutively produced protein. *Arthritis Rheum* 48: 119–133
81 Gao G, Westling J, Thompson VP, Howell TD, Gottschall PE, Sandy JD (2002) Activation of the proteolytic activity of ADAMTS4 (aggrecanase-1) by C-terminal truncation. *J Biol Chem* 277: 11034–11041
82 Chevrier A, Mort JS, Crine P, Hoemann CD, Buschmann MD (2001) Soluble recombinant neprilysin induces aggrecanase-mediated cleavage of aggrecan in cartilage explant cultures. *Arch Biochem Biophys* 396: 178–186
83 Hotoda N, Koike H, Sasagawa N, Ishiura S (2002) A secreted form of human ADAM9 has an alpha-secretase activity for APP. *Biochem Biophys Res Commun* 293: 800–805
84 Shi ZD, Xu WZ, Loechel F, Wewer UM, Murphy LJ (2000) ADAM 12, a disintegrin metalloprotease, interacts with insulin-like growth factor-binding protein-3. *J Biol Chem* 275: 18574–18580
85 Gilpin BJ, Loechel F, Mattei MG, Engvall E, Albrechtsen R, Wewer UM (1998) A novel, secreted form of human ADAM 12 (meltrin alpha) provokes myogenesis *in vivo*. *J Biol Chem* 273: 157–166
86 Harada T, Nishie A, Torigoe K, Ikezaki K, Shono T, Maehara Y, Kuwano M, Wada M (2000) The specific expression of three novel splice variant forms of human metalloprotease-like disintegrin-like cysteine-rich protein 2 gene in brain tissues and gliomas. *Jap J Cancer Res* 91: 1001–1006
87 Bevitt DJ, Mohamed J, Catterall JB, Li Z, Arris CE, Hiscott P, Sheridan C, Langton KP, Barker MD, Clarke MP, McKie N (2003) Expression of ADAMTS metalloproteinases in the retinal pigment epithelium derived cell line ARPE-19: transcriptional regulation by TNFα. *Biochim Biophys Acta* 1626: 83–91
88 Zheng XL, Chung D, Takayama TK, Majerus EM, Sadler JE, Fujikawa K (2001) Structure of von Willebrand factor-cleaving protease (ADAMTS13), a metalloprotease involved in thrombotic thrombocytopenic purpura. *J Biol Chem* 276: 41059–41063
89 Colige A, Sieron AL, Li SW, Schwarze U, Petty E, Wertelecki W, Wilcox W, Krakow D, Cohn DH, Reardon W et al (1999) Human Ehlers-Danlos syndrome type VIIC and bovine dermatosparaxis are caused by mutations in the procollagen IN-proteinase gene. *Am J Human Gen* 65: 308–317
90 Colige A, Vandenberghe I, Thiry M, Lambert CA, Van Beeumen J, Li SW, Prockop DJ, Lapiere CM, Nusgens BV (2002) Cloning and characterization of ADAMTS-14, a novel ADAMTS displaying high homology with ADAMTS-2 and ADAMTS-3. *J Biol Chem* 277: 5756–5766
91 Nagase H, Woessner JF (1993) Role of endogenous proteinases in the degradation of cartilage matrix. In: JF Woessner, DS Howell (eds): *Joint cartilage degradation*. Marcel-Dekker, New York, 159–185
92 Campbell IK, Wojta J, Novak U, Hamilton JA (1994) Cytokine modulation of plas-

minogen activator inhibitor-1 (PAI-1) production by human articular cartilage and chondrocytes. Downregulation by tumor necrosis factor alpha and upregulation by transforming growth factor-β basic fibroblast growth factor. *Biochim Biophys Acta* 1226: 277–285
93 Bryson H, Bunning RA, Feltell R, Kam CM, Kerrigan J, Powers JC, Buttle DJ (1998) A serine proteinase inactivator inhibits chondrocyte-mediated cartilage proteoglycan breakdown occurring in response to proinflammatory cytokines. *Arch Biochem Biophys* 355: 15–25
94 Ronday HK, van der Laan WH, Tak PP, de Roos JA, Bank RA, TeKoppele JM, Froelich CJ, Hack CE, Hogendoorn PC, Breedveld FC, Verheijen JH (2001) Human granzyme B mediates cartilage proteoglycan degradation and is expressed at the invasive front of the synovium in rheumatoid arthritis. *Rheumatology* 40: 55–61
95 Moldovan F, Pelletier JP, Mineau F, Dupuis M, Cloutier JM, Martel-Pelletier J (2000) Modulation of collagenase 3 in human osteoarthritic cartilage by activation of extracellular transforming growth factor beta role of furin convertase. *Arthritis Rheum* 43: 2100–2109
96 Pedrozo HA, Schwartz Z, Robinson M, Gomes R, Dean DD, Bonewal LF, Boyan BD (1999) Potential mechanisms for the plasmin-mediated release and activation of latent transforming growth factor-beta1 from the extracellular matrix of growth plate chondrocytes. *Endocrinology* 140: 5806–5816
97 Yamaguchi K, Sakiyama H, Matsumoto M, Moriya H, Sakiyama S (1990) Degradation of type I and II collagen by human activated C1-s. *FEBS Lett* 268: 206–208
98 Clemmons DR, Busby WH Jr, Garmong A, Schultz DR, Howell DS, Altman RD, Karr R (2002) Inhibition of insulin-like growth factor binding protein 5 proteolysis in articular cartilage and joint fluid results in enhanced concentrations of insulin-like growth factor 1 and is associated with improved osteoarthritis. *Arthritis Rheum* 46: 694–703
99 Fowlkes JL, Thrailkill KM, Serra DM, Suzuki K, Nagase H (1995) Matrix metalloproteinases as insulin-like growth factor binding protein-degrading proteinases. *Prog Growth Factor Res* 6: 255–263
100 Feldmann M, Brennan FM, Maini RN (1996) Role of cytokines in rheumatoid arthritis. *Annu Rev Immunol* 14: 397–440
101 Camerer E, Kolsto AB, Prydz H (1996) Cell biology of tissue factor, the principal initiator of blood coagulation. *Thromb Res* 81: 1–41
102 Salvi R, Peclat V, So A, Busso N (2000) Enhanced expression of genes involved in coagulation and fibrinolysis in murine arthritis. *Arthritis Res* 2: 504–512
103 Nakano S, Ikata T, Kinoshita I, Kanematsu J, Yasuoka S (1999) Characteristics of the protease activity in synovial fluid from patients with rheumatoid arthritis and osteoarthritis. *Clin Exp Rheumatol* 17: 161–170
104 Varisco PA, Peclat V, van Ness K, Bischof-Delaloye A, So A, Busso N (2000) Effect of thrombin inhibition on synovial inflammation in antigen induced arthritis. *Ann Rheum Dis* 59: 781–787

105 Marty I, Peclat V, Kirdaite G, Salvi R, So A, Busso N (2001) Amelioration of collagen-induced arthritis by thrombin inhibition. *J Clin Invest* 107:631–640
106 Hirano F, Kobayashi A, Hirano Y, Nomura Y, Fukawa E, Makino I (2002) Thrombin-induced expression of RANTES mRNA through protease activated receptor-1 in human synovial fibroblasts. *Ann Rheum Dis* 61: 834–837
107 Ferrell WR, Lockart JC, Kelso EB, Dunning L, Plevin R, Meek SE, Smith AJ, Hunter GD, McLean JS, McGarry F et al (2003) Essential role for proteinase-activated receptor-2 in arthritis. *J Clin Invest* 111: 35–41
108 Fernandes JC, Martel-Pelletier J, Lascau-Coman V, Moldovan F, Jovanovic D, Raynauld JP, Pelletier JP (1998) Collagenase-1 and collagenases-3 synthesis in normal and early experimental osteoarthritic canine cartilage: an immunohistochemical study. *J Rheumatol* 25: 1585–1594
109 Sandy JD, Sriratana A, Brown HLG, Lowther DA (1981) Evidence for polymorphonuclear leukocyte derived proteinases in arthritic cartilage. *Biochem J* 193: 193–202
110 Schalkwijk J, Joosten LAB, Van den Berg WB, Van de Putte LBA (1990) Antigen induced arthritis in beige mice. *Ann Rheum Dis* 49: 607–610
111 Turk V, Turk B, Turk D (2001) Lysosomal cysteine proteinases: facts and opportunities. *EMBO J* 20: 4629–4633
112 Etherington DJ (1972) The nature of the collagenolytic cathepsin of rat liver and its distribution in other rat tissues. *Biochem J* 127: 685–692
113 Burleigh MC, Barrett AJ, Lazarus GS (1974) Cathepsin B1. A lysosomal enzyme that degrades native collagen. *Biochem J* 137: 387–398
114 Delaisse J-M, Eeckhout Y, Vaes G (1980) Inhibition of bone resorption in culture by inhibitors of thiol proteinases. *Biochem J* 192: 365–368
115 Delaisse J-M, Eeckhout Y, Vaes G (1984) *In vivo* and *in vitro* evidence for the involvement of cysteine proteinases in bone resorption. *Biochem Biophys Res Commun* 125: 441–447
116 Nguyen Q, Mort JS, Roughley PJ (1990) Cartilage proteoglycan aggregate is degraded more extensively by cathepsin L than by cathepsin B. *Biochem J* 266: 569–573
117 Buttle DJ, Bramwell H, Hollander AP (1995) Proteolytic mechanisms of cartilage breakdown: a target for arthritis therapy? *Clin Pathol: Mol Pathol* 48: 167M–177M
118 Everts V, Van Der Zee E, Creemers L, Beertsen W (1996) Phagocytosis and intracellular digestion of collagen, its role in turnover and remodelling. *Histochem J* 28: 229–245
119 Everts V, Beersten W, Tigchelaar-Gutter W (1985) The digestion of phagocytosed collagen is inhibited by the proteinases inhibitors leupeptin and E-64. *Collagen Rel Res* 5: 315–336
120 Creemers LB, Hoeben KA, Jansen DC, Buttle DJ, Beertsen W, Everts V (1998) Participation of intracellular cysteine proteinases, in particular cathepsin B, in degradation of collagen in periosteal tissue explants. *Matrix Biol* 6: 575–584
121 Hill PA, Buttle DJ, Jones SJ, Boyde A, Murata M, Reynolds JJ, Meikle M (1994) Inhibition of bone resorption by selective inactivators of cysteine proteinases. *J Cell Biochem* 56: 118–130

122 Ikeda Y, Ikata T, Mishiro T, Nakano S, Ikebe M, Yasuoka S (2000) Cathepsins B and L in synovial fluids from patients with rheumatoid arthritis and the effect of cathepsin B on the activation of pro-urokinase. *J Med Invest* 47: 61–75

123 Kobayashi H, Moniwa N, Sugimura M, Shinohara H, Ohi H, Terao T (1993) Effects of membrane-associated cathepsin B on the activation of receptor bound prourokinase and subsequent invasion of reconstituted basement membranes. *Biochim Biophys Acta* 1178: 55–62

124 Inaoka T, Bilbe G, Ishibashi O, Tezuka K, Kumegawa M, Kokubo T (1995) Molecular cloning of human cDNA for cathepsin K: novel cysteine proteinase predominantly expressed in bone. *Biochem Biophys Res Commun* 206: 89–96

125 Bromme D, Okamoto K (1995) Human cathepsin O2, a novel cysteine protease highly expressed in osteoclastomas and ovary molecular cloning, sequencing and tissue distribution. *Biol Chem Hoppe Seyler* 376: 379–384

126 Kirschke H, Barrett AJ, Rawlings ND (1995) Proteinases 1: lysosomal cysteine proteinases. *Protein Profile* 2: 1581–1620

127 Bossard, MJ, Tomaszek TA, Thompson SK, Amegadzie BY, Hanning CR, Jones C, Kurdyla JT, McNulty DE, Drake FH, Gowen M (1996) Proteolytic activity of human osteoclast cathepsin K – expression, purification, activation and substrate identification. *J Biol Chem* 271: 12517–12524

128 Drake FH, Dodds RA, James IE, Connor JR, Debouck C, Richardson S (1996) Cathepsin K, but not cathepsins B, L, or S, is abundantly expressed in human osteoclasts. *J Biol Chem* 271: 12511–12516

129 Inui T, Ishibashi O, Inaoka T, Origane Y, Kumegawa M, Kokubo T, Yamamura T (1997) Cathepsin K antisense oligodeoxynucleotide inhibits osteoclastic bone resorption. *J Biol Chem* 272: 8109–8112

130 Li Z, Hou WS, Bromme D (2000) Collagenolytic activity of cathepsin K is specifically modulated by cartilage resident chondroitin sulfates. *Biochemistry* 39: 529–536

131 Hou WS, Li W, Keyszer G, Weber E, Levy R, Klein MJ, Gravallese EM, Goldring SR, Bromme D (2002) Comparison of cathepsins K and S expression within the rheumatoid and osteoarthritic synovium. *Arthritis Rheum* 46: 663–674

132 Hou WS, Li Z, Gordon RE, Chan K, Klein MJ, Levy R, Keysser M, Keyszer G, Bromme D (2001) Cathepsin K is a critical protease in synovial fibroblast-mediated collagen degradation. *Am J Pathol* 159(6): 2167–2177

133 Villadangos JA, Bryant RA, Deussing J, Driessen C, Lennon-Dumenil AM, Riese RJ, Roth W, Saftig P, Shi GP, Chapman HA et al (1999) Proteases involved in MHC class II antigen presentation. *Immunol Rev* 172: 109–120

134 Lemaire R, Huet G, Zerimech F, Grard G, Fontaine C, Duquesnoy B, Flipo RM (1997) Selective induction of the secretion of cathepsins B and L by cytokines in synovial fibroblast-like cells. *Br J Rheumatol* 36: 735–743

135 De Ceuninck F, Poiraudeau S, Pagano M, Tsagris L, Blanchard O, Willeput J and Corvol M (1995) Inhibition of chondrocyte cathepsin B and L activities by insulin-like

growth factor-II (IGF-II) and its Ser29 variant *in vitro*: possible role of the mannose 6-phosphate/IGF-II receptor. *Mol Cell Endocrinol* 113: 205–213
136 Kaneko M, Tomita T, Nakase T, Ohsawa Y, Seki H, Takeuchi E, Takano H, Shi K, Takahi K, Kominami E (2001) Expression of proteinases and inflammatory cytokines in subchondral bone regions in the destructive joint of rheumatoid arthritis. *Rheumatology* 40: 247–255
137 Kamolmatyakul S, Chen W, Li, YP (2001) Interferon-gamma downregulates gene expression of cathepsin K in osteoclasts and inhibits osteoclast formation. *J Dent Res* 80: 351–355
138 Yamada N, Niwa S, Tsujimura T, Iwasaki T, Sugihara A, Futani H, Hayashi S, Okamura H, Akedo H, Terada N (2002) Interleukin-18 and interleukin-12 synergistically inhibit osteoclastic bone-resorbing activity. *Bone* 30: 901–908
139 Uusitalo H, Hiltunen A, Soderstrom M, Aro HT, Vuorio E (2000) Expression of cathepsins B, H, K, L, and S and matrix metalloproteinases 9 and 13 during chondrocyte hypertrophy and endochondral ossification in mouse fracture callus. *Calcif Tissue Int* 67: 382–390
140 Kostoulas G, Lang A, Nagase H, Baici A (1999) Stimulation of angiogenesis through cathepsin B inactivation of the tissue inhibitors of matrix metalloproteinases. *FEBS Lett* 455: 286–290
141 Parkes C (1986) In: AJ Barrett, G Salvesen (eds): *Proteinase inhibitors*. Elsevier, Amsterdam, 571–587
142 Suzuki K, Shimizu K, Hamamoto T, Nakagawa Y, Murachi T, Yamamuro T (1992) Characterization of proteoglycan degradation by calpain. *Biochem J* 285: 857–862
143 Yasuda T, Shimizu K, Nakagawa Y, Yamamoto S, Niibayashi H, Yamamuro T (1995) m-calpain in rat growth plate chondrocyte cultures: its involvement in the matrix mineralization process. *Dev Biol* 170: 159–168
144 Szomor Z, Shimizu K, Fujimori Y, Yamamoto S, Yamamuro T (1995) Appearance of calpain correlates with arthritis and cartilage destruction in collagen induced arthritic knee joints of mice. *Ann Rheum Dis* 54: 477–483
145 Ishikawa H, Nakagawa Y, Shimizu K, Nishihara H, Matsusue Y, Nakamura T (1999) Inflammatory cytokines induced downregulation of m-calpa mRNA expression in fibroblastic synoviocytes from patients with osteoarthritis and rheumatoid arthritis. *Biochem Biophys Res Commun* 266: 341–346
146 Buttle, DJ, Saklatvala J (1992) Lysosomal cysteine endopeptidases mediate interleukin 1-stimulated cartilage proteoglycans degradation. *Biochem J* 287: 657–661
147 Adkinson AM, Raptis SZ, Kelley DG, Pham CT (2002) Dipeptidyl peptidase I activates neutrophil-derived serine proteases and regulates the development of acute experiment arthritis. *J Clin Invest* 109: 363–371
148 Wlodawer A (1995) Proteasome: a complex protease with a new fold and a distinct mechanism. *Structure* 3: 417–420
149 Sakai T, Kambe F, Mitsuyama H, Ishiguro N, Kurokouchi K, Takigawa M, Iwata H, Seo H (2001) Tumor necrosis factor alpha induces expression of genes for matrix degrada-

tion in human chondrocyte-like HCS-2/8 cells through activation of NF-kappaB: abrogation of the tumor necrosis factor alpha effect by proteasome inhibitors. *J Bone Miner* Res 16: 1272–1280
150 Gravallese EM, Goldring SR (2000) Cellular mechanisms and the role of cytokines in bone erosions in rheumatoid arthritis. *Arthritis Rheum* 43: 2143–2151
151 Everts V, Delaisse JM, Korper W, Niehof A, Vaes G, Beertsen W (1992) Degradation of collagen in the bone-resorbing compartment underlying the osteoclast involves both cysteine and matrix metalloproteinases. *J Cell Physiol* 150: 221–231
152 Kafienah N, Bromme D, Buttle DJ, Croucher LJ, Hollander AP (1998) Human cathepsin K cleaves native type I and II collagens at the N-terminal end of the triple helix. *Biochem J* 331: 727–732
153 Hummel KM, Petrow PK, Franz JK, Müller-Ladner U, Aicher WK, Gay RE, Bromme D, Gay S (1998) Cysteine cathepsin K mRNA is expressed in synovium of patients with rheumatoid arthritis and is detected at sites of synovial bone destruction. *J Rheumatol* 25: 1887–1894
154 Hollberg K, Hultenby K, Hayman A, Cox T, Andersson G (2002) Osteoclasts from mice deficient in tartrate-resistant acid phosphatase have altered ruffled borders and disturbed intracellular vesicular transport. *Exp Cell Res* 279: 227–238
155 Redlich K, Hayer S, Maier A, Dunstan CR, Tohidast-Akrad M, Lang S, Turk B, Pietschmann P, Woloszczuk W, Haralambous S et al (2002) Tumor necrosis factor α-mediated joint destruction is inhibited by targeting osteoclasts with osteoprotegerin. Arthritis Rhem 46: 785–792
156 Corisdeo S, Gyda M, Zaidi M, Moonga BS, Troen BR (2001) New insights into the regulation of cathepsin K gene expression by osteoprotegerin ligand. *Biochem Biophys Res Commun* 285: 335–339
157 Kong YY, Yoshida H, Sarosi I, Tan HL, Timms E, Capparelli C, Morony S, Oliveira-dos-Santos AJ, Van G, Itie A, Khoo W (1999) OPGL is a key regulator of osteoclastogenesis, lymphocyte development and lymph-node organogenesis. *Nature* 397: 315–323
158 Kotake S, Udagawa N, Takahashi N, Matsuzaki K, Itoh K, Ishiyama S, Saito S, Inoue K, Kamatani N, Gillespie MT et al (1999) IL-17 in synovial fluids from patients with rheumatoid arthritis is a potent stimulator of osteoclastogenesis. *J Clin Invest* 103: 1345–1352
159 Bush KA, Walker JS, Lee CS, Kirkham BW (2001) Cytokine expression and synovial pathology in the initiation and spontaneous resolution phases of adjuvant arthritis: interleukin-17 expression is upregulated in early disease. *Clin Exp Immunol* 123: 487–495
160 Chabaud M, Lubberts E, Joosten L, van den Berg W, Miossec P (2001) IL-17 derived from juxta-articular bone and synovium contributes to joint degradation in rheumatoid arthritis. *Arthritis Res* 3: 168–177
161 Van den Berg WB (2001) Uncoupling of inflammatory and destructive mechanisms in arthritis. *Semin Arthritis Rheum* 30 (5 Suppl 2): 7–16
162 Keffer J, Probert L, Cazlaris H, Georgopoulos S, Kaslaris E, Kioussis D, Kollias G

(1991) Transgenic mice expressing human tumor necrosis factor: a predictive genetic model of arthritis. *EMBO J* 13: 4025–4031
163 Ghivazzani SC, Kang R, Georgescu HI, Lechman ER, Jaffurs D, Engle JM, Watkins SC, Tindal MH, Suchanek MK, McKenzie LR et al (1997) Constitutive intra articular expression of human IL-1β following gene transfer to rabbit synovium produces all major pathologies of human RA. *J Immunol* 159: 3604–3612
164 Probert L, Plows D, Kontogeorgos G, Kollias G (1995) The type I IL-1 receptor acts in series with TNF1 to induce arthritis in TNF1 transgenic mice. *Eur J Immunol* 25: 1794–1797
165 van den Berg WB (2001) Anti-cytokine therapy in chronic destructive arthritis. *Arthritis Res* 3: 18–26
166 Van den Berg, WB (2002) Lessons from animal models of arthritis. *Curr Rheumatol Rep* 4: 232–239
167 Mori L, Iselin S, De Libero G, Lesslauer W (1996) Attenuation of collagen induced arthritis in 55 kDa TNF receptor type I (TNFR)-IgG1 treated and TNFR1 deficient mice. *J Immunol* 157: 3178–3182
168 Joosten LAB, Helsen MMA, Saxne T, van de Loo FAJ, Heinegard D, van den Berg WB (1999) IL-1αβ blockade prevents cartilage and bone destruction in murine type II collagen-induced arthritis, whereas TNF-α blockage only ameliorates joint inflammation. *J Immunol* 163: 5049–5055
169 Van den Berg WB, Joosten LAB, Kollias G, van de Loo FAJ (1999) Role of TNFα in experimental arthritis: separate activity of IL-1β in chronicity and cartilage destruction. *Ann Rheum Dis* 58 (Suppl I): S140–S148
170 Horai R, Saijo S, Tanioka H, Nakae S, Sudo K, Okahara A, Ikuse T, Asano M, Iwakura Y (2000) Development of chronic inflammatory arthropathy resembling rheumatoid arthritis in IL-1 receptor antagonist-deficient mice. *J Exp Med* 191: 313–320
171 Müller-Ladner U, Evans CH, Franklin BN, Roberts CR, Gay RE, Robbins PD, Evans CH, Gay S (1997) Gene transfer of cytokine inhibitors into human synovial fibroblasts in the SCID mouse model. *Arthritis Rheum* 42: 490–497
172 Müller-Ladner U, Roberts CR, Franklin BN, Gay RE, Robbins PD, Evans CH, Gay S (1997) Human IL-1ra gene transfer into human synovial fibroblasts is chondroprotective. *J Immunol* 158: 3492–3498
173 Joosten LAB, Heuvelmans-Jacobs M, Lubberts E, van de Loo FAJ, Bakker AC, Helsen MMA, Richards CD, van den Berg WB (2002) Local Interleukin-12 gene transfer promotes conversion of an acute arthritis to a chronic destructive arthritis. *Arthritis Rheum* 46: 1379–1389
174 Joosten LAB, Radstake TRD, Lubberts E, van den Bersselaar LAM, van Riel PLCM, van Lent PLEM, Barrera P, van den Berg WB (2003) Association of interleukin-18 expression with enhanced levels of both Interleukin-1β and tumor necrosis factor α in knee synovial tissue of patients with rheumatoid arthritis. *Arthritis Rheum* 48: 339–347
175 Plater-Zyberk C, Jooster LA, Helsen MM, Sattonnet-Roche P, Siegfried C, Alouani S, van de Loo FA, Graber P, Aloni S, Cirillo R et al (2001) Therapeutic effect of neutraliz-

ing endogenous IL-18 activity the collagen-induced model of arthritis. *J Clin Invest* 108: 1825–1832

176 Lubberts E, Joosten LA, Oppers B, van den Bersselaar L, Coenen-de Roo CJ, Kolls JK, Schwarzenberger P, van de Loo FA, Van den Berg WB (2001) IL-1-independent role of IL-17 in synovial inflammation and joint destruction during collagen-induced arthritis. *J Immunol* 167: 1004–1013

177 Lubberts E, Joosten LA, Chabaud M, van Den Bersselaar L, Oppers B, Coenen-De Roo CJ, Richards CD, Miossec P, van Den Berg WB (2000) IL-4 gene therapy for collagen arthritis suppresses synovial IL-17 and osteoprotegerin ligand and prevents bone erosion. *J Clin Invest* 105:1697–710

178 Scharstuhl A, Glansbeek HL, van Beuningen HM, Vitters EL, van der Kraan PM, Van den Berg WB (2002) Inhibition of endogenous TGFβ during experimental osteoarthritis prevents osteophyte formation and impairs cartilage repair. *J Immunol* 169: 507–514

179 van den Berg WB, van der Kraan PM, Scharstuhl A, van Beuningen HM (2001) Growth factors and cartilage repair. Clin Orthop 391: S244–250

180 Hui W, Cawston TE, Rowan AD (2003) Transforming growth factor β1 and insulin-like growth factor 1 block collagen degradation induced by oncostatin M in combination with tumour necrosis factor α from bovine cartilage. *Ann Rheum Dis* 62(2): 172–174

181 Verschure PJ, Marle JV, Joosten LA, Helsen MM, Lafeber FP, Berg WB (1996) Localization of insulin-like growth factor-1 receptor in human normal and osteoarthritic cartilage in relation to proteoglycan synthesis and content. *Br J Rheumatol* 35: 1044–1055

182 Joosten LA, Lubberts E, Helsen MM, Saxne T, Coenen-de Roo CJ, Heinegard D, van den Berg WB (1999) Protection against cartilage and bone destruction by systemic interleukin-4 treatment in established murine type II collagen induced arthritis. *Arthritis Res* 1: 81–91

183 Lubberts E, Joosten LA, Helsen MM, van den Berg WB (1998) Regulatory role of interleukin 10 in joint inflammation and cartilage destruction in murine streptococcal cell wall (SCW) arthritis. More therapeutic benefit with IL-4/IL-10 combination therapy than with IL-10 treatment alone. *Cytokine* 10: 361–369

184 Cleaver C, Rowan AD, Cawston TE (2001) Interleukin-13 blocks the release of collagen fragments from bovine nasal cartilage treated with proinflammatory cytokines by preventing pro-collagenase activation. *Annals Rheum Dis* 60: 150–157

185 Cawston TE, Curry VA, Summers CA, Clark IM, Riley GP, Life PF, Spaull JR, Goldring MB, Koshy PJT, Rowan AD, Shingleton WD (1998) The role of oncostatin M in animal and human connective tissue collagen turnover and its localization within the rheumatoid joint. *Arthritis Rheum* 41: 1760–1771

186 Müller-Ladner U, Kriegsmann J, Franklin BN, Matsumoto S, Geiler T, Gay RE, Gay S (1996) Synovial fibroblasts of patients with rheumatoid arthritis attach to and invade normal human cartilage when engrafted into SCID mice. *Am J Pathol* 149: 1607–1615

187 Pap T, Claus A, Ohtsu S, Hummel KM, Schwartz P, Drynda S, Pap G, Machner A, Stein B, George M et al (2003) Osteoclast-independent bone resorption by fibroblast-like cells. *Arthritis Res Ther* 5: R163–173

188 Blumenfeld I, Livne E (1999) The role of transforming growth factor beta, insulin like-growth factor-1 in osteoarthritis and aging of joints. *Exp Gerontol* 34: 821–829
189 Hunzinger EB (1999) Articular cartilage repair: are the intrinsic biological constraints undermining this process insuperable? *Osteoarthritis Cartilage* 7: 15–28

Receptor activator of nuclear factor κB ligand (RANKL) *versus* osteoprotegerin (OPG): the yin and yang of joint integrity

Brad Bolon[1], Victoria Shalhoub[2], Paul J. Kostenuik[2], Debra Zack[3] and Ulrich Feige[3]

[1]Department of Pathology, Amgen Inc., [2]Department of Metabolic Disorders, Amgen Inc., [3]Department of Inflammation Research, Amgen Inc., One Amgen Center Drive, M/S 29-M-B, Thousand Oaks, CA 91320-1799, USA

Introduction

"Life manifests in duality" is a philosophical tenet in many cultures. This idea is embodied clearly by the Asian concept of yin and yang, in which the complementary forces of dark (yin) and light (yang) govern all aspects of life. Yin and yang convey the interdependence of opposites. Such reciprocity is a common feature of many biological processes – including skeletal maintenance.

In the skeletal system, duality is evident at the cellular level in the interplay between osteoblasts and osteoclasts as they regulate growth, maintenance and repair during health and disease. Osteoblasts are derived from mesenchymal stem cells in bone marrow, produce bone matrix, and express receptors for most bone-regulating molecules (regardless of whether the main action is to drive bone formation or bone resorption [1]). Osteoclasts arise from phagocytic precursors of the monocyte/macrophage lineage and remove bone matrix. In normal bone remodeling, osteoblasts support and regulate osteoclast activity [2, 3], and *vice versa* [4]. However, in pathologic states, "activated" cells (e.g., leukocytes, synoviocytes) produce cytokines [5], chemokines [6] and other molecules [7] that can enhance osteoclast formation and function with minimal or no need for osteoblast participation.

Duality is also evident in the molecular regulation of bone resorption. This process is controlled by the equilibrium between two molecules discovered during the past decade: receptor activator of NF-κB (RANK) ligand (RANKL), which acts *via* the membrane-bound receptor RANK to stimulate osteoclast production, activation and survival; and osteoprotegerin (OPG), which inhibits bone resorption due to its function as a soluble, non-signaling decoy receptor for RANKL. In the inflamed joint, therefore, RANKL promotes erosions (an undesirable consequence [yin] [8]), while OPG protects skeletal architecture (the desired outcome [yang]). The balance between RANKL and OPG determines the degree to which RANK-mediated signaling is activated, and thus the extent to which bone is built or destroyed.

The present chapter outlines the cellular, molecular and physiological attributes of the OPG/RANKL/RANK signaling pathway with regard to the immune and

Cytokines and Joint Injury, edited by Wim B. van den Berg and Pierre Miossec

skeletal systems, emphasizing alterations that develop during immune-mediated arthritis.

Molecular biology of the OPG/RANKL/RANK system

The known relationships between the various extracellular components of the OPG/RANKL/RANK signaling pathway are depicted schematically in Figure 1. These activities were defined *in vitro* using osteoclast precursors (human and rodent) and confirmed *in vivo* utilizing mice in which components of the pathway were added or subtracted by either genetic engineering or administration of recombinant proteins. The core pathway consists of the common ligand (RANKL), the activating receptor (RANK) and the soluble decoy receptor (OPG). Expression of both RANKL and OPG are modulated by many other extracellular ligands. Finally, the impact of RANKL–RANK binding is transmitted by interweaving intracellular signal transduction cascades (Fig. 2) comprised of a growing host of enzymes and transcription factors.

RANKL

RANKL is one of two necessary and sufficient signals that control osteoclast differentiation, activation and survival [9–11]. [The other is macrophage-colony stimulating factor (M-CSF), which acts – at least in part – by inducing expression of

Figure 1
Schematic representation of the osteoprotegerin (OPG)/RANK ligand (RANKL)/receptor activator of NF-κB (RANK) signaling pathway in normal bone maintenance and immune-mediated arthritis
Under normal conditions (Panel A), RANKL generated by osteoblasts under the influence of many pro-resorptive signals (listed inside large white arrow) binds to its specific membrane-bound receptor RANK, thereby triggering a network of intracellular kinase cascades (Fig. 2) that promote osteoclast differentiation, activation and survival. OPG downregulates resorption of bone by acting as a soluble sink for RANKL, thereby preventing osteoclast formation and eliciting osteoclast apoptosis. In the arthritic joint (Panel B), expression of RANKL in large quantities by activated leukocytes (T lymphocytes and macrophages) and synovial fibroblasts promotes extensive osteoclast differentiation, activation and survival. OPG, while present, is not sufficient to block RANKL activity.
Abbreviations: CFU-M, colony forming unit–macrophage; IL, interleukin; LIF, leukemia inhibitory factor; M–CSF, macrophage-colony stimulating factor; PGE_2, prostaglandin E_2; PTH, parathyroid hormone; PTHrP, PTH-related protein; TNF, tumor necrosis factor.

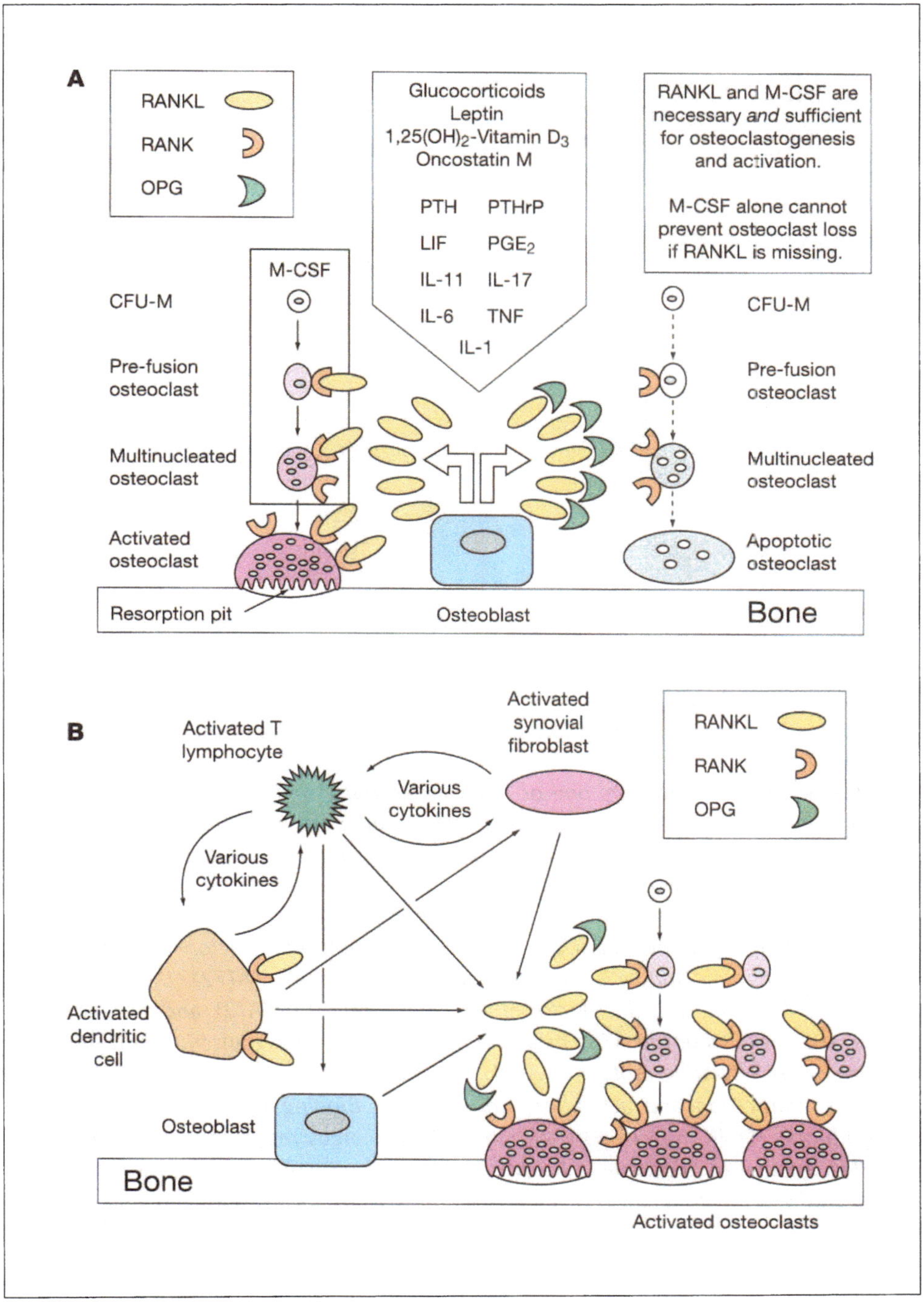
A
RANKL
RANK
OPG
Glucocorticoids
Leptin
1,25(OH)2-Vitamin D3
Oncostatin M
PTH
PTHrP
LIF
PGE2
IL-11
IL-17
IL-6
TNF
IL-1
RANKL and M-CSF are necessary *and* sufficient for osteoclastogenesis and activation.
M-CSF alone cannot prevent osteoclast loss if RANKL is missing.
M-CSF
CFU-M
Pre-fusion osteoclast
Multinucleated osteoclast
Activated osteoclast
CFU-M
Pre-fusion osteoclast
Multinucleated osteoclast
Apoptotic osteoclast
Resorption pit
Osteoblast
Bone
B
Activated T lymphocyte
Activated synovial fibroblast
RANKL
RANK
OPG
Various cytokines
Various cytokines
Activated dendritic cell
Osteoblast
Bone
Activated osteoclasts

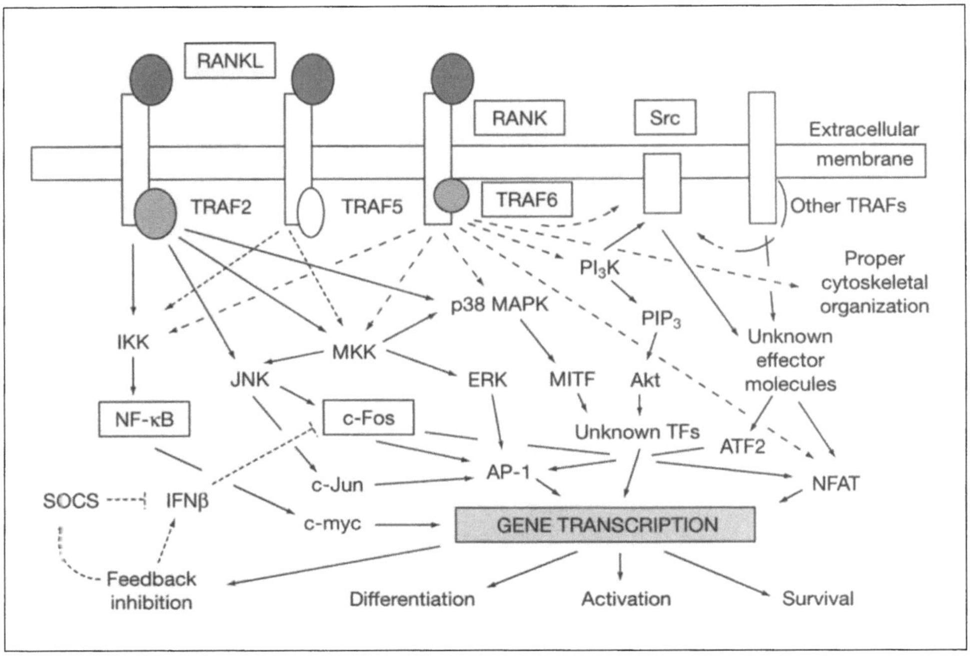

Figure 2

Osteoclast differentiation, activation and survival result from stimulation of cellular RANK by RANKL, which enables various TNF receptor-associated factors (TRAFs; ovals) to associate with RANK

Each RANK/TRAF pair triggers one or more intracellular signal transduction pathways, including the serine/threonine kinase Akt, extracellular signal-regulated protein kinase (ERK), IκB kinase (IKK), c-Jun amino terminal kinase (JNK), p38 mitogen-activated protein kinase (MAPK) systems, and the Src protein tyrosine kinase. In some cases, these kinases are enabled by the action of other enzymes, such as MAPK kinases (MKK). These cascades subsequently interact with one or more transcription factors (TFs) to initiate gene transcription, including activating transcription factor-2 (ATF2), the proto-oncogenes c-Jun and c-myc, nuclear factor of activated T cells 2 (NFAT2), and nuclear factor-κB (NF-κB). Proteins of the ATF, Fos and Jun families form the transcription factor complex known as activating protein-1 (AP-1). Gene transcription leads to osteoclast differentiation, maturation and activation, and survival. Binding of RANKL by OPG (not shown) blocks RANK-mediated signal transduction, thereby resulting in osteoclast apoptosis (Fig. 1 Panel A). Blockade of RANK signaling by null mutations of certain molecules in this pathway (names encased in boxes) inhibits osteoclast function and leads to osteopetrosis (as detailed in Tab. 2). Additional abbreviations: IFNβ, interferon-β; MITF, microphthalmia-associated transcription factor; PI_3K, phosphatidylinositol-3 kinase; PIP_3, phosphatidylinositol triphosphate; SOCS, suppressors of cytokine signaling.

RANKL and other elements of the RANK signal transduction pathway [12].] Alternative acronyms for RANKL in the literature include ODF (osteoclast differentiation factor [11]), OPGL (osteoprotegerin ligand [10]), TRANCE (TNF-related activation-induced cytokine [13]) and TNFSF11 (TNF ligand superfamily member 11, which is the accepted permanent nomenclature for this molecule [14]). The human *RANKL* gene is located on chromosome 13q14 [15]. The *RANKL* gene shares a common origin with another TNF superfamily member, CD40 ligand (CD40L), as indicated by their common structures (single-copy gene comprised of five exons [16]) and shared ability to promote function and survival of dendritic cells [13, 17]. The *RANKL* promoter contains multiple regulatory elements [16] important to transcription of genes that control bone mass, including binding sites for activating protein-1 (AP-1, a RANK-activated transcription factor complex), core-binding factor α1 (CBFA1, an osteoblast-specific transcription factor involved in osteoblast differentiation), and vitamin D_3 [18]. Expression of *RANKL* is also subject to epigenetic control based on the degree of methylation near the transcription start site [19]. The extent of methylation varies among different cell and tissue types.

Once translated, RANKL is produced by self-assembly of monomers into a non-covalently linked, stable, membrane-bound homotrimer (human protein: 317 aa, MW 45 kd after post-translational modification of the 35 kd protein) [15, 20, 21] that can be cleaved later to release a soluble homotrimer (26 kd). Both membrane-bound and soluble RANKL are potent agonists of RANK [15, 22]. Using mouse RANKL (316 aa), production of the soluble form is mediated by TNFα converting enzyme (TACE). However, the low specificity constant for RANKL (1,000-fold lower than that of the prototypic TACE substrate, TNFα) implies that an as yet undiscovered protease is the likely main catalyst for this conversion [23]. TACE-mediated cleavage of human RANKL has not been demonstrated, although extensive homology between full-length human and mouse RANKL (87% at the amino acid level [10]) suggests that TACE should have this capacity for the human protein as well. The degree of homology between RANKL and other TNF family members, such as TNFα and TRAIL (TNF-related apoptosis-inducing ligand), is approximately 20% [15].

RANKL is highly expressed in bone, intestine and peripheral lymphoid organs, and is found to a lesser degree in many other tissues including heart, lung, skeletal muscle and thyroid [10, 11, 15, 17]. In adults, the expression of RANKL in normal bone is highest near remodeling zones such as the hypertrophic chondrocytes of growth plates, the primary spongiosa, the metaphysis and the periosteal side of the diaphyseal collar [10, 24]. RANKL is produced chiefly by immature osteoblasts during normal bone remodeling [25]. Activated $CD4^+$ and $CD8^+$ T lymphocytes [15, 22], activated macrophages [26] and activated synovial fibroblasts [27] as well as immature osteoblasts are major sources of RANKL in arthritic joints.

RANKL acts *in vitro* to alter the expression of more than 600 genes engaged in osteoclast differentiation and activation [28]. Among these, RANKL induces numerous cytokines, chemokines and growth factors [28, 29] as well as the transcription factor NFAT2 (nuclear factor of activated T cells 2), which acts in concert with AP-1 to alter the transcription of many cytokine genes [30, 31]. RANKL also enhances production of proteins that activate and stabilize TNF receptor-associated factors (TRAFs), which connect RANK to many downstream effector molecules [32]. Recent reports indicate that adding molecules other than M-CSF may also augment RANKL-mediated induction of osteoclasts. For example, bone morphogenetic protein-2 (BMP-2) cross-communicates with RANKL to support osteoclast differentiation and survival [33]. Transforming growth factor-β (TGFβ), activin A (another TGFβ superfamily member), and prostaglandin E_2 (PGE_2) act synergistically with RANKL to control osteoclast differentiation [34–37]. In addition, the glycosphingolipid lactosylceramide is required before RANKL can initiate osteoclastogenesis [38]. Interestingly, RANKL also induces osteoclasts to secrete platelet-derived growth factor BB (PDGF-BB), an osteoblastogenesis inhibitory factor that appears to allow for feedback inhibition to downregulate RANKL production by osteoblasts [4].

RANK

RANK is the cellular receptor that mediates all known actions of RANKL. The accepted alternative designation for this molecule is TNFRSF11A (TNF receptor superfamily member 11A [14]). In humans, the 10 exons of the *RANK* gene map to chromosome 18q21 [17]. RANK is induced by M-CSF [29] and is expressed chiefly on osteoclasts and their precursors [39] as well as on dendritic cells and mature T lymphocytes [40]. At the amino acid level, human RANK (616 aa [MW 97 kd]) exhibits 70% homology to the mouse version [17]. The cytoplasmic tail of RANK has multiple binding sites for TRAFs, which amplify the impact of the RANKL/RANK complex *via* a series of kinase cascades [32, 41–43]. In young adult mice, RANK is expressed highly in the metaphysis on multinucleated cells (osteoclasts) distributed in the primary spongiosa adjacent to the growth plate and in the periosteum, and to a lesser extent in hypertrophic chondrocytes [39]. Expression is reduced substantially in senescent rats [24].

OPG

OPG is a soluble decoy receptor that binds RANKL before it can activate RANK-mediated signaling. Synonymous acronyms for OPG include OCIF (osteoclast inhibitory factor [44]), FDCR-1 (follicular dendritic cell receptor-1 [45]), TR1 (TNF receptor superfamily member 1 [46]), and TNFRSF11B (TNF receptor superfamily member 11B, the current preferred nomenclature for this protein [14]). Human

OPG is a single-copy gene comprised of 5 exons that spans 29 kB of chromosome 8q24 [47]. The OPG promoter harbors a recognition element for CBFA1, the binding of which leads to an increase in OPG expression [48].

Human OPG (401 aa, MW 44 kd) is found in various body fluids in both monomeric and dimeric variants (55 kd or 110 kd, respectively, after post-translational N-glycosylation [49]). The main OPG form released by cells is the disulfide-linked homodimer [49], but the monomeric variant is secreted in small amounts [49]. Additional monomer is produced in plasma by proteolysis [50], and the monomeric form is reported to be the predominant version in circulation [51]. Mean serum concentrations under normal conditions vary from study to study, and assay formats likely have a significant influence on the absolute OPG levels that are measured. For example, mean serum OPG in normal mice has been assessed at 500 pg/ml, although the range of individual animal values (200 to 1,300 pg/ml) is quite broad [49, 51]. In normal healthy humans, serum OPG levels have been reported to vary from 50 to 200 pg/ml [52–54] to as high as 1 to 2 ng/ml [55]. Both the monomer and the dimer are cleaved during processing to yield a mature protein of 380 aa [50]; the activity of the final product derived from either parent form is similar [50]. The constitution of OPG is maintained across species, with human and rat proteins sharing approximately 85% homology at the amino acid level [49].

OPG is a unique member of the TNF receptor (TNFR) superfamily in that it lacks the membrane-spanning sequence common to most TNFRs [44, 49]. The RANKL-binding site of OPG consists of four highly conserved, cysteine-rich, TNFR-like domains located near the N-terminus; OPG immobilizes RANKL in a 1:1 molar ratio. The C-terminus of OPG contains a heparin-binding domain [56], which may function to anchor soluble OPG to cell surfaces [45] and allow for its internalization and degradation (demonstrated in multiple myeloma cells [57]). The C-terminus bears two death domain homologous (DDH) regions of unknown function [56]. The extracellular location of the DDH regions suggests that they do not initiate osteoclast apoptosis.

OPG is expressed in an even wider range of mouse tissues than is RANKL, with high levels found in lung, heart, kidney and placenta [49, 51]. In humans, OPG is expressed in virtually all tissues – including bone – except for peripheral blood lymphocytes [44]. In bones of young animals, OPG immunoreactivity is localized to resorbed bone surfaces and bone matrices in cement lines as well as in osteoblasts and osteocytes [24, 58]. The molecule is also found in proliferating chondrocytes [24] and in the superficial zone of normal cartilage [59]. Expression decreases with age at all sites except the endosteal lining [24]. Interestingly, post-translational processing of OPG in bone appears to confer a different complement of sugar residues than does glycosylation in other tissues [58], suggesting that the bone- and soft tissue-specific variants might have divergent functions.

At present, OPG is not known to act in concert with other molecules to directly regulate RANKL activity. However, at least two other potential RANKL-modulating entities have been described. First, the TNF family member TRAIL can block the anti-osteoclast activities of OPG *in vitro* [60], suggesting the possibility that RANKL and TRAIL can cooperate to circumvent the regulatory effects of OPG. Under normal homeostatic conditions, however, TRAIL is not an essential means of regulating OPG *in vivo* [61]. Second, a novel calcium-dependent (C type) lectin designated murine osteoclast inhibitory lectin (mOCIL) is reported to directly reduce RANKL-mediated osteoclast differentiation in an OPG-independent manner [62]. In mouse bone marrow cultures, mOCIL and OPG share the ability to suppress osteoclast formation during early proliferation, but only OPG is available to block production during later stages [62]. Thus, for practical purposes OPG remains the chief means of damping RANKL-mediated osteoclast function during all stages of the cell's life.

Extracellular regulatory elements

RANKL serves as the ultimate ligand by which many pro-resorptive factors indirectly instigate bone degradation under both physiologic and pathologic conditions. This precept is substantiated by the induction of osteoclastogenesis and hypercalcemia in wild type mice but not in mice lacking RANK following injection of a pro-resorptive molecule – either parathyroid hormone (PTH)-related protein (PTHrP), vitamin D (as the active metabolite $1{,}25(OH)_2D_3$), interleukin-1β (IL-1β), or TNFα [63]. As shown in Table 1, many factors that induce bone resorption function by inducing RANKL and/or by suppressing OPG [64–66]. Several therapeutic agents elicit these changes in human osteoblasts, including the glucocorticoids dexamethasone [67] and hydrocortisone [68] as well as the immunosuppressive agents cyclosporin A, FK-506, and rapamycin [69]. Similarly, fibroblast growth factor-2 (FGF-2 [70]), insulin-like growth factor-I (IGF-I [71]), PGE_2 [11, 72–75], PTH [76–79], PTHrP [8], thyroid hormone [80] and vitamin D [11, 25, 76] encourage osteoclastogenesis by upregulating RANKL expression while reducing OPG production by murine osteoblasts and osteoblastic cell lines. A similar effect is engendered in human bone marrow stromal cells by vitamin D [81]. In contrast, under normal physiological conditions, bone degradation is prevented by those factors that augment OPG expression while lowering that of RANKL. For example, high calcium levels in the microenvironment of activated osteoclasts hinder additional resorption by raising local OPG expression [44, 58]. Both TGFβ [73, 82] and several topoisomerase II inhibitors [83] increase expression of OPG while downregulating that of RANKL. The hormone estrogen inhibits osteoclast function by promoting OPG expression [84, 85], an effect mediated *via* estrogen receptor type α [85] and also indirectly *via* TGFβ [66, 86]. Estrogen suppresses RANKL-induced osteoclast formation in mouse bone marrow stromal cells [87], although this action has yet to be confirmed as stemming directly from estrogen-induced inhibition of RANKL expression. Two other hormones, leptin [88] and melatonin [89], also suppress

Table 1 - Regulation of OPG and RANKL expression by selected osteotropic ligands

Treatment	OPG	RANKL	Comments	References
Cytokines				
Histamine	ND	↑ mRNA	Mu-OB	[210]
IL-1	↑ mRNA/prot.	↑ mRNA	Hu-BMSC, Hu-OB	[74, 90, 91]
			Mu-OB	[11]
IL-6	ND	↑ mRNA	Mu-BMSC, Mu-OB	[211]
IL-7	ND	↑ mRNA/prot.	Activated Hu-T cells	[212]
IL-11	↑ mRNA	↑ mRNA	Mu-OB	[11, 76]
IL-17	↑ mRNA	↑ mRNA	Mu-OB	[92]
IL-18	↑ mRNA	no change	Mu-BMSC, Mu-OB	[213]
Prostaglandin E_2 (PGE_2)	↓ mRNA/prot.	↑ mRNA	Hu-BMSC, Hu-OB	[11, 72–74]
Transforming growth factor-β (TGFβ)	↓ mRNA/prot.	↓ mRNA	Mu-BMSC, Mu-OB	[73, 82]
			Rat-OB	[214, 215]
Tumor necrosis factor-α (TNFα)	↓ mRNA/prot.	↑ mRNA	Hu-BMSC, Hu-OB	[74, 90, 91, 93]
Hormones				
Calcium	↑ mRNA	ND	Mu-BMSC, Mu-OB	[44]
5α-Dihydrotestosterone	↓ mRNA/prot.	ND	Hu-OB	[216]
17β-Estradiol	↑ mRNA/prot.	ND	Hu-OB, Mu-BMSC	[84]
Glucocorticoids	↓ mRNA/prot.	↑ mRNA	Hu-BMSC, Hu-OB	[67, 68, 74]
			Mu-OB	[73]
Insulin-like growth factor-I (IGF-I)	↓ mRNA	↑ mRNA	Mu-BMSC	[71]
Leptin	↑ mRNA/prot.	ND	Hu-PBMC, Mu-Spleen	[88]
Melatonin	↑ mRNA/prot.	↓ mRNA	Mu-OB	[89]

Table 1 (continued)

Treatment	OPG	RANKL	Comments	References
Parathyroid hormone (PTH)	↓ mRNA	↑ mRNA	Hu-OB, Mu-OB, Rat-OB	[76, 77]
			Mu-BMSC	[78, 79]
			Mu-Bone, Rat-Bone	[11]
PTH-related protein (PTHrP)	↓ mRNA	↑ mRNA	Hu-Breast cancer cells expressing PTHrP	[8]
Vitamin D (1,25$(OH)_2D_3$)	↓ mRNA	↑ mRNA	Mu-BMSC, Mu-OB	[11, 25, 44, 73, 76]
	No change	↑ mRNA	Mu-OB	[217]
	↓ mRNA	↑ mRNA	Hu-BMSC, Hu-RA-SF	[27, 81]
	↑ mRNA/prot.	ND	Hu-OB	[90]
Chemicals and drugs				
Bisphosphonates	↓ mRNA/prot.	↑ mRNA	Hu-OB	[218]
Calcium ionophores	↑ mRNA	↑ mRNA	Mu-OB	[219]
Cyclooxygenase-2 (COX-2) inhibitors	↓ mRNA	↑ mRNA	Mu-BMSC	[75]
Heparin (in synergy with IL-11)	ND	↑ mRNA	Mu-BMSC	[220]
Topoisomerase II inhibitors	↑ mRNA	↓ mRNA	Mu-BMSC	[83]
Immune activators				
αTCR	ND	↑ mRNA	Activated Hu-T cells	[15]
αCD3/IL-2/ConA	↓ mRNA/prot.	↑ mRNA	Activated Hu-T cells	[197]
αCD3	↑ mRNA	ND	Activated Mu-T cells	[22]
	ND	↑ mRNA	Activated Hu-T cells (RA)	[157]
αCD28	↑ mRNA	ND	Activated Mu-T cells	[22]
ConA	↑ mRNA	ND	Activated Mu-T cells	[22]
PHA	ND	↑ mRNA	Activated Hu-T cells	[162]

Table 1 (continued)

Treatment	**OPG**	**RANKL**	**Comments**	**References**
PMA	↑ mRNA	ND	Activated Mu-T cells	[22]
	↑ mRNA/prot.	↑ mRNA	Mu-OB, Rat-OB	[219, 221]
Immunosuppressants				
Cyclosporin A	↓ mRNA	↑ mRNA	Hu-BMSC	[69]
FK-506	↓ mRNA	↑ mRNA	Hu-BMSC	[69]
Rapamycin	↓ mRNA	↑ mRNA	Hu-BMSC	[69]
	↑ mRNA/prot.	ND	Hu-OB	[69]

Abbreviations: ND, not determined; BMSC, bone marrow stromal cells or cell line; ConA, concanavalin A, a lectin activator of T cells; Hu, human; IL, interleukin; Mu, murine; OB, osteoblastic cells or cell line; OPG, osteoprotegerin; RANKL, RANK ligand; PBMC, peripheral blood mononuclear cells; PHA, phytohemagglutinin, a lectin activator of T cells; PMA, phorbol myristoyl acetate, a protein kinase C activator; RA, rheumatoid arthritis; RA-SF, synovial fibroblasts obtained from RA patients; TCR, T cell receptor

↑ = *enhanced expression*

↓ = *reduced expression*

osteoclast formation by augmenting OPG expression. Thus, osteoclast net activity directed by numerous hormones and cytokines is regulated by their different abilities to reciprocally control RANKL and OPG production [64–66].

Paradoxically, some cytokines enhance the expression of both OPG and RANKL (Tab. 1). This pattern has been verified for the proinflammatory agents IL-1β [90, 91], IL-11 [11, 76], IL-17 [92], TNFα [63, 90, 91, 93] and TNFβ [93]. Thus, control of osteoclast net activity resides not in the absolute amounts of OPG and RANKL at a given site, but rather in their ratio. This principle is corroborated by the ability of supplemental OPG to counteract the osteolytic effects of many pro-resorptive, RANKL-inducing stimuli. For example, OPG blocks PTH-induced osteoclast differentiation [78] and bone resorption *in vitro* [64, 94] and *in vivo* [95]. Comparable results are obtained when recombinant OPG is used to neutralize PTHrP produced by tumor xenografts [96, 97]. Finally, OPG (50 ng/ml) blocks osteoclastogenesis and the formation of resorption pits mediated by RANKL-expressing cells isolated from joints of RA patients; the number of resorption pits is well correlated to the OPG:RANKL ratio [98]. Therefore, pro-resorptive agents would be anticipated to skew the OPG:RANKL ratio toward a relative excess of RANKL, even if they actually function to increase the expression of OPG as well. Interestingly, serum concentrations of both RANKL and OPG are elevated in RA patients [99].

Intracellular effector elements

Like other TNFR family members, an intracellular kinase cascade drives gene transcription initiated by RANK activation (Fig. 2). Signaling by TNFR proteins progresses by two pathways; those that contain cytoplasmic death domains and can directly induce apoptosis, and those (such as RANK) that do not. The cytoplasmic tail of RANK has unique TRAF interaction motifs (TIMs [100]) for binding TRAFs 1, 2, 3, 5 and 6 [39, 41–43, 101]. For instance, TRAFs 2 and 5 bind to the C terminus of RANK, while TRAF6 interacts with a RANK domain located closer to the inner surface of the plasma membrane [39]. TRAF2 also serves as an adaptor protein for other TNFR molecules [102], while TRAF5 [103] and TRAF6 [104] also participate in TNFα-induced osteoclastogenesis. Of the known TRAFs, only TRAF6 also transduces signals through IL-1 receptor/Toll-like receptor pathways [105–107]; TRAF6 also partakes in IL-17-mediated signaling [108]. Interestingly, RANKL but not IL-1 or TNF evokes oscillations in intracellular calcium concentrations [109], suggesting that this RANKL action is not mediated by TRAF2 or TRAF6. The various TRAFs appear to regulate different phases of osteoclast biology, with TRAF 2 and TRAF5 involved in osteoclastogenesis while TRAF6 controls both formation and activation (reviewed in [32]). TRAF6 is particularly critical in RANK-mediated signal amplification as it is the principal mediator activating gene transcription *via* NF-κB [42, 43] and it also functions at least in part by organizing the osteoclast cytoskeleton [106]. However, deletion of any single TRAF reduces but does not prevent osteoclast formation, thereby demonstrating that the differentiation programs mediated by each TRAF are redundant without being identical [106]. RANKL induces factors capable of activating and/or stabilizing TRAFs [28].

RANK/TRAF complexes amplify RANKL signaling through multiple downstream kinases. Enzymes recruited by RANK/TRAF include the Akt phosphatidylinositol-3 (PI-3) serine/threonine kinase, extracellular signal-regulated protein kinase (ERK), IκB kinase (IKK), c-Jun N-terminal kinase (JNK), p38 kinase, and the c-Src protein tyrosine kinase [32, 110–112]. In some instances, these enzymes function independently to alter osteoclast function [113]. TRAF2 appears to be a primary mediator of the JNK pathway [43], but the TRAFs tasked to direct the other kinases have yet to be defined. RANK-mediated activity of each cascade is independently regulated [114]. The various kinases promote binding of the transcription factors AP-1 (a complex of dimeric transcription factors composed of ATF [activating transcription factor], Fos, or Jun subunits that share a common DNA binding site [115]), NFATc2, and NF-κB to genes required for osteoclast formation and activity [28, 109, 116]. In the case of *NFAT*, the promoter region contains binding sites for AP-1 and NF-κB [117], which provides a mechanism by which RANKL can upregulate its expression [28]. RANKL also increases expression of c-myc, a transcription factor that controls differentiation and proliferation of osteoclasts [118]. These factors are required at various stages of osteoclast differentiation program. For example, NF-κB components p50 and p52 are not needed for genesis of progenitor cells but are indispensable for differentiation of precursors to mature osteoclasts [119]. In like manner, independent activation of both JNK and NF-κB are necessary for completion of osteoclastogenesis [113]. The complexity of the intracellular amplification cascades provides an explanation for the synergistic activities of RANKL and M-CSF in osteoclastogenesis, in that M-CSF induces RANK as well as TRAF2, PI-3 kinase, and MEKK3 (mitogen-activated protein kinase [MAP] or ERK kinase kinase 3 [29]).

In addition to direct inhibition of RANKL by OPG, RANK-regulated osteoclast activity is controlled indirectly by several means (Fig. 2). For instance, in osteoclast precursor cells RANKL elicits production of interferon-β (IFNβ), which acts to inhibit the AP-1 component c-Fos [120, 121]. However, RANKL also induces suppressors of cytokine signaling (SOCS), which block the IFNβ-mediated dampening of RANKL signaling [120]. The TRAF6 component of RANK-regulated osteoclastogenesis is modulated by the regulatory element TRAF6-inhibitory zinc finger protein (TIZ [122]) and destabilized by T lymphocyte-derived interferon-γ (IFNγ [123]), while other as yet undiscovered factors likely block interactions between other TRAFs and their downstream effectors. Importantly, the cellular receptors for proinflammatory cytokines with osteoclastogenic activity, including IL-1 and TNFα, also use TRAFs [32], thereby providing a means by which multiple pro-resorptive pathways may be coupled, or even additive [124].

A dynamic equilibrium between OPG and RANKL regulates osteoclast differentiation, activation and survival

At the cellular level, the degree of bone degradation depends on the balance between production, activation and survival of osteoclasts. RANKL promotes these three activities, while OPG blocks differentiation of precursors while simultaneously thwarting the maturation and retention of previously formed cells. With respect to differentiation, RANKL suppresses the macrophage phenotype of mouse bone marrow cells and promotes the generation of osteoclasts [125]. OPG blocks osteoclastogenesis in mouse bone marrow [94] and human peripheral blood mononuclear cells [126] cultured with RANKL and M-CSF, and it eliminates their formation in severely inflamed joints of rats with adjuvant- [127] and collagen- [128] induced arthritis. OPG impedes RANKL-induced fusion of osteoclast precursors [129] but does not lower the number of precursors [49]. Activation of mature osteoclasts is inhibited by OPG *in vitro* [10, 64, 130]. Injection of OPG into normal and hypercalcemic mice reduces circulating concentrations of calcium (which depend on bone resorption by pre-formed cells) within two hours [95, 96], a time course explicable only by direct blockade of mature osteoclasts. One explanation for this rapid reversal is the capacity of OPG to reverse RANKL-mediated osteoclast attachment to bone [131, 132]. In addition, experiments *in vitro* suggest that OPG works rapidly to downregulate expression of cathepsin K, the main protease in osteoclasts responsible for bone resorption [133]. Finally, OPG greatly lessens osteoclast survival *in vitro* by neutralizing the ability of RANKL to prevent osteoclast apoptosis [134, 135]. More mature osteoclasts are especially sensitive to OPG [134]. Significant apoptosis of osteoclasts results only if OPG is present for from one [73, 130] to three days [134, 135]. However, in mice and rats a single OPG injection *in vivo* induces substantial osteoclast apoptosis within six hours [135], leading to a profound reduction in osteoclast populations within 12 to 48 hours [134, 136]. Again, the OPG:RANKL ratio governs the extent of osteoclast differentiation, activation and survival.

Skeletal phenotypes resulting from *in vivo* genetic alterations in proteins of the OPG/RANK/RANKL pathway

Mouse phenotypes

The importance of the OPG/RANK/RANKL pathway to skeletal development and homeostasis is demonstrated amply by the plethora of bone abnormalities that occur in mice with spontaneous or engineered mutations of elements in this transduction system (Tab. 2). The two principal changes are osteopetrosis, which results from reduced RANK-mediated signaling, and osteoporosis, which reflects excessive activity in RANK-controlled kinase cascades.

Table 2 - In vivo *bone and immune system phenotypes of mice with engineered mutations in OPG, RANKL, RANK or molecules associated with the RANK signaling pathway*

Stage of osteoclast development	**Gene/function**	**Mouse phenotypes**	**Immune consequences**	**Refs.**
Precursor proliferation and survival	*Macrophage colony stimulating factor (M-CSF)* Osteoclast differentiation factor (together with RANKL)	Spontaneous knockout – osteopetrosis, resulting from an osteoclastogenesis defect (known as the op/op mouse)	None reported	[222]
Commitment and differentiation	*Receptor activator of NF-κB (RANK)* Cellular receptor for RANKL (acting to mediate osteoclast differentiation and activation)	Transgenic – osteopetrosis (for soluble RANK, which yields an osteoclastogenesis defect, but normal osteoclast precursors). Knockout – osteopetrosis (via intrinsic defects in osteoclast precursors, osteoclastogenesis, and calcium metabolism), lymph node agenesis, and mammary gland dysgenesis.	Knockout – extrinsic defect in B cell development. No effect on T cell and dendritic cell development.	[63]
Commitment and differentiation	*RANK ligand (RANKL)* (1) Osteoclast differentiation and activation factor (2) Dendritic cell survival factor	Transgenic – osteoporosis and arterial calcification. Knockout – osteopetrosis (by osteoclastogenesis defect, but normal osteoclast precursors), lymph node agenesis, and mammary gland dysgenesis.	Knockout – intrinsic defect in T cell development. Normal dendritic cells.	[137, 138]

Table 2 (continued)

Stage of osteoclast development	Gene/function	Mouse phenotypes	Immune consequences	Refs.
Commitment and differentiation	*Nuclear factor-κB (NF-κB)* RANK-activated transcription factor	Knockout (deletion of p50 and/or p52 subunits) – osteopetrosis (*via* defects in osteoclastogenesis and osteoclast maturation)	Impaired B cell maturation and macrophage function, with altered thymic and splenic architectures (if both p50 and p52 deleted)	[145]
Commitment and differentiation	*Activator protein-1 (AP-1)* RANK-activated transcription factor	Knockout (of the c-Fos component of AP-1) – osteopetrosis (via osteoclastogenesis defect)	Reduced B cell numbers (due to impaired marrow environment)	[144, 146, 223]
Regulation of commitment and differentiation (RANKL activity)	*Interferon-β (INFβ)* Inhibitor of RANKL-induced c-Fos (AP-1) expression	Knockout – osteoporosis (*via* increased osteoclastogenesis).	None reported	[121]
Regulation of commitment and differentiation (RANKL activity)	*Osteoprotegerin (OPG)* Soluble decoy receptor for RANKL	Transgenic – osteopetrosis (by osteoclastogenesis defect, but normal osteoclast precursors). Knockout – osteoporosis (*via* increased osteoclastogenesis) and arterial calcification.	Knockout – extrinsic defect in B cell development. No effect on T cell and dendritic cell development.	[49]

Table 2 (continued)

Stage of osteoclast development	Gene/function	Mouse phenotypes	Immune consequences	Refs.
Regulation of commitment and differentiation (OPG activity?)	*Klotho (kl/kl)* Unknown entity	Spontaneous knockout – osteopetrosis (*via* enhanced OPG levels, leading to fewer osteoclasts) and early death.	Spontaneous knockout – Reduced B cell numbers	[164, 175]
Maturation	*TNFR-associated factor 6 (TRAF6)* A cytoplasmic co-factor that binds to activated RANK	Knockout – osteopetrosis (*via* defects in osteoclastogenesis and osteoclast activation, and a lack of osteoclast precursors), lymph node agenesis, thymic atrophy, and periportal hepatic inflammation.	Knockout – Decreased B cell responsiveness, with defective IL-1-mediated signaling	[141, 142]
Maturation	*Src* RANK-associated cytoplasmic protein tyrosine kinase	Knockout – osteopetrosis (from a defect in osteoclast activation).	None reported	[143]
Survival	*Src homology 2-containing inositol-5-phosphatase (SHIP)*	Knockout – osteopetrosis (from osteoclast hypersensitivity to M-CSF and RANKL, and to longer osteoclast survival).	None reported	[224]

RANKL

Mice engineered to lack RANKL (*Rankl*$^{-/-}$) develop osteopetrosis with systemic bone deformities and failed tooth eruption as a consequence of essentially universal osteoclast annihilation [137, 138]. The increased bone mass is partially reversed by restoring RANKL through transgenic over-expression [138]. Interestingly, deletion of Rankl does not yield an inherent defect in osteoclasts, as *in vitro* treatment of *Rankl*$^{-/-}$ osteoclast precursors with RANKL and M-CSF results in the production of osteoclasts [137]. Instead, the defect in *Rankl*$^{-/-}$ mice is an indirect effect of impaired RANKL production by stromal cells.

RANK

Deletion of cellular RANK, or inhibition of RANK activity, also abolishes osteoclasts. Mice with null mutations of cellular RANK (*Rank*$^{-/-}$) exhibit a phenotype that is essentially identical to that of *Rankl*$^{-/-}$ animals, including severe osteopetrosis with multiple skeletal deformities and an absence of tooth eruption [63]. However, unlike their *Rankl*$^{-/-}$ counterparts, *Rank*$^{-/-}$ osteoclast precursors are inherently defective [63] as shown by their failure to support osteoclastogenesis in the presence of RANKL and M-CSF [10]. Responsiveness to RANKL and M-CSF can be restored if RANK is re-introduced into the defective *Rank*$^{-/-}$ precursors [10], and the resulting osteoclasts are functional as shown by their capacity to reverse osteopetrosis in *Rank*$^{-/-}$ mice [63]. Introduction of a soluble RANK either by injection or transgenic over-expression of RANK-Fc (a soluble, non-signaling fusion protein combining the RANKL-binding domain of RANK with the constant [Fc] region of human immunoglobulin G_1 [39]) also creates a RANKL deficiency that culminates in increased bone mass. Injection of young adult mice with RANK-Fc leads to higher bone density near sites of skeletal growth, while over-expression throughout the entire lifespan yields generalized osteopetrosis [39].

OPG

In vivo studies with genetically engineered mice have shown that excess OPG leads to profound osteoclast depletion and osteopetrosis, while an OPG deficit allows pronounced osteoclast expansion and osteoporosis. Transgenic mice that over-express rat *Opg* are markedly osteopetrotic at birth, but do not develop deformed bones or defective tooth eruption [49, 139]. The extent of osteopetrosis is highly correlated with circulating OPG concentrations; all mice are affected if serum OPG levels exceed 10 ng/ml (normal serum levels are ≤ 1–2 ng/ml [49]). In contrast, mice with null mutations of OPG (*Opg*$^{-/-}$) develop progressive osteoporosis with many fractures beginning shortly after birth due to widespread and uncontrolled osteoclast production [140]. Heterozygous mice lacking one OPG copy

(*Opg*$^{+/-}$) also loses bone mass to a significant degree over time [140]. The osteoporotic phenotype of *Opg*$^{-/-}$ mice can be rescued by administering OPG, either by crossbreeding with transgenic mice that over-express *Opg* or by injecting recombinant OPG [139]. Thus, supplemental OPG can counteract an imbalanced OPG:RANKL ratio in which a relative excess of RANKL is leading to enhanced bone loss.

Intracellular effector elements

Null mutations of many proteins located downstream of the RANKL/RANK complex also lead to osteopetrosis (Tab. 2). Factors for which this finding has been reported include TRAF6 [141, 142], the TRAF6 binding partner Src tyrosine kinase [143], and transcription factors c-Fos (a component of AP-1 [144]) and NF-κB [145]. The osteopetrotic phenotype in mice lacking c-Fos is rescued by over-expressing Fra-1, another Fos family member that is a normal constituent of AP-1 [146].

Human genetic diseases

Three of 11 genetic defects known to induce sclerosing bone dysplasias in humans have been attributed to mutations in elements of the OPG/RANK/RANKL pathway [147]. Activating mutations in *RANK* have been demonstrated in kindreds with two autosomal dominant, metabolic diseases of bone: familial expansile osteolysis [148] and expansile skeletal hyperphosphatasia [149]. Paget's disease of bone has been associated with activating mutations in *RANK* (in a single Japanese family [148]) or *p62* [150], a sequestosome component that supports NF-κB activation through the RANKL/RANK pathway [151]. However, while present in many patients, polymorphisms in *OPG* and *RANK* have not been linked to disease induction in most cases of sporadic and familial Paget's disease [152, 153]. An inactivating mutation of *OPG* has recently been identified as the cause of idiopathic hyperphosphatasia, an autosomal recessive metabolic bone disease described in one Australian family [154]. Alterations in the *RANKL* gene have yet to be linked with human disease.

The role of the OPG/RANK/RANKL pathway in bone pathophysiology

Numerous reports implicate an altered OPG:RANKL equilibrium as a causative factor in diseases associated with excessive bone resorption. Two conditions for which a burgeoning literature exists are rheumatoid arthritis (RA) and osteoporosis.

The OPG/RANK/RANKL pathway in rheumatoid arthritis and experimental arthritides

Several lines of evidence indicate that a RANKL-mediated increase in osteoclast net activity contributes significantly to the joint erosions that are characteristic of RA as well as animal models of arthritis [32, 98]. First, osteoclast populations are expanded in RA patients [155, 156] and in animals in which immune-mediated arthritis has been induced [22, 127, 136]. Second, RANKL is expressed prominently in proximity to bone erosions in both RA patients [26, 98, 157, 158] and rodents [159–161]. Third, osteoclast activity in RA and experimental arthritis in rats is controlled chiefly by RANKL derived from activated $CD4^{+}$ T cells [22, 26, 162], macrophages [26] and synovial fibroblasts [27], although RANKL from activated B cells [163, 164], osteoblasts and bone marrow stromal cells may also play a role [11, 165]. RANKL levels in human arthritis patients are highest in those individuals with extensive inflammation (e.g., active RA) relative to those with a minimal leukocyte infiltrate (e.g., osteoarthritis) [26]; interestingly, active RA also is characterized by elevated OPG concentrations in serum and synovial fluid, although the OPG levels are not correlated with disease severity [166]. In addition to the direct production of RANKL, activated T cells from RA patients emit IL-1β [91], IL-17 [92, 158], and TNFα [91], all of which induce RANKL expression. Inhibition of these and other proinflammatory cytokines (granulocyte/macrophage-colony stimulating factor [GM-CSF], interferon-γ [IFNγ], IL-6) in the absence of RANKL blockade does not hinder T cell-induced osteoclastogenesis *in vitro* [22, 92]. Finally, the severity of skeletal destruction is lessened in *Rankl*-deficient mice with serum transfer-induced arthritis [167], indicating that RANKL can modulate the severity of an antibody-driven inflammatory process.

The OPG/RANK/RANKL pathway in inflammation-associated osteoporosis

Local (peri-articular [155]) and systemic [168] osteoporosis is a common concurrent finding in chronic RA (reviewed in [169]). Increased osteoclast function is the principal mechanism by which secondary osteoporosis develops in RA patients [170], with activation of existing cells being of more importance than differentiation of precursors [171]. The excess RANKL production typical of the arthritic joint provides a reasonable explanation for this finding. Support for this hypothesis is provided by evaluation of mice with collagen-induced arthritis, which exhibit increased RANKL expression both within the inflamed joints and at distant sites [161]. Further substantiation for this premise is the ability of supplemental OPG to ameliorate bone loss in mice with null mutations of cytotoxic T-lymphocyte-associated protein 4 (*ctla4*), which exhibit widespread inflammation secondary to constitutive T cell activation [22]. These data implicate an imbalanced OPG:RANKL ratio as the likely pathogenesis for osteoporosis that accompanies immune-mediated disease.

The OPG/RANK/RANKL pathway in immune system function

The role of OPG, RANK and RANKL in immune system function recently has been reviewed [40]. Growing evidence exists to support RANKL activities in both immune system development and in immune responsiveness. The major implications are summarized briefly below. However, the role of the OPG/RANK/RANKL pathway in inflammation requires further exploration as prior studies in experimental models of immune-mediated arthritis have demonstrated no capacity for RANKL inhibitors to modify established inflammation [127, 128, 136].

Development

Molecules of the OPG/RANK/RANKL pathway are widely distributed throughout elements of the immune system during both development and maturity. RANKL is expressed broadly in lymphoid organs, including lymph nodes, spleen, thymus (including immature $CD4^+CD8^+$ thymocytes), and gut-associated lymphoid tissues (Peyer's patches) [10, 17].

RANKL/RANK signaling obviously is required for the genesis of lymph nodes as these organs are absent in mice with null mutations of *Rankl* [137], *Rank* [63, 172], and *Traf6* [142]. Interestingly, splenic and Peyer's patch architectures in *Rankl*$^{-/-}$ and *Rank*$^{-/-}$ animals are essentially normal, despite some modest alterations in organ size. Selective ablation of lymph nodes is not a consequence of *Rankl* and *Rank* deletion as these two molecules also are expressed in spleen and Peyer's patches [40]. Aberrant lymph node formation also cannot be attributed to defective homing of *Rankl*$^{-/-}$ lymphocytes [137]. Taken together, these data suggest that the specific and essential role of RANKL and RANK in specifying lymphoid organ architecture is limited to lymph node organogenesis, likely as a growth and/or survival factor on lymph node organizing cells during embryonic development. However, RANKL may also play a cooperative role with TNFα in specifying Peyer's patch development as mice with targeted removal of either *Rankl* [137] or TNF receptor type I (*TnfrI*) have small Peyer's patches [173], while animals deficient in both *Rankl* and *TnfrI* lack Peyer's patches altogether [40]. The exact cellular and molecular mechanisms of RANKL/RANK regulated lymph node morphogenesis and the link between lymph node and Peyer's patch formation have yet to be elicited.

RANKL is involved in regulating early differentiation and expansion of T lymphocytes. RANKL is found on $CD4^-CD8^-$ early thymocyte precursors [17], which also express the pre T cell receptor (TCR). In *Rankl*$^{-/-}$ mice, $CD4^-CD8^-CD44^-$

CD25$^+$ precursors do not progress to become CD4$^-$CD8$^-$CD44$^-$CD25$^-$ thymocytes [137], a necessary stage before genesis of immature CD4$^+$CD8$^+$ cells that ultimately can undergo positive selection to become mature CD4$^+$ or CD8$^+$ T cells. This developmental defect is intrinsic to the bone marrow-derived precursors and not to anomalies in the thymic microenvironment during differentiation [40]. Thymocyte differentiation *in vitro* is normal for cells derived from *Rank*$^{-/-}$ mice [63, 172] but is altered in cells obtained from *Rankl*$^{-/-}$ animals [40]. The difference in thymocyte differentiation, which is the only discernible distinction between the *Rankl*$^{-/-}$ and *Rank*$^{-/-}$ phenotypes, suggests that RANKL might act on another receptor during early thymocyte development.

The OPG/RANKL/RANK pathway also contributes to B cell development and function. *Rankl*$^{-/-}$ and *Rank*$^{-/-}$ mice have reduced numbers of mature B220$^+$ B cells in the spleen and lymph nodes, and slightly disorganized B cell areas in primary splenic follicles [137, 172]. RANKL regulates early B cell differentiation from the B220$^+$CD43$^+$CD25$^-$ pro-B cell to the B220$^+$CD43$^-$CD25$^+$ pre-B cell [137]. Interestingly, B cells in bone marrow can produce RANKL [164], and RANK is expressed at low levels on activated B cells [174]. Immature B cell stages can differentiate *in vitro* to form osteoclasts when stimulated with RANKL and M-CSF [164]. The functional importance of B cells in osteoclast function is confirmed by assessment of *klotho* (*kl/kl*) mice, which are characterized by osteopetrosis and death. The *kl/kl* phenotype, which results from persistent elevation of OPG with fewer osteoclasts [175], occurs in conjunction with greatly reduced B cell numbers [164]. Finally, OPG appears to be required for efficient generation of antibodies [176].

Communication between immune cells

Molecules of the OPG/RANK/RANKL pathway mediate interactions among many immune effector cells, particularly those between dendritic cells (DCs) and T lymphocytes (Fig. 3). Immature DCs are dispersed in tissues to capture pathogens and process their antigens. Mature DCs with captured antigen migrate to the T cell zones of secondary lymphoid organs, where they present the antigen to T cells during initiation of the adaptive immune response [177]. DCs and mature T cells express RANK, while activated CD4$^+$ and CD8$^+$ T cells produce both soluble and membrane-bound RANKL [15, 22, 174], suggesting that RANKL regulates DC and T cell exchanges in paracrine and autocrine manners, respectively. RANKL/RANK signaling on DCs promotes DC survival *in vitro* and *in vivo* [13, 17, 178]. For this purpose, RANKL can work in conjunction with CD40L (another TNF ligand family member found on activated T cells) or TNFα [174]; RANKL appears to control later events associated with producing immune system memory, while CD40L seems to specify initial priming in response to novel antigens [40]. The response of DCs to RANKL depends on their anatomic location, as indicated by different RANKL-

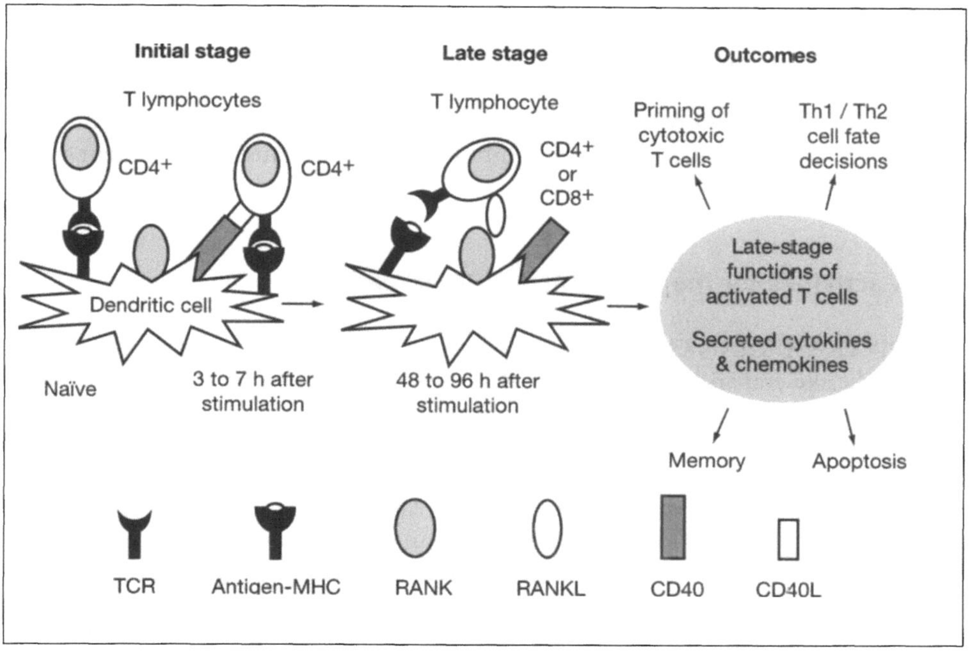

Figure 3
Proposed RANKL–RANK interactions during immune responses
At the initial stage, dendritic cells (DCs) are primed by CD4⁺ T cells that express CD40L and generate lymphokines. CD40L is rapidly upregulated (expression maximal at 3–7 hrs and gone by 24 hrs). Later responses (e.g., Th1/Th2 cell fate decisions, antigen-specific memory) are evoked by subsequent RANKL production at high levels (expression peaks at 48 hrs and is sustained through 96 hrs) on both CD4⁺ and CD8⁺ T cells.
Abbreviations: CD40L, CD40 ligand; DC, dendritic cell; MHC, major histocompatability complex; TCR, T cell receptor. This figure is adapted (with the publisher's permission) from [40].

mediated functions in mucosal DCs *versus* splenic DCs [179]. Thus, RANKL/RANK interactions can regulate DC functions, T cell activation and memory induction, and T cell/DC communication – and do so in a tissue-specific fashion.

OPG reasonably might be presumed to control these RANKL-regulated immune cell interactions. OPG is produced by and binds to DCs, the latter property apparently mediated by attachment of hyaluronic-binding domains of OPG to the DC surface [45]. Serum OPG concentrations are increased significantly in disorders characterized by persistent immune system activation [180]. OPG binds both RANKL [10, 11] and TRAIL [60], suggesting that the dynamic equilibrium between this trio of proteins might govern the fate of DCs – particularly at sites of active inflammation. However, *in vivo* interactions of consequence only exist for RANKL with OPG, and not for TRAIL with OPG [61]. Furthermore, the outcome of OPG-mediated RANKL inhibition *in vivo* is limited to preserving the structural integrity of arthritic joints, and not to reducing inflammation [127, 128, 136]. Thus, the anti-erosive effects provided by RANKL inhibition will dwarf any improvement in the inflammatory component of arthritis.

Other activities for the OPG/RANKL/RANK pathway

The OPG/RANK/RANKL pathway appears to act mainly in regulating the functions of the skeletal and immune systems. However, molecules of this signaling pathway also have fundamental roles in mammary gland development and vascular biology.

RANKL is an essential morphogen for the mammary gland (reviewed in [181]). Epithelium in the mammary gland expresses RANK as the lobulo-alveolar conduits are assembled [182]. A likely biological explanation for this arrangement is that RANKL-mediated increases in osteoclast activity are necessary to mobilize maternal calcium from bone during lactation. Pregnant mice lacking *Rankl* or *Rank* do not form mammary glands, which results in neonatal death by starvation [182]. If one of these two molecules is missing, epithelium in budding glands undergoes reduced proliferation in conjunction with enhanced apoptosis. Administration of exogenous RANKL to pregnant *Rankl*$^{-/-}$ animals reverses mammary gland agenesis.

Surprisingly, OPG is an important factor for maintaining vascular health. OPG is an NF-κB-dependent survival factor for endothelial cells [183]. In addition, OPG is required for arterial wall integrity as indicated by buildup of calcified plaques in large muscular arteries of *Opg*$^{-/-}$ mice [140]. These lesions occur in the tunica media, a site at which OPG normally is expressed [49]. Vascular expression of OPG occurs in smooth muscle cells, not endothelial cells [184], and can be regulated in both positive [184] and negative [185] fashions. In contrast, RANKL and RANK normally are absent at this site, though these two molecules are found in conjunction with the calcified foci [139]. The vascular mineralization can be prevented by over-expression of OPG throughout development in *Opg*$^{-/-}$ mice, but administration of OPG after lesions are formed cannot rescue the arterial walls [139]. Interestingly, bone resorption is linked to vascular calcification in warfarin-treated rats [186] and postmenopausal women [187, 188]. However, women with osteoporosis usually develop calcification of the tunica intima, often adjacent to atherosclerotic plaques. The relationship, if any, between RANKL-mediated lesions in *Opg*$^{-/-}$ mice and arterial calcification during postmenopausal osteoporosis remains to be determined. Nevertheless, OPG clearly is associated with altered cardiovascular function in humans, as the presence and severity of coronary artery disease is correlated significantly with serum OPG concentrations [189]. OPG and RANKL are expressed in early and advanced atherosclerotic lesions [190], and certain polymorphisms in the promoter region of the *OPG* gene are associated with decreased arterial blood flow in healthy people [191]. Further research will be required to ascertain whether or not OPG plays a role in the pathogenesis of reduced arterial elasticity in patients with RA [192].

Therapeutic promise of RANKL/RANK inhibition

Preclinical experience with RANKL blockade in arthritic animals

Bone turnover is substantially diminished when circulating OPG concentrations exceed 10 ng/ml (in OPG-transgenic mice [49]) to 30 ng/ml (in normal cynomolgus monkeys [193]). As shown in Table 3, administration of a RANKL-binding molecule (OPG) effectively bars skeletal resorption in the face of severe, chronic inflammation. In animals with experimental arthritis, recombinant OPG completely prevents loss of bone mineral density and bone erosion in arthritic joints, but does not impact inflammation [22, 127, 128, 136] (Fig. 4). Articular cartilage is also sustained by OPG, although this effect is an indirect result of protecting subchondral bone [22, 127, 128, 167]. Treatment with OPG does not alter expression of RANKL or OPG in the inflamed joints [128]. Maintenance of bone in arthritic joints by OPG treatment depends on both the dose [22, 127] and treatment schedule [127], with better preservation afforded by early intervention (near disease onset). Furthermore, a seven-day OPG course beginning at disease onset prevents erosions for three weeks [127], while a single OPG dose at disease onset is effective for only four days [136]. Plausible explanations for this result are that one OPG dose removes only mature osteoclasts, while multiple doses also ablate newly differentiated cells, or that so much RANKL is produced in arthritic joints that a single OPG dose cannot completely inhibit RANKL activity. Therefore, the best therapeutic efficacy in active arthritis likely will require multiple doses of a RANKL inhibitor given intermittently. Taken together, the implication of these studies is that RANKL blockade should be initiated as early as possible to reduce the degree of bone and cartilage destruction in arthritic joints.

Administration of a RANKL inhibitor in combination with agents that inhibit IL-1 and TNFα may enhance anti-erosive efficacy. Several facts support this contention:

- First, administration of such agents can normalize serum RANKL levels [99].
- Second, IL-1β and TNFα are potent pro-resorptive cytokines that direct destruction of bone and cartilage in RA [194], with TNFα appearing to exert primary control over osteoclast differentiation while IL-1 mainly regulates osteoclast function [195, 196]. However, IL-1 does augment osteoclast differentiation in the pres-

Table 3 - In vivo *efficacy of recombinant osteoprotegerin (OPG)[1], a direct-acting RANKL inhibitor, in animal models of immune-mediated arthritis or osteoporosis and in human clinical trials*

In vivo system	Regimen	Skeletal effect	Refs.
Rodents			
Adjuvant-induced arthritis in rats	≥ 1 mg/kg/day SC for 7 or 14 days	Prevents joint erosion but not inflammation	[22, 127, 136]
Collagen-induced arthritis in rats	≥ 0.06 mg/kg/day SC for 7 days	Prevents joint erosion but not inflammation	[127, 128]
Chronic polyarthritis in TNFα-transgenic mice	15 mg/kg IV every 3rd day for 35 days	Reduces bone erosions	[209]
Osteoporosis (marked) in *Opg* knockout mice	50 mg/kg/day IV 3x/week for 4 weeks	Reverses osteoporosis	[139]
Osteoporosis in ovariectomized rats	10 mg/kg SC, 3x/week for 5.5 months	Reverses chronic loss of bone mass	[225]
Osteoporosis in ovariectomized mice	Adenoviral gene therapy with OPG, once i.v.	Prevents loss of bone mass	[226]
Humans			
Phase I clinical trial in postmenopausal women (healthy)	0.1 to 3.0 mg/kg SC once	Dose-dependent reduction in bone turnover (for weeks)	[207]
Phase I clinical trial in patients with metastatic mammary carcinoma or multiple myeloma (with osteolysis at multiple sites)	0.1 to 3.0 mg/kg SC once	Dose-dependent reduction in bone turnover (for weeks)	[208]

[1]*Fusion of the constant (Fc) region of human immunoglobulin to the RANKL-binding region of OPG substantially extends the circulating half-life*

ence of RANKL-producing activated T cells [197], apparently through direct activation of osteoclasts *via* IL-1 receptor types I and II [198].
- Third, the effects of IL-1 and TNF on osteoclasts are mediated chiefly by RANKL [63].
- Fourth, both cytokines induce RANKL expression in vascular endothelial cells [199], suggesting that new capillaries in the pannus tissue that erodes bone are likely an important additional source of pro-resorptive ligands.
- Fifth, molecules that inhibit IL-1 and TNF effectively preserve the architecture of arthritic joints in their own right [200].

Both IL-1 and TNF are reported to kindle osteoclast production through RANKL-independent pathways [195, 201–204]. However, in mice the capacity of TNF to manipulate osteoclasts in a RANKL-independent manner depends on strain-specific differences in cytokine production; furthermore, the efficiency of osteoclastogenesis mediated by TNF is much lower than that elicited by RANKL [205]. A recent report suggests that TNF does not induce osteoclastogenesis, but rather targets macrophages that have been primed with minuscule amounts of RANKL [206]. Neither IL-1 nor TNF induces osteoclast formation in *Rank*$^{-/-}$ mice [63], indicating that bone preservation by IL-1 and TNF inhibitors likely is an indirect result of quenching joint inflammation and the downregulation of RANKL production by activated leukocytes.

Human clinical experience with RANKL inhibition

As in animal models of disease, RANKL inhibition also is an effective means of inhibiting bone turnover in humans (Tab. 3). A single subcutaneous injection of recombinant OPG into healthy postmenopausal women [207] or patients with advanced osteolytic neoplasia involving multiple bones [208] yields rapid and profound reductions in bone turnover for weeks. Thus, administration of OPG to block RANKL-mediated bone resorption is efficacious in humans, even in the face of aggressive bone-eroding skeletal diseases. RANKL elaborated by activated T cells and synovial fibroblasts regulates osteoclast differentiation in joints of RA patients [22, 98, 157], suggesting that RANKL blockade should be an effective therapeutic strategy for sparing joint architecture in this disease. However, clinical trials of RANKL inhibitors have yet to be initiated in RA.

Summary

The OPG/RANK/RANKL pathway obviously plays a critical role in defining the extent of skeletal destruction in severely inflamed joints. In this setting, RANKL

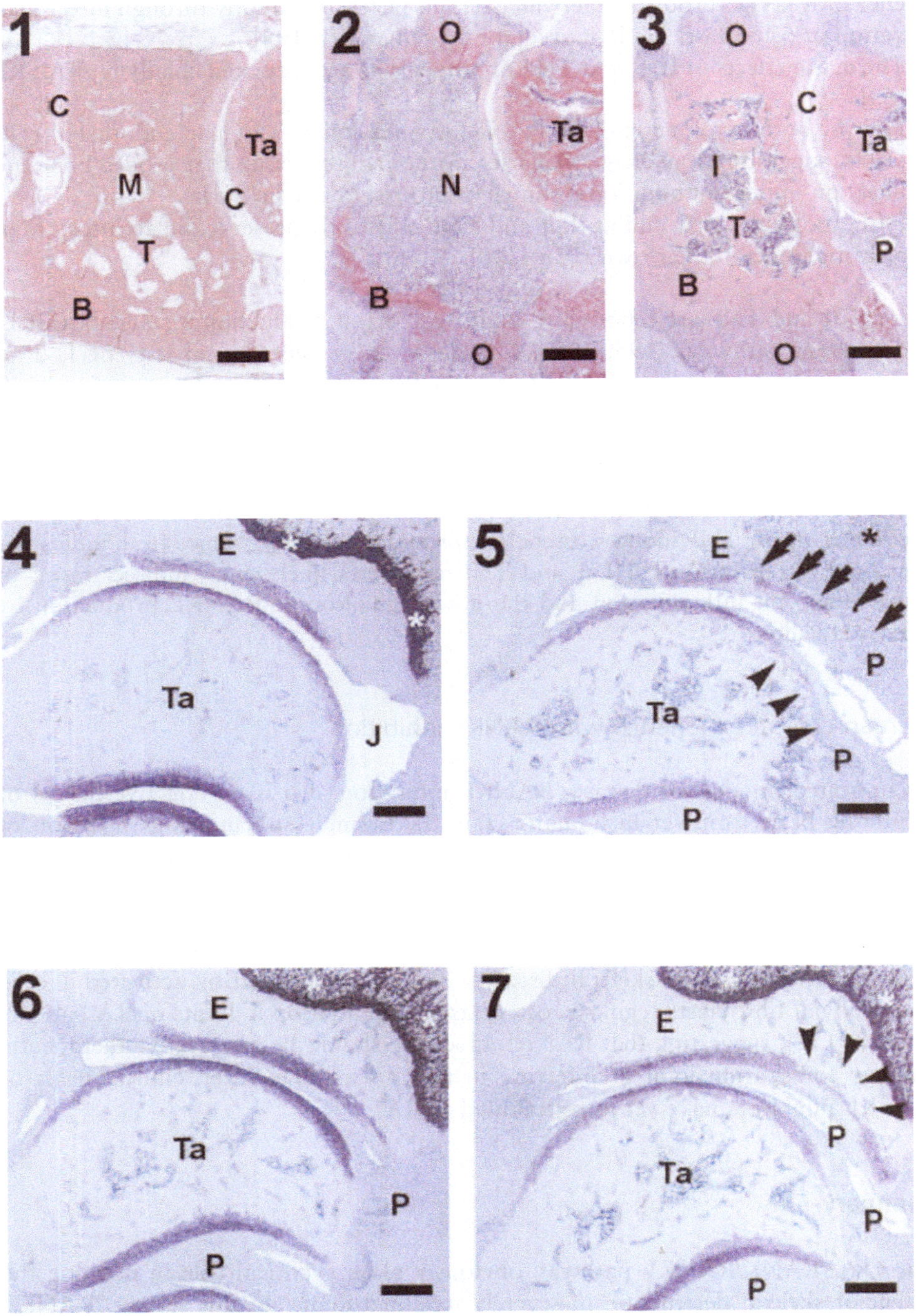
1
C
Ta
M
C
T
B
2
O
Ta
N
B
O
3
O
C
Ta
I
T
P
B
O
4
E
Ta
J
5
E
P
Ta
P
P
6
E
Ta
P
P
7
E
P
Ta
P
P

derived from activated leukocytes and synovial fibroblasts promotes osteoclastogenesis, osteoclast activation, and osteoclast survival. This pro-erosive impetus is countered by OPG, which serves as a RANKL sink and thereby prevents osteoclast expansion, activation, and persistence. RANKL serves as the final common mediator for many pro-resorptive ligands – including numerous cytokines – and, therefore, serves as a final common signal to promote bone destruction. Thus, joint integrity in arthritis will depend chiefly on the balance between OPG and RANKL, with skeletal dissolution occurring when the OPG:RANKL ratio is low (i.e., the quantity of RANKL exceeds the amount of OPG available to sequester it).

Figure 4
Even in the presence of severe inflammation, OPG directly prevents bone erosion and indirectly inhibits cartilage destruction in rats with adjuvant-induced arthritis (AdA)
Panels 1–3: Navicular tarsal bone, hind paw. Normal animals (1) have thick cortical bone (B) with smooth periosteal and endosteal bone surfaces, thick bony trabeculae (T) dividing the fat-filled marrow cavity (M), and intact joint cartilages (C) resting on subchondral bone plates. Relative to normal controls, animals with untreated AdA (2) feature marked reduction and loss of cortical and trabecular bone, extensive osteophyte production (O), and replacement of the fatty marrow by necrotic inflammatory cells (N) at the peak of clinical arthritis (seven days after disease onset). Injection of AdA rats with OPG (3) at doses ≥ 1 mg/kg (by subcutaneous bolus for seven days beginning at disease onset) precludes bone erosion; thus, cartilage architecture is preserved even though inflammatory cells fill the marrow (I) and the expanded synovial tissue (pannus, P).
Panels 4–7: Tibiotarsal joint, hind paw. In normal animals (4), cartilage lining the joint cavities (J) and comprising the growth plate of the distal tibia (white asterisks) have intact matrix proteoglycans (as indicated by uniform toluidine blue staining), and cortical and trabecular bone is intact. Rats with untreated AdA (5) are characterized by extensive inflammation surrounding and within the joint cavity – including extension of pannus (P) along cartilage surfaces; loss of cartilage proteoglycans is indicated by multifocal absence of toluidine blue staining due to penetration of the subchondral bone (arrows) and/or overgrowth of pannus (arrowheads). The growth plate (normal position indicated by black asterisk) and associated bone trabeculae of the distal tibia are absent. Even while severe inflammation is ongoing, administration of OPG (as described above) to AdA rats protects bone and cartilage matrix integrity by preserving subchondral bone (6), although some proteoglycan loss may occur at the periphery of joints in which invasion by pannus (P) is extensive (7, arrowheads). The primary spongiosa of OPG-treated animals (white asterisks) are thickened relative to those of normal rats.
Other abbreviations: E, epiphysis of the distal tibia; Ta, talus.
Panels 1–3: Hematoxylin and eosin (for general joint architecture), bars = 400 mm.
Panels 4–7: Tibiotarsal joint, toluidine blue (for cartilage matrix integrity), bars = 250 mm.

Inhibition of RANKL is a proven means of reducing joint erosions in experimental models of immune-mediated arthritis [22, 127, 128, 209]. Because the pathogenesis of skeletal damage in these models is similar to that reported in human patients with RA, RANKL blockade should prove to be a suitable therapeutic strategy in people once suitable agents have been tested. The degree of bone protection will depend on the treatment schedule, with early and aggressive intervention yielding more complete preservation, and periodic re-treatment required to control osteoclasts and prevent erosions over time. Despite the apparent importance of RANKL as a mediator in dendritic cell–T cell communication, RANKL blockade has little impact on the severity of inflammation in the arthritic joint. Thus, long-term treatment of joint inflammation likely will require combination therapy with both a RANKL inhibitor (to provide maximal protection of skeletal integrity) with one or more anti-inflammatory agents (to quench inflammation, thereby removing the irritant that promotes continued osteoclast net activity).

References

1 Rodan GA, Martin TJ (1981) Role of osteoblasts in hormonal control of bone resorption – a hypothesis. *Calcif Tissue Int* 33: 349–351

2 Jimi E, Nakamura I, Amano H, Taguchi Y, Tsurukai T, Tamura M, Takahashi N, Suda T (1996) Osteoclast function is activated by osteoblastic cells through a mechanism involving cell-to-cell contact. *Endocrinology* 137: 2187–2190

3 Katagiri T, Takahashi N (2002) Regulatory mechanisms of osteoblast and osteoclast differentiation. *Oral Dis* 8: 147–159

4 Kubota K, Sakikawa C, Katsumata M, Nakamura T, Wakabayashi K (2002) Platelet-derived growth factor BB secreted from osteoclasts acts as an osteoblastogenesis inhibitory factor. *J Bone Min Res* 17: 257–265

5 Keyszer GM, Heer AH, Gay S (1994) Cytokines and oncogenes in cellular interactions of rheumatoid arthritis. *Stem Cells* 12: 75–86

6 Taubman MA, Kawai T (2001) Involvement of T-lymphocytes in periodontal disease and in direct and indirect induction of bone resorption. *Crit Rev Oral Biol Med* 12: 125–135

7 Lerner UH (1994) Regulation of bone metabolism by the kallikrein-kinin system, the coagulation cascade, and the acute-phase reactants. *Oral Surg Oral Med Oral Pathol* 78: 481–493

8 Thomas RJ, Guise TA, Yin JJ, Elliott J, Horwood NJ, Martin TJ, Gillespie MT (1999) Breast cancer cells interact with osteoblasts to support osteoclast formation. *Endocrinology* 140: 4451–4458

9 Fuller K, Wong B, Fox S, Choi Y, Chambers TJ (1998) TRANCE is necessary and sufficient for osteoblast-mediated activation of bone resorption in osteoclasts. *J Exp Med* 188: 997–1001

10 Lacey DL, Timms E, Tan HL, Kelley MJ, Dunstan CR, Burgess T, Elliott R, Colombero A, Elliott G, Scully S et al (1998) Osteoprotegerin ligand is a cytokine that regulates osteoclast differentiation and activation. *Cell* 93: 165–176

11 Yasuda H, Shima N, Nakagawa N, Yamaguchi K, Kinosaki M, Mochizuki S, Tomoyasu A, Yano K, Goto M, Murakami A et al (1998) Osteoclast differentiation factor is a ligand for osteoprotegerin / osteoclastogenesis-inhibitory factor and is identical to TRANCE / RANKL. *Proc Natl Acad Sci USA* 95: 3597–3602

12 Teitelbaum SL (2000) Bone resorption by osteoclasts. *Science* 289: 1504–1508

13 Wong BR, Josien R, Lee SY, Sauter B, Li HL, Steinman RM, Choi Y (1997) TRANCE (tumor necrosis factor [TNF]-related activation-induced cytokine), a new TNF family member predominantly expressed in T cells, is a dendritic cell-specific survival factor. *J Exp Med* 186: 2075–2080

14 HUGO Gene Nomenclature Committee (1998) The new TNF nomenclature scheme. http://www.gene.ucl.ac.uk/nomenclature/genefamily/tnftop.html. (Last accessed March 18, 2003)

15 Wong BR, Rho J, Arron J, Robinson E, Orlinick J, Chao M, Kalachikov S, E. C, Bartlett FSI, Frankel WN et al (1997) TRANCE is a novel ligand of the tumor necrosis receptor family that activates c-Jun N-terminal kinase in T cells. *J Biol Chem* 272: 25190–25194

16 Kodaira K, Mizuno A, Yasuda H, Shima N, Murakami A, Ueda M, Higashio K (1999) Cloning and characterization of the gene encoding mouse osteoclast differentiation factor. *Gene* 230: 121–127

17 Anderson DM, Maraskovsky E, Billingsley WL, Dougall WC, Tometsko ME, Roux ER, Teepe MC, DuBose RF, Cosman D, Galibert L (1997) A homologue of the TNF receptor and its ligand enhance T cell growth and dendritic cell function. *Nature* 390: 175–179

18 Kitazawa R, Kitazawa S (2002) Vitamin D_3 augments osteoclastogenesis *via* vitamin D-responsive element of mouse RANKL gene promoter. *Biochem Biophys Res Comm* 290: 650–655

19 Kitazawa S, Kitazawa R (2002) Epigenetic control of mouse receptor activator of NF-κB ligand gene expression. *Biochem Biophys Res Comm* 293: 126–131

20 Lam J, Nelson CA, Ross PF, Teitelbaum SL, Fremont DH (2001) Crystal structure of the TRANCE / RANKL cytokine reveals determinants of receptor-ligand specificity. *J Clin Invest* 108: 971–979

21 Ito S, Wakabayashi K, Ubukata O, Hayashi S, Okada F, Hata T (2002) Crystal structure of the extracellular domain of mouse RANK ligand at 2.2-Å resolution. *J Biol Chem* 277: 6631–6636

22 Kong YY, Feige U, Sarosi I, Bolon B, Tafuri A, Morony S, Capparelli C, Li J, Elliott R, McCabe S et al (1999) Activated T cells regulate bone loss and joint destruction in adjuvant arthritis through osteoprotegerin ligand. *Nature* 402: 304–309

23 Lum L, Wong BR, Josien R, Becherer JD, Erdjument-Bromage H, Schlondorff J, Tempst P, Choi Y, Blobel CP (1999) Evidence for a role of tumor necrosis factor-α (TNF-α)-con-

verting enzyme-like protease in shedding of TRANCE, a TNF family member involved in osteoclastogenesis and dendritic cell survival. *J Biol Chem* 274: 13613–13618

24 Ikeda T, Utsuyama M, Hirokawa K (2001) Expression profiles of receptor activator of nuclear factor κB ligand, receptor activator of nuclear factor κB, and osteoprotegerin messenger RNA in aged and ovariectomized rat bones. *J Bone Min Res* 16: 1416–1425

25 Nagai M, Sato N (1999) Reciprocal gene expression of osteoclastogenesis inhibitory factor and osteoclast differentiation factor regulates osteoclast formation. *Biochem Biophys Res Comm* 257: 719–723

26 Crotti TN, Smith MD, Weedon H, Ahern MJ, Findlay DM, Kraan M, Tak PP, Haynes DR (2002) Receptor activator of NF-κB ligand (RANKL) expression in synovial tissue from patients with rheumatoid arthritis, spondyloarthropathy, osteoarthritis, and from normal patients: semiquantitative and quantitative analysis. *Ann Rheum Dis* 61: 1047–1054

27 Takayanagi H, Iizuka H, Juji T, Nakagawa T, Yamamoto A, Miyazaki T, Koshihara Y, Oda H, Nakamura K, Tanaka S (2000) Involvement of receptor activator of nuclear factor κB ligand / osteoclast differentiation factor in osteoclastogenesis from synoviocytes in rheumatoid arthritis. *Arthritis Rheum* 43: 259–269

28 Ishida N, Hayashi K, Hoshijima M, Ogawa T, Koga S, Miyatake Y, Kumegawa M, Kimura T, Takeya T (2002) Large scale gene expression analysis of osteoclastogenesis *in vitro* and elucidation of NFAT2 as a key regulator. *J Biol Chem* 277: 41147–41156

29 Cappellen D, Luong-Nguyen N-H, Bongiovanni S, Grenet O, Wanke C, Susa M (2002) Transcriptional program of mouse osteoclast differentiation governed by the macrophage colony-stimulating factor and the ligand for the receptor activator of NF-κB. *J Biol Chem* 277: 21971–21982

30 Cyert MS (2001) Regulation of nuclear localization during signaling. *J Biol Chem* 276: 20805–20808

31 Macian F, Lopez-Rodriguez C, Rao A (2001) Partners in transcription: NFAT and AP-1. *Oncogene* 20: 2476–2489

32 Gravallese EM, Galson DL, Goldring SR, Auron PE (2001) The role of TNF-receptor family members and other TRAF-dependent receptors in bone resorption. *Arthritis Res* 3: 6–12

33 Itoh K, Udagawa N, Katagiri T, Iemura S, Ueno N, Yasuda H, Higashio K, Quinn JMW, Gillespie MT, Martin TJ et al (2001) Bone morphogenetic protein-2 stimulates osteoclast differentiation and survival supported by receptor activator of nuclear factor-κB ligand. *Endocrinology* 142: 3656–3662

34 Wani MR, Fuller K, Kim NS, Choi Y, Chambers TJ (1999) Prostaglandin E_2 cooperates with TRANCE in osteoclast induction from hemopoietic precursors: synergistic activation of differentiation, cell spreading, and fusion. *Endocrinology* 140: 1927–1935

35 Fuller K, Lean JM, Bayley KE, Wani MR, Chambers TJ (2000) A role for TGFβ1 in osteoclast differentiation and survival. *J Cell Sci* 113: 2445–2463

36 Fuller K, Bayley KE, Chambers TJ (2000) Activin A is an essential cofactor for osteoclast induction. *Biochem Biophys Res Comm* 268: 2–7

37 Murase Y, Okahashi N, Koseki T, Itoh K, Udagawa N, Hashimoto O, Sugino H, Noguchi T, Nishihara T (2001) Possible involvement of protein kinases and Smad2 signaling pathways on osteoclast differentiation enhanced by activin A. *J Cell Physiol* 188: 236–242

38 Iwamoto T, Fukumoto S, Kanaoka K, Sakai E, Shibata M, Fukumoto E, Inokuchi J-I, Takamiya K, Furukawa K, Furukawa K et al (2001) Lactosylceramide is essential for the osteoclastogenesis mediated by macrophage-colony stimulating factor and receptor activator of nuclear factor-κB ligand. *J Biol Chem* 276: 46031–46038

39 Hsu H, Lacey DL, Dunstan CR, Solovyev I, Colombero A, Timms E, Tan HL, Elliott G, Kelley MJ, Sarosi I et al (1999) Tumor necrosis factor receptor family member RANK mediates osteoclast differentiation and activation induced by osteoprotegerin ligand. *Proc Natl Acad Sci USA* 96: 3540–3545

40 Kong YY, Boyle WJ, Penninger JM (2000) Osteoprotegerin ligand: a regulator of immune response and bone physiology. *Immunol Today* 21: 495–502

41 Darnay BG, Haridas V, Ni J, Moore PA, Aggarwal BB (1998) Characterization of the intracellular domain of receptor activator of NF-κB (RANK) – interaction with tumor necrosis factor receptor-associated factors and activation of NF-κB and c-Jun N-terminal kinase. *J Biol Chem* 273: 20551–20555

42 Galibert L, Tometsko ME, Anderson DM, Cosman D, Dougall WC (1998) The involvement of multiple tumor necrosis factor receptor (TNFR)-associated factors in the signaling mechanisms of receptor activator of NF-κB, a member of the TNFR superfamily. *J Biol Chem* 273: 34120–34127

43 Darnay BG, Ni J, Moore PA, Aggarwal BB (1999) Activation of NF-κB by RANK requires tumor necrosis factor receptor-associated factor (TRAF) 6 and NF-κB-inducing kinase – Identification of a novel TRAF6 interaction motif. *J Biol Chem* 274: 7224–7731

44 Yasuda H, Shima N, Nakagawa N, Mochizuki SI, Yano K, Fujise N, Sato Y, Goto M, Yamaguchi K, Kuriyama M et al (1998) Identity of osteoclastogenesis inhibitory factor (OCIF) and osteoprotegerin (OPG): a mechanism by which OPG / OCIF inhibits osteoclastogenesis *in vitro*. *Endocrinology* 139: 1329–1337

45 Yun TJ, Chaudhary PM, Shu GL, Frazer JK, Ewings MK, Schwartz SM, Pascual V, Hood LE, Clark EA (1998) OPG / FDCR-1, a TNF receptor family member, is expressed in lymphoid cells and is upregulated by ligating CD40. *J Immunol* 161: 6113–6121

46 Kwon BS, Wang S, Udagawa N, Haridas V, Lee ZH, Kim KK, Oh KO, Greene J, Li Y, Su J et al (1998) TR1, a new member of the tumor necrosis factor receptor superfamily, induces fibroblast proliferation and inhibits osteoclastogenesis and bone resorption. *FASEB J* 12: 845–854

47 Morinaga T, Nakagawa N, Yasuda H, Tsuda E, Higashio K (1998) Cloning and characterization of the gene encoding human osteoprotegerin / osteoclastogenesis-inhibitory factor. *Eur J Biochem* 254: 685–691

48 Thirunavukkarasu K, Halladay DL, Miles RR, Yang XH, Galvin RJS, Chandrasekhar S, Martin TJ, Onyia JE (2000) The osteoblast-specific transcription factor Cbfa1 con-

tributes to the expression of osteoprotegerin, a potent inhibitor of osteoclast differentiation and function. *J Biol Chem* 275: 25163–25172
49 Simonet WS, Lacey DL, Dunstan CR, Kelley M, Chang MS, Luethy R, Nguyen HQ, Wooden S, Bennett L, Boone T et al (1997) Osteoprotegerin: a novel secreted protein involved in the regulation of bone density. *Cell* 89: 309–319
50 Tomoyasu A, Goto M, Fujise N, Mochizuki S, Yasuda H, Morinaga T, Tsuda E, Higashio K (1998) Characterization of monomeric and homodimeric forms of osteoclastogenesis inhibitory factor. *Biochem Biophys Res Comm* 245: 382–387
51 Yano K, Shibata O, Mizuno A, Kobayashi F, Higashio K, Morinaga T, Tsuda E (2001) Immunological study on circulating murine osteoprotegerin / osteoclastogenesis inhibitory factor (OPG / OCIF): possible role of OPG / OCIF in the prevention of osteoporosis in pregnancy. *Biochem Biophys Res Comm* 288: 217–224
52 Szulc P, Hofbauer LC, Heufelder AE, Roth S, Delmas PD (2001) Osteoprotegerin serum levels in men: correlation with age, estrogen, and testosterone status. *J Clin Endocrinol Metab* 86: 3162–3165
53 Khosla S, Arrighi HM, Melton LJI, Atkinson EJ, O'Fallon WM, Dunstan C, Riggs BL (2002) Correlates of osteoprotegerin levels in women and men. *Osteoporosis Intl* 13: 394–399
54 Fahrleitner A, Prenner G, Leb G, Tscheliessnigg KH, Piswanger-Solkner C, Obermayer-Pietsch B, Portugaller HR, Berghold A, Dobnig H (2003) Serum osteoprotegerin is a major determinant of bone density development and prevalent vertebral fracture status following cardiac transplantation. *Bone* 32: 96–106
55 Yano K, Tsuda E, Washida N, Kobayashi F, Goto M, Harada A, Ikeda K, Higashio K, Yamada Y (1999) Immunological characterization of circulating osteoprotegerin / osteoclastogenesis inhibitory factor: Increased serum concentrations in postmenopausal women with osteoporosis. *J Bone Miner Res* 14: 518–527
56 Yamaguchi K, Kinosaki M, Goto M, Kobayashi F, Tsuda E, Morinaga T, Higashio K (1998) Characterization of structural domains of human osteoclastogenesis inhibitory factor. *J Biol Chem* 273: 5117–5123
57 Standal T, Seidel C, Hjertner O, Plesner T, Sanderson RD, Waage A, Borset M, Sundan A (2002) Osteoprotegerin is bound, internalized, and degraded by multiple myeloma cells. *Blood* 100: 3002–3007
58 Nakamura H, Tsuji T, Hirata A, Yamamoto T (2002) Localization of osteoprotegerin (OPG) on bone surfaces and cement lines in rat tibia. *J Histochem Cytochem* 50: 945–953
59 Komuro H, Olee T, Kuhn K, Quach J, Brinson DC, Shikhman A, Valbracht J, Creighton-Achermann L, Lotz M (2001) The osteoprotegerin / receptor activator of nuclear factor κB / receptor activator of nuclear factor κB ligand system in cartilage. *Arthritis Rheum* 44: 2768–2776
60 Emery JG, Mcdonnell P, Burke MB, Deen KC, Lyn S, Silverman C, Dul E, Appelbaum ER, Eichman C, Diprinzio R et al (1998) Osteoprotegerin is a receptor for the cytotoxic ligand TRAIL. *J Biol Chem* 273: 14363–14367

61 Sedger LM, Glaccum MB, Schuh JCL, Kanaly ST, Williamson E, Kayagaki N, Yun T, Smolak P, Le T, Goodwin R et al (2002) Characterization of the *in vivo* function of TNF-α-related apoptosis-inducing ligand, TRAIL/Apo2L, using TRAIL/Apo2L gene-deficient mice. *Eur J Immunol* 32: 2246–2254

62 Zhou H, Kartsogiannis V, Hu YS, Elliott J, Quinn JMW, McKinstry WJ, Gillespie MT, Ng KW (2001) A novel osteoblast-derived C-type lectin that inhibits osteoclast formation. *J Biol Chem* 276: 14916–14923

63 Li J, Sarosi I, Yan X-Q, Morony S, Capparelli C, Tan HL, McCabe S, Elliott R, Scully S, Van G et al (2000) RANK is the intrinsic hematopoietic cell surface receptor that controls osteoclastogenesis and regulation of bone mass and calcium metabolism. *Proc Natl Acad Sci USA* 97: 1566–1571

64 Tsukii K, Shima N, Mochizuki S, Yamaguchi K, Kinosaki M, Yano K, Shibata O, Udagawa N, Yasuda H, Suda T et al (1998) Osteoclast differentiation factor mediates an essential signal for bone resorption induced by 1α,25-dihydroxyvitamin D_3, prostaglandin E2, or parathyroid hormone in the microenvironment of bone. *Biochem Biophys Res Comm* 246: 337–341

65 Morony S, Capparelli C, Lee R, Shimamoto G, Boone T, Lacey DL, Dunstan CR (1999) A chimeric form of osteoprotegerin inhibits hypercalcemia and bone resorption induced by IL-1β, TNF-α, PTH, PTHrP, and $1,25(OH)_2D_3$. *J Bone Min Res* 14: 1478–1485

66 Hofbauer LC, Khosla S, Dunstan CR, Lacey DL, Boyle WJ, Riggs BL (2000) The roles of osteoprotegerin and osteoprotegerin ligand in the paracrine regulation of bone resorption. *J Bone Miner Res* 15: 2–12

67 Hofbauer LC, Gori F, Riggs BL, Lacey DL, Dunstan CR, Spelsberg TC, Khosla S (1999) Stimulation of osteoprotegerin ligand and inhibition of osteoprotegerin production by glucocorticoids in human osteoblastic lineage cells: potential paracrine mechanisms of glucocorticoid-induced osteoporosis. *Endocrinology* 140: 4382–4389

68 Vidal NOA, Brändström H, Jonsson KB, Ohlsson C (1998) Osteoprotegerin mRNA is expressed in primary human osteoblast-like cells: downregulation by glucocorticoids. *J Endocrinol* 159: 191–195

69 Hofbauer LC, Shui C, Riggs BL, Dunstan CR, Spelsberg TC, O'Brien T, Khosla S (2001) Effects of immunosuppressants on receptor activator of NF-κB ligand and osteoprotegerin production by human osteoblastic and coronary artery smooth muscle cells. *Biochem Biophys Res Comm* 280: 334–339

70 Chikazu D, Katagiri M, Ogasawara T, Ogata N, Shimoaka T, Takato T, Nakamura K, Kawaguchi H (2001) Regulation of osteoclast differentiation by fibroblast growth factor-2: stimulation of receptor activator of nuclear factor κB ligand / osteoclast differentiation factor expression in osteoblasts and inhibition of macrophage colony-stimulating factor function in osteoclast precursors. *J Bone Min Res* 16: 2074–2081

71 Rubin J, Ackert-Bicknell CL, Zhu L, Fan X, Murphy TC, Nanes MS, Marcus R, Holloway L, Beamer WG, Rosen CJ (2002) IGF-I regulates osteoprotegerin (OPG) and receptor activator of nuclear factor-κB ligand *in vitro* and OPG *in vivo*. *J Clin Endocrinol Metab* 87: 4273–4279

72 Brändström H, Jonsson KB, Ohlsson C, Vidal O, Ljunghall S, Ljunggren Ö (1998) Regulation of osteoprotegerin mRNA levels by prostaglandin E_2 in human bone marrow stromal cells. *Biochem Biophys Res Comm* 247: 338–341

73 Murakami T, Yamamoto M, Ono K, Nishikawa M, Nagata N, Motoyoshi K, Akatsu T (1998) Transforming growth factor-β1 increases mRNA levels of osteoclastogenesis inhibitory factor in osteoblastic / stromal cells and inhibits the survival of murine osteoclast-like cells. *Biochem Biophys Res Comm* 252: 747–752

74 Brändström H, Björkman T, Ljunggren Ö (2001) Regulation of osteoprotegerin secretion from primary cultures of human bone marrow stromal cells. *Biochem Biophys Res Comm* 280: 831–835

75 Gruber R, Karreth F, Fischer MB, Watzek G (2002) Platelet-released supernatants stimulate formation of osteoclast-like cells through a prostaglandin / RANKL-dependent mechanism. *Bone* 30: 726–732

76 Horwood NJ, Elliott J, Martin TJ, Gillespie MT (1998) Osteotropic agents regulate the expression of osteoclast differentiation factor and osteoprotegerin in osteoblastic stromal cells. *Endocrinology* 139: 4743–4746

77 Lee SK, Lorenzo JA (1999) Parathyroid hormone stimulates TRANCE and inhibits osteoprotegerin messenger ribonucleic acid expression in murine bone marrow cultures: correlation with osteoclast-like cell formation. *Endorinology* 140: 3552–3561

78 Itoh K, Udagawa N, Matsuzaki K, Takami M, Amano H, Shinki T, Ueno Y, Takahashi N, Suda T (2000) Importance of membrane- or matrix-associated forms of M-CSF and RANKL / ODF in osteoclastogenesis supported by SaOS-4/3 cells expressing recombinant PTH/PTHrP receptors. *J Bone Min Res* 15: 1766–1775

79 Ma YL, Cain RL, Halladay DL, Yang X, Zeng Q, Miles RR, Chandrasekhar S, Martin TJ, Onyia JE (2001) Catabolic effects of continuous human PTH (1-38) *in vivo* is associated with sustained stimulation of RANKL and inhibition of osteoprotegerin and gene-associated bone formation. *Endocrinology* 142: 4047–4054

80 Miura M, Tanaka K, Komatsu Y, Suda M, Yasoda A, Sakuma Y, Ozasa A, Nakao K (2002) A novel interaction between thyroid hormones and $1,25(OH)_2D_3$ in osteoclast formation. *Biochem Biophys Res Comm* 291: 987–994

81 Quinn JM, Horwood NJ, Elliott J, Gillespie MT, Martin TJ (2000) Fibroblastic stromal cells express receptor activator of NF-κB ligand and support osteoclast differentiation. *J Bone Min Res* 15: 1459–1466

82 Takai H, Kanematsu M, Yano K, Tsuda E, Higashio K, Ikeda K, Watanabe K, Yamada Y (1998) Transforming growth factor-β stimulates the production of osteoprotegerin / osteoclastogenesis inhibitory factor by bone marrow stromal cells. *J Biol Chem* 273: 27091–27096

83 Yamagishi T, Otsuka E, Hagiwara H (2001) Reciprocal control of expression of mRNAs for osteoclast differentiation factor and OPG in osteogenic stromal cells by genistein: evidence for the involvement of topoisomerase II in osteoclastogenesis. *Endocrinology* 142: 3632–3637

84 Hofbauer LC, Khosla S, Dunstan CR, Lacey DL, Spelsberg TC, Riggs BL (1999) Estro-

gen stimulates gene expression and protein production of osteoprotegerin in human osteoblastic cells. *Endocrinology* 140: 4367–4370

85 Lindberg MK, Erlandsson M, Alatalo SL, Windahl S, Andersson G, Halleen JM, Carlsten H, Gustafsson J-Å, Ohlsson C (2001) Estrogen receptor α, but not estrogen receptor β, is involved in the regulation of the OPG/RANKL (osteoprotegerin / receptor activator of NF-κB ligand) ratio and serum interleukin-6 in male mice. *J Endocrinol* 171: 425–433

86 Srivasta S, Toraldo G, Weitzmann MN, Cenci S, Ross FP, Pacifici R (2001) Estrogen decreases osteoclast formation by downregulating receptor activator of NF-κB ligand (RANKL)-induced JNK activation. *J Biol Chem* 276: 8836–8840

87 Shevde NK, Bendixen AC, Dienger KM, Pike JW (2000) Estrogens suppress RANK ligand-induced osteoclast differentiation *via* a stromal cell independent mechanism involving c-Jun repression. *Proc Natl Acad Sci USA* 97: 7829–7834

88 Holloway WR, Collier FM, Aitken CJ, Myers DE, Hodge JM, Malakellis M, Gough T, J., Collier GR, Nicholson GC (2002) Leptin inhibits osteoclast generation. *J Bone Min Res* 17: 200–209

89 Koyama H, Nakade O, Takada Y, Kaku T, Lau K-HW (2002) Melatonin at pharmacologic doses increases bone mass by suppressing resorption through downregulation of the RANKL-mediated osteoclast formation and activation. *J Bone Min Res* 17: 1219–1229

90 Hofbauer LC, Dunstan CR, Spelsberg TC, Riggs BL, Khosla S (1998) Osteoprotegerin production by human osteoblast lineage cells is stimulated by vitamin D, bone morphogenetic protein-2, and cytokines. *Biochem Biophys Res Comm* 250: 776–781

91 Hofbauer LC, Lacey DL, Dunstan CR, Spelsberg TC, Riggs BL, Khosla S (1999) Interleukin-1β and tumor necrosis factor-α, but not interleukin-6, stimulate osteoprotegerin ligand gene expression in human osteoblastic cells. *Bone* 25: 255–259

92 Kotake S, Udagawa N, Takahashi N, Matsuzaki K, Itoh K, Ishiyama S, Saito S, Inoue K, Kamatani N, Gillespie MT et al (1999) IL-17 in synovial fluids from patients with rheumatoid arthritis is a potent stimulator of osteoclastogenesis. *J Clin Invest* 103: 1345–1352

93 Brändström H, Jonsson KB, Vidal O, Ljunghall S, Ohlsson C, Ljunggren Ö (1998) Tumor necrosis factor-α and -β upregulate the levels of osteoprotegerin mRNA in human osteosarcoma MG-63 cells. *Biochem Biophys Res Comm* 248: 454–457

94 Udagawa N, Takahashi N, Yasuda H, Mizuno A, Itoh K, Ueno Y, Shinki T, Gillespie MT, Martin TJ, Higashio K et al (2000) Osteoprotegerin produced by osteoblasts is an important regulator in osteoclast development and function. *Endocrinology* 141: 3478–3484

95 Yamamoto M, Murakami T, Nishikawa M, Tsuda E, Mochizuki S, Higashio K, Akatsu T, Motoyoshi K, Nagata N (1998) Hypocalcemic effect of osteoclastogenesis inhibitory factor / osteoprotegerin in the thyroparathyroidectomized rat. *Endocrinology* 139: 4012–4015

96 Akatsu T, Murakami T, Ono K, Nishikawa M, Tsuda E, Mochizuki SI, Fujise N,

Higashio K, Motoyoshi K, Yamamoto M et al (1998) Osteoclastogenesis inhibitory factor exhibits hypocalcemic effects in normal mice and in hypercalcemic nude mice carrying tumors associated with humoral hypercalcemia of malignancy. *Bone* 23: 495–498
97 Capparelli C, Kostenuik PJ, Morony S, Starnes C, Weimann B, Van G, Scully S, Qi MY, Lacey DL, Dunstan CR (2000) Osteoprotegerin prevents and reverses hypercalcemia in a murine model of humoral hypercalcemia of malignancy. *Cancer Res* 60: 783–787
98 Haynes DR, Crotti TN, Loric M, Bain GI, Atkins GJ, Findlay DM (2001) Osteoprotegerin and receptor activator of nuclear factor κB ligand (RANKL) regulate osteoclast formation by cells in the human rheumatoid arthritic joint. *Rheumatology* 40: 623–630
99 Ziolkowska M, Kurowska M, Radzikowska A, Luszczykiewicz G, Wiland P, Dziewczopolski W, Filipowicz-Sosnowska A, Pazdur J, Szechinski J, Kowalczewski J et al (2002) High levels of osteoprotegerin and soluble receptor activator of nuclear factor κB ligand in serum of rheumatoid arthritis patients and their normalization after anti-tumor necrosis factor a treatment. *Arthritis Rheum* 46: 1744–1753
100 Arch RH, Gedrich RW, Thompson CB (1998) Tumor necrosis factor receptor-associated factors (TRAFs) – a family of adaptor proteins that regulate life and death. *Genes Dev* 12: 2821–2830
101 Kim HH, Lee DE, Shin JN, Lee YS, Jeon YM, Chung CH, Ni JA, Kwon BS, Lee ZH (1999) Receptor activator of NF-κB recruits multiple TRAF family adaptors and activates c-Jun N-terminal kinase. *FEBS Lett* 443: 297–302
102 Nguyen LT, Duncan GS, Mirtsos C, Ng M, Speiser DE, Shahinian A, Marino MW, Mak TW, Ohashi PS, Yeh WC (1999) TRAF2 deficiency results in hyperactivity of certain TNFR1 signals and impairment of CD40-mediated responses. *Immunity* 11: 379–389
103 Kanazawa K, Azuma Y, Nakano H, Kudo A (2003) TRAF5 functions in both RANKL- and TNF α-induced osteoclastogenesis. *J Bone Min Res* 18: 443–450
104 Kaji K, Katogi R, Azuma Y, Naito A, Inoue JI, Kudo A (2001) Tumor necrosis factor α-induced osteoclastogenesis requires tumor necrosis factor receptor-associated factor 6. *J Bone Min Res* 16: 1593–1599
105 Cao Z, Xiong J, Takeuchi M, Kurama T, Goeddel DV (1996) TRAF6 is a signal transducer for interleukin-1. *Nature* 383: 443–446
106 Armstrong AP, Tometsko ME, Glaccum M, Sutherland CL, Cosman D, Dougall WC (2002) A RANK / TRAF6-dependent signal transduction pathway is essential for osteoclast cytoskeletal organization and resorptive function. *J Biol Chem* 277: 44347–44356
107 Ye H, Arron JR, Lamothe B, Cirilli M, Kobayashi T, Shevde NK, Segal D, Dzivenu OK, Vologodskaia M, Yim M et al (2002) Distinct molecular mechanism for initiating TRAF6 signalling. *Nature* 418: 443–447
108 Schwandner R, Yamaguchi K, Cao ZD (2000) Requirement of tumor necrosis factor receptor-associated factor (TRAF) 6 in interleukin-17 signal transduction. *J Exp Med* 191: 1233–1239
109 Takayanagi H, Kim S, Koga T, Nishina H, Isshiki M, Yoshida H, Saiura A, Isobe M, Yokochi T, Inoue J et al (2002) Induction and activation of the transcription factor

NFATc1 (NFAT2) integrate RANKL signaling in terminal differentiation of osteoclasts. *Develop Cell* 3: 889–901
110 Wong BR, Besser D, Kim N, Arron JR, Vologodskaia M, Hanafusa H, Choi Y (1999) TRANCE, a TNF family member, activates Akt/PKB through a signaling complex involving TRAF6 and c-Src. *Mol Cell* 4: 1041–1049
111 Lee SE, Chung WJ, Kwak HB, Chung CH, Kwack K, Lee ZH, Kim HH (2001) Tumor necrosis factor-α supports the survival of osteoclasts through the activation of Akt and ERK. *J Biol Chem* 276: 49343–49349
112 Lee SE, Woo KM, Kim SY, Kim H-M, Kwack K, Lee ZH, Kim H-H (2002) The phosphatidylinositol 3-kinase, p38, and extracellular signal-regulated kinase pathways are involved in osteoclast differentiation. *Bone* 30: 70–77
113 Yamamoto A, Miyazaki T, Kadono Y, Takayanagi H, Miura T, Nishina H, Katada T, Wakabayashi K, Oda H, Nakamura K et al (2002) Possible involvement of IκB kinase 2 and MKK7 in osteoclastogenesis induced by receptor activator of nuclear factor κB ligand. *J Bone Min Res* 17: 612–621
114 Li X, Udagawa N, Itoh K, Suda K, Murase Y, Nishihara T, Suda T, Takahashi N (2002) p38 MAPK-mediated signals are required for inducing osteoclast differentiation but not for osteoclast function. *Endocrinology* 143: 3105–3113
115 Karin M, Liu Z, Zandi E (1997) AP-1 function and regulation. *Curr Opin Cell Biol* 9: 240–246
116 Lee ZH, Kwack K, Kim KK, Lee SH, Kim HH (2000) Activation of c-Jun N-terminal kinase and activator protein-1 by receptor activator of nuclear factor κB. *Mol Pharmacol* 58: 1536–1545
117 Zhou B, Cron RQ, Wu BR, Genin A, Wang ZL, Liu S, Robson P, Baldwin HS (2002) Regulation of the murine Nfatc1 gene by NFATc2. *J Biol Chem* 277: 10704–10711
118 Battaglino R, Kim D, Fu J, Vaage B, Fu X-Y, Stashenko P (2002) c-myc is required for osteoclast differentiation. *J Bone Min Res* 17: 763–773
119 Xing L, Bushnell TP, Carlson L, Tai Z, Tondravi M, Siebenlist U, Young F, Boyce BF (2002) NF-κB p50 and p52 expression is not required for RANK-expressing osteoclast progenitor formation but is essential for RANK- and cytokine-mediated osteoclastogenesis. *J Bone Min Res* 17: 1200–1210
120 Hayashi T, Kaneda T, Toyama Y, Kumegawa M, Hakeda Y (2002) Regulation of receptor activator of NF-κB ligand-induced osteoclastogenesis by endogenous interferon-β (INF-β) and suppressors of cytokine signaling (SOCS). The possible counteracting role of SOCSs in IFN-β-inhibited osteoclast formation. *J Biol Chem* 277: 27880–27886
121 Takayanagi H, Kim S, Matsuo K, Suzuki H, Suzuki T, Sato K, Yokochi T, Oda H, Nakamura K, Ida N et al (2002) RANKL maintains bone homeostasis through c-Fos-dependent induction of interferon-β. *Nature* 416: 744–749
122 Shin JN, Kim I, Lee JS, Koh GY, Lee ZH, Kim HH (2002) A novel zinc finger protein that inhibits osteoclastogenesis and the function of tumor necrosis factor receptor-associated factor 6. *J Biol Chem* 277: 8346–8353
123 Takayanagi H, Ogasawara K, Hida S, Chiba T, Murata S, Sato K, Takaoka A, Yokochi

T, Oda H, Tanaka K et al (2000) T cell-mediated regulation of osteoclastogenesis by signalling cross-talk between RANKL and IFN-γ. *Nature* 408: 600–605
124 Zhang YH, Heulsmann A, Tondravi MM, Mukherjee A, Abu-Amer Y (2001) Tumor necrosis factor-α (TNF) stimulates RANKL-induced osteoclastogenesis *via* coupling of TNF type 1 receptor and RANK signaling pathways. *J Biol Chem* 276: 563–568
125 Lean JM, Matsuo K, Fox SW, Fuller K, Gibson FM, Draycott G, Wani MR, Bayley KE, Wong BR, Choi Y et al (2000) Osteoclast lineage commitment of bone marrow precursors through expression of membrane-bound TRANCE. *Bone* 27: 29–40
126 Shalhoub V, Faust J, Boyle WJ, Dunstan CR, Kelley M, Kaufman S, Scully S, Van G, Lacey DL (1999) Osteoprotegerin and osteoprotegerin ligand effects on osteoclast formation from human peripheral blood mononuclear cell precursors. *J Cell Biochem* 72: 251–261
127 Campagnuolo G, Bolon B, Feige U (2002) Kinetics of bone protection by recombinant osteoprotegerin therapy in Lewis rats with adjuvant arthritis. *Arthritis Rheum* 46: 1926–1936
128 Romas E, Sims NA, Hards D, Lindsay M, Quinn JWM, Ryan PFJ, Dunstan CR, Martin TJ, Gillespie MT (2002) Osteoprotegerin reduces osteoclast numbers and prevents bone erosion in collagen-induced arthritis. *Am J Pathol* 161: 1419–1427
129 Jimi E, Akiyama S, Tsurukai T, Okahashi N, Kobayashi K, Udagawa N, T. N, Takahashi N, Suda T (1999) Osteoclast differentiation factor acts as a multifunctional regulator in murine osteoclast differentiation and function. *J Immunol* 163: 434–442
130 Hakeda Y, Kobayashi Y, Yamaguchi K, Yasuda H, Tsuda E, Higashio K, Miyata T, Kumegawa M (1998) Osteoclastogenesis inhibitory factor (OCIF) directly inhibits bone-resorbing activity of isolated mature osteoclasts. *Biochem Biophys Res Comm* 251: 796–801
131 O'Brien EA, Williams JHH, Marshall MJ (2000) Osteoprotegerin ligand regulates osteoclast adherence to the bone surface in mouse calvaria. *Biochem Biophys Res Comm* 274: 281–290
132 O'Brien EA, Williams JHH, Marshall MJ (2001) Osteoprotegerin is produced when prostaglandin synthesis is inhibited causing osteoclasts to detach from the surface of mouse parietal bone and attach to the endocranial membrane. *Bone* 28: 208–214
133 Wittrant Y, Couillaud S, Theoleyre S, Dunstan C, Heymann D, Rédini F (2002) Osteoprotegerin differentially regulates protease expression in osteoclast cultures. *Biochem Biophys Res Comm* 293: 38–44
134 Akatsu T, Murakami T, Nishikawa M, Ono K, Shinomiya N, Tsuda E, Mochizuki S, Yamaguchi K, Kinosaki M, Higashio K et al (1998) Osteoclastogenesis inhibitory factor suppresses osteoclast survival by interfering in the interaction of stromal cells with osteoclasts. *Biochem Biophys Res Comm* 250: 229–234
135 Lacey DL, Tan HL, Lu J, Kaufman S, Van G, Qiu WR, Rattan A, Scully S, Fletcher F, Juan T et al (2000) Osteoprotegerin ligand modulates murine osteoclast survival *in vitro* and *in vivo*. *Am J Pathol* 157: 435–448
136 Bolon B, Campagnuolo G, Feige U (2002) Duration of bone protection afforded by a

single injection of recombinant osteoprotegerin (OPG) in male Lewis rats with adjuvant arthritis. *Cell Mol Life Sci* 59: 1569–1576

137 Kong YY, Yoshida H, Sarosi I, Tan HL, Timms E, Capparelli C, Morony S, Oliveira dos Santos AJ, Van G, Itie A et al (1999) OPGL is a key regulator of osteoclastogenesis, lymphocyte development and lymph node organogenesis. *Nature* 397: 315–323

138 Kim N, Odgren PR, Kim DK, Marks SCJ, Choi Y (2000) Diverse roles of the tumor necrosis factor family member TRANCE in skeletal physiology revealed by TRANCE deficiency and partial rescue by a lymphocyte-expressed TRANCE transgene. *Proc Natl Acad Sci USA* 97: 10905–10910

139 Min H, Morony S, Sarosi I, Dunstan CR, Capparelli C, Scully S, Van G, Kaufman S, Kostenuik PJ, Lacey DL et al (2000) Osteoprotegerin reverses osteoporosis by inhibiting endosteal osteoclasts and prevents vascular calcification by blocking a process resembling osteoclastogenesis. *J Exp Med* 192: 463–474

140 Bucay N, Sarosi I, Dunstan CR, Morony S, Tarpley J, Capparelli C, Scully S, Tan HL, Xu W, Lacey DL et al (1998) Osteoprotegerin-deficient mice develop early onset osteoporosis and arterial calcification. *Genes Dev* 12: 1260–1268

141 Lomaga MA, Yeh WC, Sarosi I, Duncan GS, Furlonger C, Ho A, Morony S, Capparelli C, Van G, Kaufman S et al (1999) TRAF6 deficiency results in osteopetrosis and defective interleukin-1, CD40, and LPS signaling. *Genes Dev* 13: 1015–1024

142 Naito A, Azuma S, Tanaka S, Miyazaki T, Takaki S, Takatsu K, Nakao K, Nakamura K, Katsuki M, Yamamoto T et al (1999) Severe osteopetrosis, defective interleukin-1 signaling and lymph node organogenesis in TRAF6-deficient mice. *Genes to Cells* 4: 353–362

143 Boyce BF, Hughes DE, Wright KR, Xing LP, Dai AH (1999) Recent advances in bone biology provide insight into the pathogenesis of bone diseases. *Lab Invest* 79: 83–94

144 Wang ZQ, Ovitt C, Grigoriadis A, Mohle-Ateinlein U, Ruther U, Wagner EF (1992) Bone and haematopoeitic defects in mice lacking c-fos. *Nature* 360: 741–745

145 Franzoso G, Carlson L, Xing L, Poljak L, Shores EW, Brown KD, Leonardi A, Tran T, Boyce BF, Siebenlist U (1997) Requirement for NF-κB in osteoclast and B-cell development. *Genes Dev* 11: 3482–3496

146 Matsuo K, Owens JM, Tonko M, Elliot C, Chambers TJ, Wagner EF (2000) Fosl1 is a transcriptional target of c-Fos during osteoclast differentiation. *Nat Genet* 24: 184–187

147 Janssens K, Van Hul W (2002) Molecular genetics of too much bone. *Hum Mol Genet* 11: 2385–2393

148 Hughes AE, Ralston SH, Marken J, Bell C, MacPherson H, Wallace RGH, van Hul W, Whyte MP, Nakatsuka K, Hovy L et al (2000) Mutations in TNFRSF11A, affecting the signal peptide of RANK, cause familial expansile osteolysis. *Nat Genet* 24: 45–48

149 Whyte MP, Hughes AE (2002) Expansile skeletal hyperphosphatasia is caused by a 15-base pair tandem duplication in TNFRSF11A encoding RANK and is allelic to familial expansile osteolysis. *J Bone Min Res* 17: 26–29

150 Laurin N, Brown JP, Morissette J, Raymond V (2002) Recurrent mutation of the gene

encoding sequestosome 1 (SQSTM1 / p62) in Paget disease of bone. *Am J Hum Genet* 70: 1582–1588
151 Sanz L, Sanchez P, Lallena MJ, Diaz M, Moscat J (1999) The interaction of p62 with RIP links the atypical PKCs to NF-κB activation. *EMBO J* 18: 3044–3053
152 Sparks AB, Peterson SN, Bell C, Loftus BJ, Hocking L, Cahill DP, Frassica FJ, Streeten EA, Levine MA, Fraser CM et al (2001) Mutation screening of the TNFRSF11A gene encoding receptor activator of NFκB (RANK) in familial and sporadic Paget's disease of bone and osteosarcoma. *Calcif Tissue Int* 68: 151–155
153 Wuyts W, van Wesenbeeck L, Morales-Piga A, Ralston S, Hocking L, Vanhoenacker F, Westhovens R, Verbruggen L, Anderson D, Hughes A et al (2001) Evaluation of the role of RANK and OPG genes in Paget's disease of bone. *Bone* 28: 104–107
154 Cundy T, Hegde M, Naot D, Chong B, King A, Wallace R, Love DR, Seidel J, Fawkner M, Banovic T et al (2002) A mutation in the gene TNFRSF11B encoding osteoprotegerin causes an idiopathic hyperphosphatasia phenotype. *Hum Mol Genet* 11: 2119–2127
155 Shimizu S, Shiozawa S, Shiozawa K, Imura S, Fujita T (1985) Quantitative histologic studies on the pathogenesis of periarticular osteoporosis in rheumatoid arthritis. *Arthritis Rheum* 28: 25–31
156 Gravallese EM, Harada Y, Wang JT, Gorn AH, Thornhill TS, Goldring SR (1998) Identification of cell types responsible for bone resorption in rheumatoid arthritis and juvenile rheumatoid arthritis. *Am J Pathol* 152: 943–951
157 Gravallese EM, Manning C, Tsay A, Naito A, Pan C, Amento E, Goldring SR (2000) Synovial tissue in rheumatoid arthritis is a source of osteoclast differentiation factor. *Arthritis Rheum* 43: 250–258
158 Toraldo G, Roggia C, Qian WP, Pacifici R, Weitzmann MN (2003) IL-7 induces bone loss *in vivo* by induction of receptor activator of nuclear factor κB and tumor necrosis factor-α from T cells. *Proc Natl Acad Sci USA* 100: 125–130
159 Romas E, Bakharevski O, Hards DK, Kartsogiannis V, Quinn JM, Ryan PF, Martin TJ, Gillespie MT (2000) Expression of osteoclast differentiation factor at sites of bone erosion in collagen-induced arthritis. *Arthritis Rheum* 43: 821–826
160 Lubberts E, Oppers-Walgreen B, Pettit AR, van den Bersselaar L, Joosten LAB, Goldring SR, Gravallese EM, van den Berg WB (2002) Increase in expression of receptor activator of nuclear factor κB at sites of bone erosion correlates with progression of inflammation in evolving collagen-induced arthritis. *Arthritis Rheum* 46: 3055–3064
161 Mori H, Kitazawa R, Mizuki S, Nose M, Maeda S, Kitazawa S (2002) RANK ligand, RANK, and OPG expression in type II collagen-induced arthritis mouse. *Histochem Cell* Biol 117: 283–292
162 Kotake S, Udagawa N, Hakoda M, Mogi M, Yano K, Tsuda E, Takahashi K, Furuya T, Ishiyama S, Kim K-J et al (2001) Activated human T cells directly induce osteoclastogenesis from human monocytes. *Arthritis Rheum* 44: 1003–1012
163 Choi Y, Woo KM, Ko S-H, Lee YJ, Park S-J, Kim H-M, Kwon BS (2001) Osteoclasto-

genesis is enhanced by activated B cells but suppressed by activated CD8+ T cells. Eur *J Immunol* 31: 2179–2188

164 Manabe N, Kawaguchi H, Chikuda H, Miyaura C, Inada M, Nagai R, Nabeshima Y, Nakamura K, Sinclair A, Scheuermann RH et al (2001) Connection between B lymphocyte and osteoclast differentiation pathways. *J Immunol* 167: 2625–2631

165 Tsurukai T, Udagawa N, Matsuzaki K, Takahashi N, Suda T (2000) Roles of macrophage-colony stimulating factor and osteoclast differentiation factor in osteoclastogenesis. *J Bone Min Res* 18: 177–184

166 Feuerherm AJ, Børset M, Seidel C, Sundan A, Leistad L, Østensen M, Faxvaag A (2001) Elevated levels of osteoprotegerin (OPG) and hepatocyte growth factor (HGF) in rheumatoid arthritis. *Scand J Rheumatol* 30: 229–234

167 Pettit AR, Ji H, von Stechow D, Müller R, Goldring SR, Choi Y, Benoist C, Gravallese EM (2001) TRANCE / RANKL knockout mice are protected from bone erosion in a serum transfer model of arthritis. *Am J Pathol* 159: 1689–1699

168 Toritsuka Y, Nakamura N, Lee SB, Hashimoto J, Yasui N, Shino K, Ochi T (1997) Osteoclastogenesis in iliac bone marrow of patients with rheumatoid arthritis. *J Rheumatol* 24: 1690–1696

169 Romas E, Gillespie MT, Martin TJ (2002) Involvement of receptor activator of NFκB ligand and tumor necrosis factor-α in bone destruction in rheumatoid arthritis. *Bone* 30: 340–346

170 Gough A, Sambrook P, Devlin J, Huisson A, Njeh C, Robbins P et al (1998) Osteoclastic activation is the principal mechanism leading to secondary osteoporosis in rheumatoid arthritis. *J Rheumatol* 25: 1282–1289

171 Hirayama T, Danks L, Sabokbar A, Athanasou NA (2002) Osteoclast formation and activity in the pathogenesis of osteoporosis in rheumatoid arthritis. *Rheumatology* 41: 1232–1239

172 Dougall WC, Glaccum M, Charrier K, Rohrbach K, Brasel K, De Smedt T, Daro E, Smith J, Tometsko ME, Maliszewski CR et al (1999) RANK is essential for osteoclast and lymph node development. *Genes Dev* 13: 2412–2424

173 Matsumoto M, al. e (1996) Role of lymphotoxin and the type I TNF receptor in the formation of germinal centers. *Science* 271: 1289–1291

174 Josien R, Wong BR, Li HL, Steinman RM, Choi Y (1999) TRANCE, a TNF family member, is differentially expressed on T cell subsets and induces cytokine production in dendritic cells. *J Immunol* 162: 2562–2568

175 Yamashita T, Okada S, Higashio K, Nabeshima Y-I, Noda M (2002) Double mutations in klotho and osteoprotegerin gene loci rescued osteopetrotic phenotype. *Endocrinology* 143: 4711–4717

176 Yun TJ, Tallquist MD, Aicher A, Rafferty KL, Marshall AJ, Moon JJ, Ewings MK, Mohaupt M, Herring SW, Clark EA (2001) Osteoprotegerin, a crucial regulator of bone metabolism, also regulates B cell development and function. *J Immunol* 166: 1482–1491

177 Banchereau J, Steinman RM (1998) Dendritic cells and the control of immunity. *Nature* 392: 245–252

178 Josien R, Li HL, Ingulli E, Sarma S, Wong BR, Vologodskaia M, Steinman RM, Choi Y (2000) TRANCE, a tumor necrosis factor family member, enhances the longevity and adjuvant properties of dendritic cells *in vivo*. *J Exp Med* 191: 495–502
179 Williamson E, Bilsborough JM, Viney JL (2002) Regulation of mucosal dendritic cell function by receptor activator of NF-κB (RANK) / RANK ligand interactions: impact on tolerance induction. *J Immunol* 169: 3606–3612
180 Ueland T, Bollerslev J, Godang K, Müller F, Frøland SS, Aukrust P (2001) Increased serum osteoprotegerin in disorders characterized by persistent immune activation or glucocorticoid excess: possible role in bone homeostasis. *Eur J Endocrinol* 145: 685–690
181 Martin TJ, Gillespie MT (2001) Receptor activator of nuclear factor κB ligand (RANKL): another link between breast and bone. *Tr Endocrinol Metab* 12: 2–4
182 Fata JE, Kong YY, Li J, Sasaki T, Irie-Sasaki J, Moorehead RA, Elliott R, Scully S, Voura EB, Lacey DL et al (2000) The osteoclast differentiation factor osteoprotegerin-ligand is essential for mammary gland development. *Cell* 103: 41–50
183 Malyankar UM, Scatena M, Suchland KL, Yun TJ, Clark EA, Giachelli CM (2000) Osteoprotegerin is an $\alpha_v\beta_3$-induced, NF-κB-dependent survival factor for endothelial cells. *J Biol Chem* 275: 20959–20962
184 Zhang J, Fu M, Myles D, Zhu X, Du J, Cao X, Chen YE (2002) PDGF induces osteoprotegerin expression in vascular smooth muscle cells by multiple signal pathways. *FEBS Lett* 521: 180–184
185 Fu MG, Zhang JF, Lin YM, Zhu XJ, Willson TM, Chen YQE (2002) Activation of peroxisome proliferator-activated receptor γ inhibits osteoprotegerin gene expression in human aortic smooth muscle cefls. *Biochem Biophys Res Comm* 294: 597–601
186 Price PA, June HH, Buckley JR, Williamson MK (2001) Osteoprotegerin inhibits artery calcification induced by warfarin and vitamin D. *Arterioscler Thromb Vasc Biol* 21: 1610–1616
187 Parhami F, Demer LL (1997) Arterial calcification in face of osteoporosis in aging: can we blame oxidized lipids? *Curr Opin Lipidol* 8: 312–314
188 Hak AE, Pols HA, van Hemert AM, Hofman A, Witteman JC (2000) Progression of aortic calcification is associated with metacarpal bone loss during menopause: a population-based longitudinal study. *Arterioscler Thromb Vasc Biol* 20: 1926–1931
189 Jono S, Ikari Y, Shioi A, Mori K, Miki T, Hara K, Nishizawa Y (2002) Serum osteoprotegerin levels are associated with the presence and severity of coronary artery disease. *Circulation* 106: 1192–1194
190 Dhore CR, Cleutjens JPM, Lutgens E, Cleutjens KBJM, Geusens PPM, Kitslaar PJEHM, Tordoir JHM, Spronk HMH, Vermeer C, Daemen MJAP (2001) Differential expression of bone matrix regulatory proteins in human atherosclerotic plaques. *Arterioclero Thromb Vasc Biol* 21: 1998–2003
191 Brändström H, Stiger F, Lind L, Kahan T, Melhus H, Kindmark A (2002) A single nucleotide polymorphism in the promoter region of the human gene for osteoprotegerin is related to vascular morphology and function. *Biochem Biophys Res Comm* 293: 13–17

192 Wong M, Toh L, Wilson A, Rowley K, Karschimkus C, Prior D, Romas E, Clemens L, Dragicevic G, Harianto H et al (2003) Reduced arterial elasticity in rheumatoid arthritis and the relationship to vascular disease risk factors and inflammation. *Arthritis Rheum* 48: 81–89

193 Martin SW, Sommers JS, Watson AM, Young JD, Dunstan CR (1999) The pharmacokinetics and pharmacodynamics of OPG following single dose administration in cynomolgus monkeys. *J Bone Miner Res* 14 (Suppl 1): S399

194 Feldmann M, Brennan FM, Maini RN (1996) Rheumatoid arthritis. Cell 85: 307–310

195 Kobayashi K, Takahashi N, Jimi E, Udagawa N, Takami M, Kotake S, Nakagawa N, Kinosaki M, Yamaguchi K, Shima N et al (2000) Tumor necrosis factor a stimulates osteoclast differentiation by a mechanism independent of the ODF / RANKL-RANK interaction. *J Exp Med* 191: 275–286

196 Suda T, Kobayashi K, Jimi E, Udagawa N, Takahashi N (2001) The molecular basis of osteoclast differentiation and activation. *Novartis Found Symp* 232: 235–247 (discussion 247–250)

197 Horwood NJ, Kartsogiannis V, Quinn JM, Romas E, Martin TJ, Gillespie MT (1999) Activated T lymphocytes support osteoclast formation *in vitro*. *Biochem Biophys Res Comm* 265: 144–150

198 Xu LX, Kukita T, Nakano Y, Yu H, Hotokebuchi T, Kuratani T, Iijima T, Koga T (1996) Osteoclasts in normal and adjuvant arthritis bone tissues express the mRNA for both type I and II interleukin-1 receptors. *Lab Invest* 75: 677–687

199 Collin-Osdoby P, Rothe L, Anderson F, Nelson M, Maloney W, Osdoby P (2001) Receptor activator of NF-κB and osteoprotegerin expression by human microvascular endothelial cells, regulation by inflammatory cytokines, and role in human osteoclastogenesis. *J Biol Chem* 276: 20659–20672

200 Feige U, Hu Y-L, Gasser J, Campagnuolo G, Munyakazi L, Bolon B (2000) Anti-interleukin-1 and anti-tumor necrosis factor-α synergistically inhibit adjuvant arthritis in Lewis rats. *Cell Mol Life Sci* 57: 1457–1470

201 Kimble RB, Srivastava S, Ross FP, Matayoshi A, Pacifici R (1996) Estrogen deficiency increases the ability of stromal cells to support murine osteoclastogenesis *via* an interleukin-1 and tumor necrosis factor-mediated stimulation of macrophage colony-stimulating factor production. *J Biol Chem* 271: 28890–28897

202 Fox SW, Fuller K, Chambers TJ (2000) Activation of osteoclasts by interleukin-1: divergent responsiveness in osteoclasts formed *in vivo* and *in vitro*. *J Cell Physiol* 184: 334–340

203 Fuller K, Murphy C, Kirstein B, Fox SW, Chambers TJ (2002) TNF-α potently activates osteoclasts, through a direct action independent of and strongly synergistic with RANKL. *Endocrinology* 143: 1108–1118

204 Kudo O, Fujikawa Y, Itonaga I, Sabokbar A, Torisu T, Athanasou NA (2002) Proinflammatory cytokine (TNFα / IL-1α) induction of human osteoclast formation. *J Pathol* 198: 220–227

205 Zou W, Hakim I, Tschoep K, Endres S, Bar-Shavit Z (2001) Tumor necrosis factor-α

mediates RANK ligand stimulation of osteoclast differentiation by an autocrine mechanism. *J Cell Biochem* 83: 70–83

206 Lam J, Takeshita S, Barker JE, Kanagawa O, Ross F, Teitelbaum SL (2000) TNF-α induces osteoclastogenesis by direct stimulation of macrophages exposed to permissive levels of RANK ligand. *J Clin Invest* 106: 1481–1488

207 Bekker PJ, Holloway D, Nakanishi A, Arrighi HM, Leese PT, Dunstan CR (2001) The effect of a single dose of osteoprotegerin in postmenopausal women. *J Bone Miner Res* 16: 348–360

208 Body JJ, Greipp P, Coleman RE, Facon T, Geurs F, Fermand JP, Harousseau JL, Lipton A, Mariette X, Williams CD et al (2003) A phase I study of AMGN-0007, a recombinant osteoprotegerin construct, in patients with multiple myeloma or breast carcinoma related bone metastases. *Cancer* 97 (3 Suppl S): 887–892

209 Redlich K, Hayer S, Maier A, Dunstan CR, Tohidast-Akrad M, Lang S, Türk B, Pietschmann P, Woloszczuk W, Haralambous S et al (2002) Tumor necrosis factor α-mediated joint destruction is inhibited by targeting osteoclasts with osteoprotegerin. *Arthritis Rheum* 46: 785–792

210 Deyama Y, Kikuiri T, Ohnishi G-I, Feng Y-G, Takeyama S, Hatta M, Yoshimura Y, Suzuki K (2002) Histamine stimulates production of osteoclast differentiation factor / receptor activator of nuclear factor-κB ligand by osteoblasts. *Biochem Biophys Res Comm* 298: 240–246

211 O'Brien CA, Gubrij I, Lin SC, Saylors RL, Manolagas SC (1999) STAT3 activation in stromal / osteoblastic cells is required for induction of the receptor activator of NF-κB ligand and stimulation of osteoclastogenesis by gp130-utilizing cytokines or interleukin-1 but not 1,25-dihydroxyvitamin D_3 or parathyroid hormone. *J Biol Chem* 274: 19301–19308

212 Weitzmann MN, Cenci S, Rifas L, Brown C, Pacifici R (2000) Interleukin-7 stimulates osteoclast formation by upregulating the T cell production of soluble osteoclastogenic cytokines. *Blood* 96: 1873–1878

213 Makiishi-Shimobayashi C, Tsujimura T, Iwasaki T, Yamada N, Sugihara A, Okamura H, Hayashi S, Terada N (2001) Interleukin-18 upregulates osteoprotegerin expression in stromal/osteoblastic cells. *Biochem Biophys Res Comm* 281: 361–366

214 Quinn JM, Itoh K, Udagawa N, Häusler K, Yasuda H, Shima N, Mizuno A, Higashio K, Takahashi N, Suda T et al (2001) Transforming growth factor-β affects osteoclast differentiation *via* direct and indirect actions. *J Bone Min Res* 16: 1787–1794

215 Thirunavukkarasu K, Miles RR, Halladay DL, Yang X, Galvin RJS, Chandrasekhar S, Martin TJ, Onyia JE (2001) Stimulation of osteoprotegerin (OPG) gene expression by transforming growth factor-β (TGF-β). *J Biol Chem* 276: 36241–36250

216 Hofbauer LC, Hicok KC, Chen D, Khosla S (2002) Regulation of osteoprotegerin production by androgens and anti-androgens in human osteoblastic lineage cells. *Eur J Endocrinol* 147: 269–273

217 Thomas GP, Baker SUK, Eisman JA, Gardiner EM (2001) Changing RANKL / OPG

mRNA expression in differentiating murine primary osteoblasts. *J Endocrinol* 170: 451–460
218 Viereck V, Emons G, Lauck V, Frosch K-H, Blaschke S, Gründker C, Hofbauer LC (2002) Bisphosphonates pamidronate and zoledronic acid stimulate osteoprotegerin production by primary human osteoblasts. *Biochem Biophys Res Comm* 291: 680–686
219 Takami M, Takahashi N, Udagawa N, Miyaura C, Suda K, Woo J, Martin T, Nagai K, Suda T (2000) Intracellular calcium and protein kinase C mediate expression of receptor activator of nuclear factor-κB ligand and osteoprotegerin in osteoblasts. *Endocrinology* 141: 4711–4719
220 Walton KJ, Duncan JM, Deschamps P, Shaughnessy SG (2002) Heparin acts synergistically with interleukin-11 to induce STAT3 activation and *in vitro* osteoclast formation. *Blood* 100: 2530–2536
221 Yang X, Halladay D, Onyia JE, Martin TJ, Chandrasekhar S (2002) Protein kinase C is a mediator of the synthesis and secretion of osteoprotegerin in osteoblast-like cells. *Biochem Biophys Res Comm* 290: 42–46
222 Yoshida H, Hayashi S, Kunisada T, Ogawa M, Nishikawa S, Okamura H, Sudo T, Shultz L, Nishikawa S (1990) The murine mutation osteopetrosis is in the coding region of the macrophage colony stimulating factor gene. *Nature* 345: 442–444
223 Okada S, Wang ZQ, Grigoriadis AE, Wagner EF, Vonruden T (1994) Mice lacking c-Fos have normal hematopoietic stem cells but exhibit altered B cell differentiation due to an impaired bone marrow environment. *Mol Cell Biol* 14: 382–390
224 Takeshita S, Namba N, Zhao JJ, Jiang Y, Genant HK, Silva MJ, Brodt MD, Helgason CD, Kalesnikoff J, Rauh MJ et al (2002) SHIP-deficient mice are severely osteoporotic due to increased numbers of hyper-resorptive osteoclasts. *Nature Med* 8: 943–949
225 Kostenuik PJ, Capparelli C, Morony S, Adamu S, Shimamoto G, Shen V, Lacey DL, Dunstan CR (2001) OPG and PTH (1-34) have additive effects on bone density and mechanical strength in osteopenic ovariectomized rats. *Endocrinology* 142: 4295–4304
226 Bolon B, Carter C, Daris M, Morony S, Capparelli C, Hsieh A, Mao M, Kostenuik P, Dunstan CR, Lacey DL et al (2001) Adenoviral delivery of osteoprotegerin ameliorates bone resorption in a mouse ovariectomy model of osteoporosis. *Mol Ther* 3: 197–205

Cytokine regulation of bone destruction

Erik Lubberts and Wim B. van den Berg

Rheumatology Research and Advanced Therapeutics, University Medical Center Nijmegen, 189 Geert Grooteplein 26–28, 6500 HB Nijmegen, The Netherlands

Introduction

Bone is a dynamic tissue, composed of cells, collagenous matrix and inorganic elements. Per year, it has been estimated that 10% of the total bone mass in humans is being remodeled [1]. The growth, development and maintenance of bone is a highly regulated process [2]. Morphogenesis and remodeling of bone involve coordinate regulation of bone-forming cells (osteoblasts) and bone-resorbing cells (osteoclasts) [3, 4]. Osteoblasts and osteoclasts arise from distinct cell lineages by physiologically controlled processes that involve cytokine, growth factors and hormones [4, 5]. Osteoblasts arise from mesenchymal stem cells, whereas osteoclasts differentiate from hematopoietic monocyte/macrophage precursors. Osteoclasts are rarely seen under normal conditions. Imbalance between osteoclast and osteoblast activities can arise from a wide variety of hormonal changes or perturbations of inflammatory and growth factors, resulting in skeletal abnormalities characterized by decreased (osteoporosis) or increased (osteopetrosis) bone mass. Increased osteoclast activity is also seen in many pathological disorders, including Paget's disease, lytic bone metastases, postmenopausal osteoporosis, or rheumatoid arthritis (RA), leading to net loss of bone. The discovery of the RANKL/RANK/OPG pathway was a major breakthrough in the understanding of the molecular mechanism of bone erosion. RANKL, RANK and OPG are three key molecules that regulate osteoclast recruitment and function (Fig. 1).

RANKL/RANK/OPG system

RANKL was cloned simultaneously by four independent groups [6–9]. The RANKL gene encodes a TNF superfamily molecule of 316 amino acids (38 kDa), and three RANKL subunits assemble to form the functional trimeric molecule. This trimeric RANKL is initially made as a membrane anchored molecule. Metalloprotease disintegrin TNFα convertase (TACE) and TACE-like proteases are able to release the cell surface RANKL as a soluble homotrimeric molecule by proteolytic cleavage

Cytokines and Joint Injury, edited by Wim B. van den Berg and Pierre Miossec

[10]. RANKL is an absolute requirement for osteoclast development, and it plays an important role in T cell differentiation as well. RANKL deficient (RANKL$^{-/-}$) mice exhibit typical osteopetrosis with total occlusion of bone marrow space within endosteal bone. RANKL$^{-/-}$ mice lack osteoclasts but have normal osteoclast progenitors that can differentiate into functionally active osteoclasts when co-cultured with normal osteoblasts/stromal cells. In addition, RANKL$^{-/-}$ mice completely lack lymph nodes and have a defect in B and T cell maturation and thymocyte differentiation [11]. The receptor for RANKL is RANK, a member of the TNF receptor superfamily. RANK is expressed as a transmembrane heterotrimer on the surface of hematopoietic osteoclast progenitors, mature osteoclasts, chondrocytes and mammary gland epithelial cells [6, 12]. RANK failed to bind to other members of the TNF ligand family such as Fas ligand, CD27 ligand, CD30 ligand, CD40 ligand, TNFα or TRAIL [13]. RANK deficient (RANK$^{-/-}$) mice showed a similar osteopetrotic phenotype as the RANKL$^{-/-}$ and RANK$^{-/-}$ mice exhibit a deficiency of B cells [14, 15].

OPG is a soluble decoy receptor that binds RANKL, preventing RANKL from binding to its receptor RANK. OPG transgenic mice also showed an osteopetrotic phenotype, although there appears to be no defect in lymphoid tissue development [16]. In contrast, the OPG deficient (OPG$^{-/-}$) mice develop severe osteoporosis, likely due to unopposed osteoclast-inducing activity of endogenous RANKL [17].

Osteoclast differentiation and activation by osteoblasts

Osteoblasts/stromal cells are crucially involved in osteoclast development and in the activation of osteoclasts function. Cell-to-cell contact between cells of the osteoblast lineage and hematopoietic cells is necessary for inducing differentiation of osteoclasts [13]. Development of osteoclasts proceeds within the local microenvironment of bone [4, 18–20]. This process can be replicated *ex vivo* using the co-culture of mouse calvarial osteoblasts and spleen cells [5, 21–24]. Macrophage colony-stimulating factor (M-CSF, also known as CSF-1) is crucial for osteoclast formation [25]. Osteoclasts are derived from cells of the monocyte/macrophage lineage. However, the mechanism by which osteoclast progenitors enter and leave the circulation is not fully understood. Cytokines such as IL-6, IL-11, oncostatin M, and leukemia inhibitory factor (LIF) stimulate osteoclast formation in co-cultures of osteoblasts/stromal cells and hematopoietic cells in which osteoblasts/stromal cells are the target cells [13]. These cytokines induce RANKL as a membrane-associated factor in osteoblasts/stromal cells. Osteoclast progenitors of the monocyte/macrophage lineage recognize RANKL in osteoblasts/stromal cells through cell-to-cell interaction, and then differentiate into osteoclasts. M-CSF, produced by osteoblasts/stromal cells, is a prerequisite for both proliferation and differentiation of osteoclast progenitors [13].

Osteoblasts/stromal cells can now be replaced with RANKL and M-CSF in dealing with the whole life of osteoclasts [13]. Both RANKL and IL-1 lead to enhanced osteoclast function even in the absence of osteoblasts/stromal cells (Fig. 1). RANKL can be replaced with IL-1 to induce survival, fusion and activation of osteoclasts. However, IL-1 could not support differentiation of osteoclast precursors into prefusion osteoclasts (pOCs) even in the presence of M-CSF when osteoblasts/stromal cells were absent. This suggests that RANKL is involved in physiological bone resorption, whereas IL-1 is involved in pathological bone resorption such as RA [13, 26].

It has been shown that TNFα can induce osteoclast formation *in vitro via* a mechanism that is independent of the RANKL/RANK signaling pathway [26]. In this *in vitro* system, osteoclast progenitors not only express RANK and c-Fms (M-CSF receptor), but also TNF receptors type I (p55) and II (p75). Osteoclast formation induced by RANKL was completely inhibited by adding OPG to the cultures. However, osteoclastogenesis induced by TNF was not blocked by OPG, but was blocked after adding antibodies against TNFR1 and TNFRII. These TNF antibodies had no effect on the RANKL induced osteoclastogenesis [26, 27]. TRAF2-mediated signals seem to play an important role in osteoclast differentiation induced by TNFα [26]. However, TRAF5 [28] and TRAF6 [29] also participate in TNFα induced osteoclastogenesis. On the other hand, it has been shown that TNFα alone does not induce osteoclastogenesis [30]. TNFα induces osteoclastogenesis *in vitro* and *in vivo* by directly targeting macrophages within a stromal environment that expresses permissive levels of RANKL [30]. Interestingly, although TNFα may stimulate differentiation of osteoclasts it did not activate osteoclasts to resorb bone. In contrast, IL-1 does not induce differentiation of osteoclasts in macrophage cultures without osteoblasts/stromal cells. However, IL-1 does stimulate pit-forming activity of the osteoclasts formed [13]. The IL-1/IL-1 receptor signaling pathway involves TRAF6 but not TRAF2 [27, 31]. This indicates that various TRAFs regulate different phases of osteoclast biology with TRAF2 and TRAF5 involved in osteoclastogenesis while TRAF6 controls both formation and activation [31].

Involvement of the inflamed synovial infiltrate in osteoclast differentiation and activation

RA is a chronic inflammatory joint disease, and bone erosion is a major complication. In areas of pannus infiltration, erosion of calcified cartilage and subchondral bone is common, leading to characteristic marginal erosions seen radiographically in this disease. The area of contact between pannus and cartilage/bone represents an erosive front, and the role of osteoclasts in erosion of bone in RA and in animal models of arthritis has been documented [32–37]. It was demonstrated that osteoclast formation can occur in the absence of any bone or bone marrow-derived cells [38]. Furthermore, it has been shown that synovial macrophages are capable of dif-

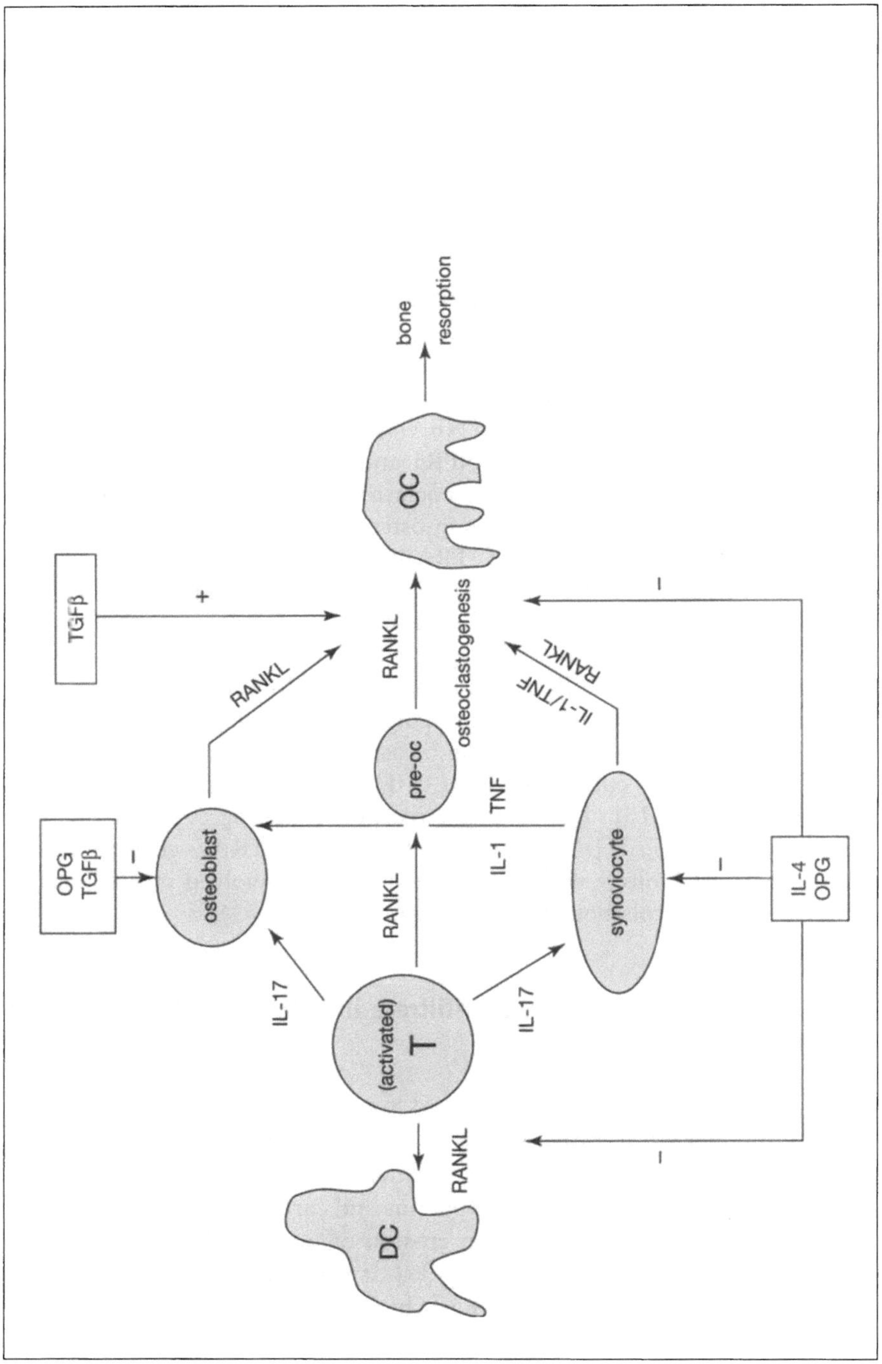

Figure 1
Schematic overview of different cell types and cytokines in the regulation of the RANKL/OPG balance and bone resorption

ferentiating into osteoclasts in the presence of rheumatoid synovial fibroblasts [38], revealing that rheumatoid synovial cells contain both osteoclast progenitors and stromal cells supporting their differentiation [38]. RANKL may play an important role in this differentiation pathway, as RANKL is considered a novel key regulator of osteoclastogenesis [11]. Many known cytokines that stimulate bone resorption act through the upregulation of this novel and essential factor. The synovial lining is normally a thin layer, one to two cells in thickness, which consist of type A and type B cells, showing macrophages and fibroblast-like properties. During (chronic) joint inflammation, considerable thickening of the lining is a characteristic event, and the subsynovial tissue becomes infiltrated by numerous macrophages and lymphocytes. Therefore, osteoclasts differentiated from monocyte/macrophage precursors in the synovial membrane are probably involved in bone destruction in arthritis. Interestingly, RANKL expression has been found in synovium of RA patients and in experimental arthritis models [11, 36, 39–42]. Recently, a marked increase in the number of cells expressing RANK (the receptor for RANKL) was found in the synovium of mice with collagen-induced arthritis (CIA) [42]. RANK expression in CIA correlated with the progression of synovial inflammation and clinical severity in evolving CIA (Fig. 2, A and B). Interestingly, RANK expression demonstrated a gradient pattern with increased numbers of RANK-positive cells within the synovial infiltrate in areas closer to periosteum and cortical bone (Fig. 2, C and D). In areas where RANK-positive cells were abundant, TRAP-positive, multinucleated osteoclast-like cells were also present at sites of focal bone erosion, suggesting differentiation of synovially derived RANK-positive osteoclast precursor cells into osteoclasts. In addition, TRAP-and cathepsin K-double-positive osteoclasts-like cells were detected on the synovial side of cortical bone at sites of early and advanced cortical bone erosion. Sites of RANK expression also correlated well with sites of RANKL expression, and there was a close correlation of the temporal expression of the receptor-ligand pair [42]. It would be of great interest to identify the lineage and origin of RANK-positive osteoclast precursor cells in the synovium. It has been shown, that RA synovial macrophages are capable of differentiating into osteoclastic bone-resorbing cells *in vitro* in the presence of RANKL and macrophage colony-stimulating factor [43]. The exact identity and stage of differentiation of RANK-expressing cells in the synovium is not yet known, but it is hypothesized that these are cells of the monocyte/macrophage lineage that can differentiate into osteoclasts in the presence of RANKL *in vivo*. The identities of the local factors within the bone microenvironment which are responsible for the upregulation of RANK and RANKL are also not known. In previous studies, using fragments of devitalized bone or polymeric material implanted subcutaneously in rats it was shown that bone fragments induced the differentiation of osteoclasts-like cells from precursors within the foreign-body granuloma [44, 45]. These findings suggest that the bone substrate itself can directly regulate the attachment and differentiation of osteoclast precursors into osteoclasts. Further support of synovial derived osteoclastogenesis and the involve-

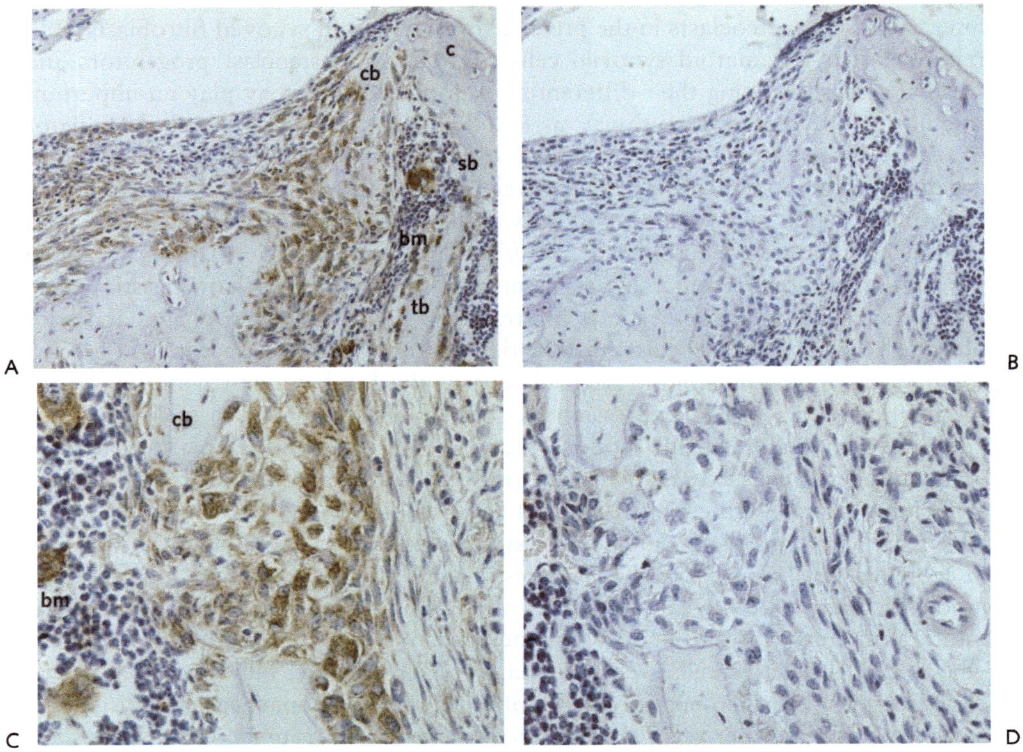

Figure 2
Expression of receptor activator of nuclear factor κB (RANK) at sites of focal bone erosion DBA-1 mice were immunized with type II collagen and knee joints were taken for histology at the time of established collagen arthritis.
A, C – RANK-positive cells at site of focal bone erosion.
B, D – No RANK-positive cells were detected in serial sections of the area using a control rabbit IgG antibody.
C, cartilage; sb, subchondral bone; cb, cortical bone; tb, trabecular bone; bm, bone marrow; s, synovitis. (Original magnification × 200 in A and B; × 400 in C and D). This figure is partly adapted from [42].

ment of these osteoclasts in bone erosion were reported for psoriatic arthritis [46]. For this form of arthritis it was shown that osteoclast precursors arise from TNFα-activated PBMCs that migrate to the inflamed synovium and subchondral bone, where they are exposed to unopposed RANKL and TNFα. This leads to osteoclastogenesis at the erosion front and in subchondral bone, resulting in a bidirectional assault on psoriatic bone [46].

Neutralizing RANKL (OPG treatment) in experimental models of arthritis

RANKL is a crucial cytokine in osteoclastogenesis and bone resorption and all factors that inhibit or enhance bone resorption, *via* osteoclasts, act *via* regulation of the RANKL/OPG balance (Fig. 1). Therefore, RANKL seems an interesting target for therapeutic intervention.

Several groups using different experimental arthritis models have shown the success of preventing bone erosion after neutralizing RANKL by the decoy receptor OPG. Studies in adjuvant arthritis (AdA) suggest that RANKL is involved in bone erosion and cartilage destruction, since systemic OPG treatment at the onset of disease prevents bone and cartilage destruction but not inflammation [47]. Furthermore, OPG reduces osteoclast numbers and prevents bone erosion in collagen-induced arthritis [48]. Although cartilage loss was also reduced by systemic OPG treatment, the effect was less striking than that on bone [48]. Uncoupling of cartilage and bone erosion has been shown using the serum transfer model in RANKL deficient mice [49]. Significant reduction of bone erosion and complete absence of multinucleated TRAP positive osteoclast-like cells were noted under these conditions. Interestingly, cartilage erosion was present in both the arthritic RANKL knockout mice and in arthritis control littermates [49]. These data support the requirement for RANKL in osteoclast-mediated bone erosion but also shows that cartilage destruction can take place independent of RANKL. Local OPG gene therapy in CIA suppressed the number of TRAP-positive and cathepsin K positive cells in the synovium and prevented bone erosion and cartilage destruction [50]. Moreover, OPG alone or in combination with biphosphonates is an effective therapeutic tool for the prevention of TNFα-mediated destruction of bone by reducing the number of bone-resorbing cells in the inflammatory tissue [51]. Furthermore, OPG protected against generalized bone loss in TNFtg mice [52]. These data strongly indicate that OPG may represent a potent tool to prevent bone loss in chronic inflammatory disorders. However, OPG does not control the inflammatory aspects and direct protective effect on cartilage is still not fully identified [53]. Therefore, further understanding of the cytokine network identifying how the crucial cytokines interact in the stimulation of RANKL and OPG is of great importance for the development of therapeutic strategies for destructive arthritis.

The influence of activated T cells on osteoclastogenesis and bone erosion

Both synovial T and non-T cell populations from RA patients expressed RANKL, but not OPG. Fibroblasts isolated from the same adult RA joints failed to express RANKL [1]. Interestingly, activated T cells can directly trigger osteoclastogenesis through RANKL [47]. Activated human T cells can directly induce osteoclastogenesis from human monocytes [54]. Furthermore, both soluble and membrane-bound

RANKL is produced by activated $CD4^+$ and $CD8^+$ T cells [8, 47]. Double immunostaining of synovial tissue of RA patients revealed that RANKL-positive cells were detected in a subset of $CD3^+$ and $CD4^+$ cells [54]. Anti-CD3 induced RANKL expression in T cell hybridoma cells and splenic T cells and RANKL expressed on T cells could effectively induce osteoclast formation from the whole population of murine splenocytes. Antigen receptor engagement induces RANKL expression which is dependent on TCR activation-induced Ca^{2+} mobilization since its expression can be blocked by cyclosporine A and TMB-8, a Ca^{2+} mobilization inhibitor [55]. Moreover, it has been shown that RANKL is regulated by calcineurin, ERK1/ERK-2 and PKC-regulated signaling pathways [8, 47].

$CTLA4^{-/-}$ deficient mice have spontaneously activated T cells and displayed severe osteoporosis compared with heterozygote littermates which could be blocked by OPG [47]. Interestingly, transfer of $CTLA4^{-/-}$ bone marrow cells into lymphocyte-deficient $RAG\text{-}1^{-/-}$ mice caused a significant decrease in total and trabecular bone mineral densities compared with control $CTLA4^{+}/\text{-}RAG1^{-/-}$ chimaeric mice. Similar loss in bone was observed after transfer of purified $CTLA4^{-/-}$ T cells into $RAG1^{-/-}$ or $RANKL^{-/-}$ mice, implying that activated T cells can directly cause bone loss *in vivo* [47]. Moreover, it has been shown that transgenic over-expression of RANKL in T cells restores osteoclastogenesis in a $RANKL^{-/-}$ background and partially restores normal bone marrow cavities [56]. Thus, systemic activation of T cells leads to bone loss, indicating that T cells through RANKL are crucial mediators of bone loss *in vivo* (Fig. 1). Moreover, RANKL expression on activated T cells may provide a bridge between bone biology *versus* immune system [57].

Cytokine regulation of the RANKL/OPG balance and bone erosion in inflammatory diseases

A large number of cytokines are able to regulate the RANKL/OPG balance, such as IL-1, TNFα, IL-6 family of cytokines [13], IL-4, IFNγ, IL-17, OPN [58], IL-7 [59, 60], IL-12, IL-18 [61], IL-15 [62]. A selection of these cytokines will be discussed in more detail with a focus on the contribution of how T cell cytokines regulate the RANKL/OPG balance.

IL-4

The anti-inflammatory cytokine IL-4 is a strong inhibitor of macrophage activity and their secreted cytokines (Fig. 1). In an *ex vivo* model of bone resorption using juxta-articular samples of bone obtained during joint surgery, IL-4 treatment prevents osteoclast formation [63]. Previously, we reported that local IL-4 gene therapy in collagen-induced arthritis prevents osteoclastogenesis and bone erosion

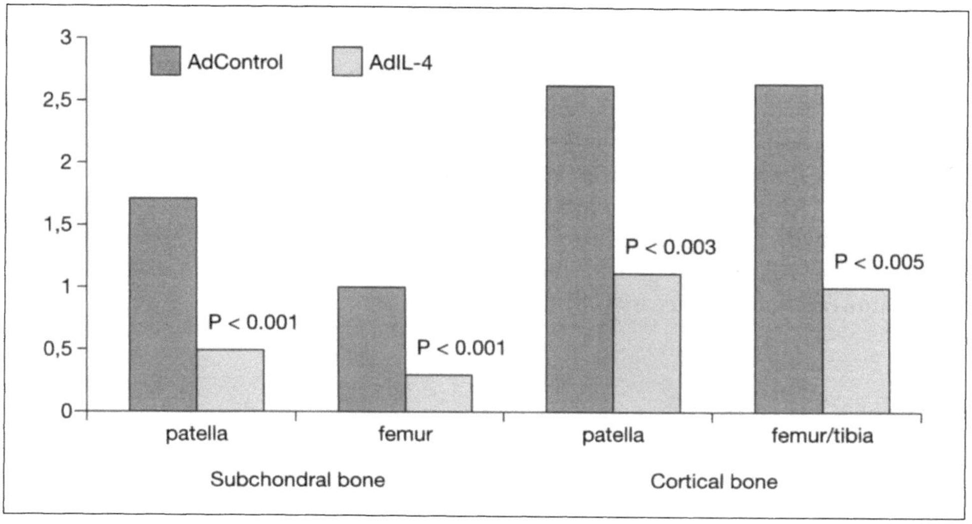

Figure 3
Histological analysis of bone erosion in CIA after local IL-4 gene therapy
DBA-1 mice were immunized with type II collagen and just before expected onset of CIA intra-articularly injected in the knee joint with 10^7 PFU of an adenoviral vector expressing IL-4 (AdIL-4) or control virus (AdControl). Seven days later, mice were sacrificed by cerval dislocation, and the knee joints were taken for histology. No differences in severity were found in the infiltrate and exudates between AdIL-4 and AdControl group (data not shown). Loss of subchondral and cortical bone in the patella and femur/tibia region was separately scored on a scale of 0–3. Results are the mean ± SD of four separate experiments with at least eight mice per group per experiment. $P < 0.001$; $P < 0.003$; $P < 0.005$ versus *control group, by Mann-Whitney rank sum test. This figure is partly adapted from [41].*

(Fig. 3) [41]. IL-4 over-expression in the knee joint of mice with collagen arthritis greatly prevented joint damage and bone erosion, although severe inflammation remained unchanged. Significantly decreased numbers of TRAP positive cells were noted after local IL-4 gene therapy, indicating inhibition of the formation of osteoclast-like cells. Furthermore, synovial T cell IL-17 mRNA expression and RANKL protein expression was markedly suppressed by local IL-4. Of interest, IL-4 suppressed type I collagen breakdown *in vitro* and enhanced synthesis of type I procollagen, suggesting that it promoted tissue repair [41]. Interestingly, local IL-4 expression in mice with collagen arthritis enhanced the mRNA expression of synovial OPG (Lubberts et al, unpublished observations). PPARγ1 seems to play an important role in the suppression of osteoclasts formation by IL-4 [64]. IL-4 inhibits osteoclast differentiation by antagonizing NF-κB activation in a STAT6-dependent manner and by blockade of the JNK, p38 and ERK mitogen-activated protein kinase pathway [65–67]. IL-4 reduces NF-κB nuclear translocation by inhibiting IκB phosphorylation, thus markedly inhibiting NF-κB DNA binding activity and blocking osteoclastogenesis entirely [65]. IL-4 fails to block osteoclastogenesis in $STAT6^{-/-}$ deficient mice but this blockade could be restored with addition of exogenous STAT6 [65]. Recently, two distinct actions of IL-4 to inhibit osteoclast formation were identified, including a direct action on myelomonocytic progenitors (from which osteoclasts derive) and an indirect action through T cells that may involve novel anti-osteoclastic factors [68].

IFNγ

In autoimmune arthritis, enhanced osteoclastic bone resorption causes severe bone damage leading to progressive joint destruction in which T cell expression of RANKL may have a critical function [47]. In fact, activated T cells promote osteoclastogenesis through RANKL expression *in vitro* [47, 69]. Exacerbation of osteoclast formation and bone destruction has been found in mice lacking one of the interferon-γ (IFNγ) receptor components [70–73]. Using an established model of endotoxin-induced bone resorption, in which the essential involvement of T cells has been documented, an exacerbation of osteoclast formation and bone destruction was observed in mice lacking one of the IFNγ receptor components [70]. This is in line with the observation of enhanced severity in the collagen-induced arthritis model in IFNγ receptor deficient mice [72, 73]. These data indicate that IFNγ may be a critical factor in maintaining bone integrity through a T cell mediated process [70]. Furthermore, it was shown *in vitro* that T cell production of IFNγ strongly suppresses osteoclastogenesis by interfering with the RANKL/RANK signaling pathway [70]. Moreover, IFNγ induces rapid degradation of the RANK adapter protein, TRAF6, which results in strong inhibition of the RANKL-induced activation of the transcription factor NF-κB and JNK [70]. IFNγ alone had no effect on TRAF6 expression, suggesting that accelerated TRAF6 degradation by IFNγ requires RANKL signaling [70]. RANKL induces the interferon-β (IFNβ) gene in osteoclast precursor cells, and the IFNβ inhibits osteoclasts differentiation by interfering with the RANKL-induced expression of c-Fos, which is an essential transcription factor for the formation of osteoclasts [74]. Interestingly, the IFNβ gene induction mechanism is dependent on c-Fos itself, indicating that the RANKL-induced c-Fos induces its own inhibitor [74]. On the other hand, local over-expression of IFNγ in the knee joint of mice with immune complex arthritis aggravated bone erosion (Lubberts, Van Lent, Nabbe, Van den Berg, unpublished observations). This suggests a dual role of IFNγ in the bone erosion process *in vivo* in which the present of T cells or T cell derived factor(s) will determine the net effect of IFNγ.

TGFβ

The role of transforming growth factor beta (TGFβ) on osteoclastogenesis is complex, with both positive and negative effects on osteoclastogenesis and bone resorption. The reason for this seems to be, in part, due to differential effects on bone marrow stromal cells that support osteoclastogenesis *versus* direct effects on osteoclastic precursor cells (Fig. 1). It has been shown that TGFβ negatively regulates osteoclastogenesis, at least in part, through the induction of OPG by osteoblasts as well as in osteoblastic/stromal cell lines [75–77]. The effect of TGFβ was mimicked by TGFβ2 and β3 but not by BMP-4, suggesting a TGFβ signal specific effect [77]. Furthermore, mutation of a Cbfa1-binding element and/or Smad-binding element in the OPG promoter region resulted in a reduction in baseline expression and in responsiveness to TGFβ and Cbfa1. This suggests the involvement and possible interaction of Cbfa1 and Smad proteins in mediating the effects of TGFβ on the OPG promoter [77]. In contrast, in the absence of osteoblast/stromal cells, TGFβ greatly increased osteoclast formation in recombinant RANKL or TNFα stimulated cultures of hematopoietic cells or RAW 264.7 macrophage-like cells, indicating that TGFβ may increase osteoclast formation *via* action on osteoclast precursors [78]. Moreover, the macrophage deactivator TGFβ1 strongly promoted TNFα induced osteoclast-like cell formation from immature bone marrow macrophages that could be abolished by IFNγ [79]. Another way that TGFβ promotes osteoclastogenesis and bone resorption is by increasing the expression of RANK, the receptor for RANKL [80]. Induction of RANK expression by TGFβ was observed in both undifferentiated RAW cells as well as in cells that had been induced to differentiate into osteoclasts by seven days treatment with M-CSF and RANKL [80]. Thus, TGFβ may act *via* more than one mechanism and although RANKL (or TNFα) is essential for osteoclast formation, factors such as TGFβ may powerfully modify the impact of osteoclastogenic stimuli [78].

IL-17

Interleukin-17 (IL-17) is a proinflammatory cytokine produced only by cells of the immune system, which contribute to joint pathology associated with RA [81]. IL-17 induces the production of proinflammatory mediators such as IL-1 and TNFα from several joint cells [82–85]. In addition, IL-17 is a potent stimulator of osteoclastogenesis (Fig. 1). Strong additive/synergistic responses between IL-17 and IL-1/TNF have been documented [86]. However, IL-17 itself has IL-1-independent catabolic capacity as well [87, 88]. Promotion of type I collagen degradation in synovium and bone by IL-17 has been demonstrated, and when combined with IL-1, a marked synergistic release of collagen was noted [89]. IL-17 in combination with TNFα increased osteoclastic resorption *in vitro* [90]. Furthermore, IL-17 induced the

A B C D

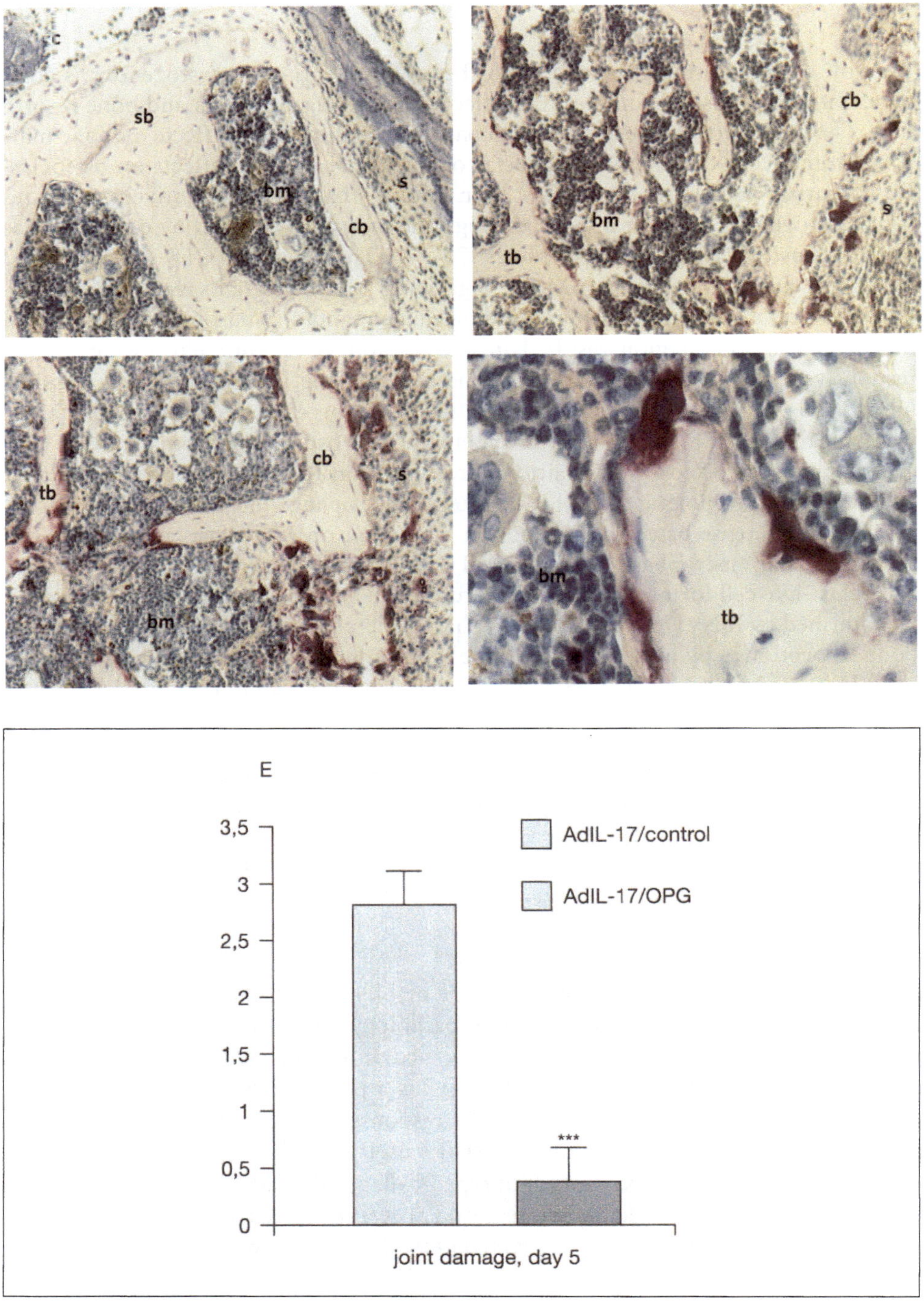

c
sb
bm
cb
s
tb
E
3,5
3
2,5
2
1,5
1
0,5
0
AdIL-17/control
AdIL-17/OPG
***
joint damage, day 5

expression of RANKL in cultures of osteoblasts [91]. IL-17 greatly stimulated osteoclast formation *via* a cell-to-cell interaction between osteoclast progenitors and osteoblasts [91]. IL-17 increased PGE_2 synthesis in cultures of osteoblasts. In addition, IL-17 induced the expression of RANKL mRNA in osteoblasts. Interestingly, OPG or a selective inhibitor of COX-2 can both completely inhibit the IL-17-induced osteoclast differentiation in co-cultures of mouse bone marrow cells and osteoblasts [91]. Early neutralization of IL-17 using sIL-17R:Fc fusion protein given systemically starting before arthritis expression in experimental arthritis prevented bone erosion [87, 92]. When anti-IL-17 antibody therapy was given after the onset of CIA, still significant prevention of bone erosion was observed [97]. Neutralizing IL-17 during this stage of CIA resulted in less RANKL positive cells in the synovium. Furthermore, IL-17 production from activated T cells seems to be required for the spontaneous development of destructive arthritis in mice deficient in IL-1 receptor antagonist [93]. Of interest, a critical role for the synovial IL-17 receptor signaling pathway in the development of bone erosion was noted using IL-17R$^{-/-}$ mice (Lubberts E et al, unpublished observation). On the other hand, local IL-17 over-expression in the knee joint of type II collagen immunized mice results in promotion of collagen arthritis and aggravates joint destruction (Fig. 4, A–D) [94]. In the collagen-induced arthritis model it was shown that IL-17 promoted bone erosion through loss of the RANKL/OPG balance [95]. Systemic OPG treatment prevented joint damage induced by local IL-17 gene transfer in type II collagen-immunized mice (Fig. 4, E). In addition, over-expression of IL-17 in the knee joints of naïve mice results in joint inflammation and bone erosion. IL-17 strongly induces synovial

Figure 4
Local over-expression of IL-17 in the knee joint of collagen type II immunized DBA-1 mice promotes the formation of TRAP-positive osteoclast-like cells in areas of subchondral, trabecular and cortical bone erosions
A – Arthritic knee joint of a mouse five days after intra-articular injection of 10^7 PFU AdControl.
B – Arthritic knee joint of a full-blown collagen-arthritic mouse with approximately the same degree of joint inflammation as the IL-17 group, showing TRAP activity in areas of trabecular and cortical bone.
C – Arthritic knee joint of a mouse five days after intra-articular injection of 10^7 PFU AdIL-17, showing TRAP activity in areas of cortical and trabecular bone.
D – Higher power images of TRAP+ cells in trabecular bone in the IL-17 group. Note the enhanced TRAP activity induced by IL-17 compared with the control groups.
E – Systemic OPG treatment prevented local IL-17-induced bone erosion in CIA.
C, cartilage; sb, subchondral bone; cb, cortical bone; tb, trabecular bone; bm, bone marrow; s, synovitis. (Original magnification × 200 in A–C; × 400 in D). This figure is partly adapted from [95].

RANKL expression which is at least partly independent of IL-1 (Lubberts E et al, unpublished observation). Systemic OPG treatment under these conditions also results in protection against bone damage without any effect on joint inflammation. These data strongly suggest that IL-17 is a potent inducer of RANKL and that the IL-17-induced promotion of bone erosion is strongly mediated by RANKL. Furthermore, it underscores the role of IL-17 in enhancement of the joint destructive process and makes IL-17 an attractive target for the treatment of destructive arthritis [96].

Concluding remarks

The bone erosion process is strongly regulated by the balance between osteoblast and osteoclast activity. In addition to the importance of osteoblast/osteoclast interaction in the induction/regulation of bone erosion within the bone microenvironment, osteoclast formation through differentiation of monocyte/macrophage precursors in the inflamed synovium may strongly contribute to the bone erosion process resulting in a bidirectional assault of bone in RA patients.

The balance between RANKL and OPG strongly regulates the bone erosion process. Several cytokines are able to influence this balance through direct or indirect effect on RANKL or OPG expression. Although OPG treatment is very efficient in preventing bone erosion, it is not anti-inflammatory or directly chondroprotective. In contrast, suppression of joint inflammation and protection of joint damage including bone and cartilage destruction were made using anti-IL-17 and anti-IL-1 therapy, suggesting anti-IL-17/anti-IL-1 as a more appropriate therapy for destructive arthritis. In addition, IL-1 and IL-17 have strong RANKL inducing and replacing capacities in osteoclast activation. Therefore, in addition to regulation of RANKL by OPG, controlling IL-17 and IL-1/TNF expression will be of great therapeutic value in the prevention of joint loss.

Acknowledgements

This work was supported by a Veni Fellowship of the Netherlands Organization for Scientific Research (NWO) grant 906-02-038 (E. Lubberts).

References

1 Theill LE, Boyle WJ, Penninger JM (2002) RANKL and RANK: T cells, bone loss, and mammalian evolution. *Annu Rev Immunol* 20: 795–823

2 Nijweide PJ, Burger EH, Feyen JHM (1986) Cells of the bone: proliferation, differentiation, and hormonal regulation. *Physiol Rev* 66: 855–886

3 Manolagas SC, Jilka RL (1995) Bone marrow, cytokines, and bone remodeling: emerging insights into the pathophysiology of osteoporosis. *N Engl J Med* 332: 305–311

4 Roodman GD (1996) Advances in bone biology: the osteoclast. *Endocr Rev* 17: 308–332

5 Suda T, Takahashi N, Martin TJ (1992) Modulation of osteoclast differentiation. *Endocr Rev* 13: 66–80

6 Anderson DM, Maraskovsky E, Billingsley WL, Dougall WC, Tometsko ME, Roux ER, Teepe MC, DuBose RF, Cosman D, Galiberts L (1997) A homologue of the TNF receptor and its ligand enhance T-cell growth and dendritic-cell function. *Nature* 390: 175–179

7 Lacey DL, Timms E, Tan HL, Kelley MJ, Dunstan CR, Burgess T, Elliott R, Colombero A, Elliot G, Scully S et al (1998) Osteoprotegerin ligand is a cytokine that regulates osteoclast differentiation and activation. *Cell* 93: 165–176

8 Wong BR, Rho J, Arron J, Robinson E, Orlinick J, Chao M, Kalachikov S, Cayani E, Bartlett FS, Frankel WN et al (1997) TRANCE is a novel ligand of the tumor necrosis factor receptor family that activates c-Jun N-terminal kinase in T cells. *J Biol Chem* 272: 25190–25194

9 Yasuda H, Shima N, Nakagawa N, Yamaguchi K, Kinosaki M, Mochizuki S, Tomoyasu A, Yano K, Goto M, Murakami A et al (1998) Osteoclast differentiation factor is a ligand for osteoprotegerin/osteoclastogenesis-inhibitory factor and is identical to TRANCE/RANKL. *Proc Natl Acad Sci USA* 95:3597–3602

10 Lum L, Wong BR, Josien R, Becherer JD, Erdjument-Bromage H, Schlondorff J, Tempst P, Choi Y, Blobel CP (1999) Evidence for a role of a tumor necrosis factor alpha (TNFα)-converting enzyme-like protease in shedding of TRANCE, a TNF family member involved in osteoclastogenesis and dendritic cell survival. *J Biol Chem* 274: 13613–13618

11 Kong YY, Yoshida H, Sarosi I, Tan HL, Timms E, Capparelli C, Morony S, Oliveira-dos-Santos AJ, Van G, Itie A et al (1999) OPGL is a key regulator of osteoclastogenesis, lymphocyte development and lymph-node organogenesis. *Nature* 397: 315–323

12 Hsu H, Lacey DL, Dunstan CR, Solovyev I, Colombero A, Timms E, Tan HL, Elliott G, Kelley MJ, Sarosi I et al (1999) Tumor necrosis factor receptor family member RANK mediates osteoclast differentiation and activation induced by osteoprotegerin ligand. *Proc Natl Acad Sci USA* 96: 3540–3545

13 Suda T, Takahashi N, Udagawa N, Jimi E, Gillespie MT, Martin TJ (1999) Modulation of osteoclast differentiation and function by the new members of the tumor necrosis factor receptor and ligand families. *Endocr Rev* 20: 345–357

14 Li J, Sarosi I, Yan XQ, Morony S, Cappartelli C, Tan HL, McCabe S, Elliot R, Scully S, Van G et al (2000) RANK is the intrinsic hematopoietic cell surface receptor that controls osteoclastogenesis and regulation of bone mass and calcium metabolism. *Proc Natl Acad Sci USA* 97: 1566–1571

15 Dougall WC, Glaccum M, Charrier K, Rohrbach K, Brasel K, De Smedt T, Daro E, Smith J, Tometsko ME, Maliszewski CR et al (1999) RANK is essential for osteoclast and lymph node development. *Genes Dev* 13: 2412–2424

16 Simonet WS, Lacey DL, Dunstan CR, Kelley M, Chang MS, Luthy R, Nguyen HQ, Wooden S, Bennett L, Boone T et al (1997) Osteoprotegerin: a novel secreted protein involved in the regulation of bone density. *Cell* 89: 309–319

17 Bucay N, Sarosi I, Dunstan CR, Morony S, Tarpley J, Capparelli C, Scully S, Tan HL, Xu W, Lacey DL et al (1998) Osteoprotegerin-deficient mice develop early onset osteoporosis and arterial calcification. *Genes Dev* 12: 1260–1268

18 Rodan GA, Martin TJ (1981) Role of osteoblasts in hormonal control of bone resorption – a hypothesis. *Calcif Tissue Int* 33: 349–351

19 Chambers TJ (1992) Regulation of osteoclast development and function. In: Rifkin BR, Gay CV (eds) *Biology and physiology of the osteoclast.* CRC Press, Boca Raton, FL, 105–128

20 Athanasou NA (1996) Cellular biology of bone-resorbing cells. *J Bone Joint Surg Am* 78: 1096–1112

21 Suda T, Udagawa N, Takahashi N (1996) Osteoclast generation. In: Raisz LG, Rodan GA, Bilezikian JP (eds) *Principles of bone biology.* Academic Press, San Diego, CA, 87–102

22 Martin TJ, Ng KW (1994) Mechanism by which cells of the osteoblast lineage control osteoclast formation and activity. *J Cell Biochem* 56: 357–366

23 Martin TJ, Udagawa N (1998) Hormonal regulation of osteoclast function. *Trends Endocrinol Metab* 9: 6–12

24 Martin TJ, Romas E, Gillespie MT (1998) Interleukins in the control of osteoclast differentiation. *Crit Rev Eukaryot Gene Expr* 8: 107–123

25 Biskobing DM, Fan X, Rubin J (1995) Characterization of M-CSF-induced proliferation nad subsequent osteoclast formation in murine marrow culture. *J Bone Miner Res* 10: 1025–1032

26 Kobayashi K, Takahashi N, Jimi E, Udagawa N, Takami M, Kotake S, Nakagawa N, Kinosaki M, Yamaguchi K, Shima N et al (2000) Tumor necrosis factor alpha stimulates osteoclast differentiation by a mechanism independent of the ODF/RANKL-RANK interaction. *J Exp Med* 191: 275–286

27 Udagawa N, Kotake S, Kamatani N, Takahashi N, Suda T (2002) The molecular mechanism of osteoclastogenesis in rheumatoid arthritis. *Arthritis Res* 4: 281–289

28 Kanazawa K, Azuma Y, Nakano H, Kudo A (2003) TRAF5 functions in both RANKL and TNFα induced osteoclastogenesis. *J Bone Min Res* 18: 443–450

29 Kaji K, Katogi R, Azuma Y, Naito A, Inoue JI, Kudo A (2001) Tumor necrosis factor-alpha-induced osteoclastogenesis requires tumor necrosis factor receptor-associated factor 6. *J Bone Min Res* 16: 1593–1599

30 Lam J, Takeshita S, Barker JE, Kanagawa O, Ross FP, Teitelbaum SL (2000) TNFα induces osteoclastogenesis by direct stimulation of macrophages exposed to permissive levels of RANK ligand. *J Clin Invest* 106: 1481–1488

31 Gravallese EM, Galson DL, Goldring SR, Auron PE (2001) The role of TNF-receptor family members and other TRAF-dependent receptors in bone resorption. *Arthritis Res* 3: 6–12

32 Bromley M, Woolley DE (1984) Chondroclasts and osteoclasts at subchondral sites of erosion in the rheumatoid joint. *Arthritis Rheum* 27: 968–975

33 Leisen JCC, Duncan H, Riddle JM, Pitchford WC (1988) The erosive front: a topographic study of the junction between the pannus and the subchondral plate in the macerated rheumatoid metacarpal head. *J Rheumatol* 15: 17–22

34 Gravallese EM, Harada Y, Wang JT, Gorn AH, Thornhill TS, Goldring SR (1998) Identification of cell types responsible for bone resorption in rheumatoid arthritis and juvenile rheumatoid arthritis. *Am J Pathol* 152: 943–951

35 Suzuki Y, Nishikaku F, Nakatuka M, Koga Y (1998) Osteoclast-like cells in murine collagen induced arthritis. *J Rheumatol* 25: 1154–1160

36 Romas E, Bakharevski O, Hards DK, Kartsogiannis V, Quinn JM, Ryan PF, Martin TJ, Gillespie MT (2000) Expression of osteoclast differentiation factor at sites of bone erosion in collagen-induced arthritis. *Arthritis Rheum* 43: 821–826

37 Redlich K, Hayer S, Ricci R, David JP, Tohidast-Akrad M, Kollias G, Steiner G, Smolen JS, Wagner EF, Schett G (2002) Osteoclasts are essential for TNFα-mediated joint destruction. *J Clin Invest* 110: 1419–1427

38 Takayanagi H, Oda H, Yamamotot S, Kawaguchi H, Tanaka S, Nishikawa T, Koshihara Y (1997) A new mechanism of bone destruction in rheumatoid arthritis: synovial fibroblast induce osteoclastogenesis. *Biochem Biophys Res Commun* 240: 279–286

39 Gravallese EM, Manning C, Tsay A, Naito A, Pan C, Amento E, Goldring SR (2000) Synovial tissue in rheumatoid arthritis is a source of osteoclast differentiation factor. *Arthritis Rheum* 43: 250–258

40 Takayanagi H, Iizuka H, Juji T, Nakagawa T, Yamamoto A, Miyazaki T, Koshihara Y, Oda H, Nakamura K, Tanaka S (2000) Involvement of rceptor activator of nuclear factor κB ligand/osteoclast differentiation factor in osteoclastogenesis from synoviocytes in rheumatoid arthritis. *Arthritis Rheum* 43: 259–269

41 Lubberts E, Joosten LAB, Chabaud M, Van den Bersselaar L, Oppers B, Coenen-de Roo CJJ, Richards CD, Miossec P, van den Berg WB (2000) IL-4 gene therapy for collagen arthritis suppresses synovial IL-17 and osteoprotegerin ligand and prevents bone erosion. *J Clin Invest* 105: 1697–1710

42 Lubberts E, Oppers-Walgreen B, Pettit AR, Van den Bersselaar L, Joosten LAB, Goldring SR, Gravallese E, van den Berg WB (2002) Increase in expression of receptor activator of nuclear factor κB at sites of bone erosion correlates with progression of inflammation in evolving collagen-induced arthritis. *Arthritis Rheum* 46: 3055–3064

43 Itonaga I, Fujikawa Y, Sabokbar A, Murray DW, Athanasou NA (2000) Rheumatoid arthritis synovial macrophage-osteoclast differentiation is osteoprotegerin ligand-dependent. *J Pathol* 192: 97–104

44 Glowacki J, Jasty M, Goldring S (1986) Comparison of multinucleatd cells elicited in rats by particulate bone, polyethylene, or ploymethylmethacrylate. *J Bone Miner Res* 1: 327–331

45 Goldring SR, Roelke M, Glowacki J (1988) Multinucleated cells elicited in response to

implants of devitalized bone particles possess receptors for clacitonin. *J Bone Miner Res* 3: 117–120
46 Ritchlin CT, Haas-Smith SA, Li P, Hicks DG, Schwarz EM (2003) Mechanism of TNFα- and RANKL-mediated osteoclastogenesis and bone resorption in psoariatic arthritis. *J Clin Invest* 111: 821–831
47 Kong YY, Feige U, Sarosi I, Bolon B, Tafuri A, Morony S, Capparelli C, Li J, Elliott R, McCabe S et al (1999) Activated T cells regulate bone loss and joint destruction in adjuvant arthritis through osteoprotegerin ligand. *Nature* 402: 304–309
48 Romas E, Sims NA, Hards DK, Lindsay M, Quinn JWM, Ryan PFJ, Dunstan CR, Martin TJ, Gillespie MT (2002) Osteoprotegerin reduces osteoclast numbers and prevents bone erosion in collagen-induced arthritis. *Am J Pathol* 161: 1419–1427
49 Pettit AR, Ji H, Von Stechow D, Muller R, Goldring SR, Choi Y, Benoist C, Gravallese EM (2001) TRANCE/RANKL knockout mice are protected from bone erosion in a serum transfer model of arthritis. *Am J Pathol* 159: 1689–1699
50 Lubberts E, Koenders MK, Joosten LAB, Schwarzenberger P, Coenen-de Roo T, Kolls JK, van den Berg WB (2002) Local OPG gene therapy prevents joint damage in collagen-induced arthritis. *Arthritis Rheum* 46: S586 (abstract 1575)
51 Redlich K, Hayer S, Maier A, Dunstan CR, Tohidast-Akrad M, Lang S, Turk B, Pietschmann P, Woloszczuk W, Haralambous S et al (2002) Tumor necrosis factor alpha-mediated joint destruction is inhibited by targeting osteoclasts with osteoprotegerin. *Arthritis Rheum* 46: 785–792
52 Schett G, Redlich K, Hayer S, Zwerina J, Bolon B, Dunstan C, Gortz B, Schulz A, Bergmeister H, Kollias G et al (2003) Osteoprotegerin protects against generalized bone loss in tumor necrosis factor-transgenic mice. *Arthritis Rheum* 48: 2042–2051
53 Komuro H, Olee T, Kuhn K, Quach J, Brinson DC, Shikhman A, Valbracht J, Creighton-Achtermann, Lotz M (2001) The osteoprotegerin/receptor activator of nuclear factor κB/receptor activator of nuclear factor κB ligand system in cartilage. *Arthritis Rheum* 44: 2768–2776
54 Kotake S, Udagawa N, Hakoda M, Mogi M, Yano K, Tsuda E, Takahashi K, Furuya T, Ishiyama S, Kim KJ et al (2001) Activated human T cells directly induce osteoclastogenesis from human monocytes. *Arthritis Rheum* 44: 1003–1012
55 Wang R, Zhang L, Zhang X, Moreno J, Celluzzi C, Tondravi M, Shi Y (2002) Regulation of activation-induced receptor activator of NF-κB ligand (RANKL) expression in T cells. *Eur J Immunol* 32: 1090–1098
56 Kim N, Odgren PR, Kim DK, Marks SC, Choi Y (2000) Diverse role of the tumor necrosis factor family member TRANCE in skeletal physiology revealed by TRANCE deficiency and partial rescue by a lymphocyte-expressed TRANCE transgene. *Proc Natl Acad Sci USA* 97: 10905–10910
57 Arron JR, Choi Y (2000) Bone *versus* immune system. *Nature* 408: 535–536
58 Yamamoto N, Sakai F, Kon S, Morimoto J, Kimura C, Yamazaki H, Okazaki I, Seki N, Fujii T, Uede T (2003) Essential role of the cryptic epitope SLAYGLR within osteopontin in a murine model of rheumatoid arthritis. *J Clin Invest* 112: 181–188

59 Weitzmann MN, Cenci S, Rifas L, Brown C, Pacifici R (2000) Interleukin-7 stimulates osteoclast formation by upregulating the T-cell production of soluble osteoclastogenic cytokines. *Blood* 96: 1873–1878

60 Toraldo G, Roggia C, Qian WP, Pacifici R, Weitzmann MN (2003) IL-7 induces bone loss *in vivo* by induction of receptor activator of nuclear factor κB ligand and tumor necrosis factor alpha from T cells. *Proc Natl Acad Sci USA* 100: 125–130

61 Horwood NJ, Elliott J, Martin TJ, Gillespie MT (2001) IL-12 alone and in synergy with IL-18 inhibits osteoclast formation *in vitro*. *J Immunol* 166: 4915–4921

62 Ogata Y, Kukita A, Kukita T, Komine M, Miyahara A, Miyazaki S, Kohashi O (1999) A novel role of IL-15 in the development of osteoclasts: inability to replace its activity with IL-2. *J Immunol* 162: 2754–2760

63 Miossec P, Chomarat P, Dechanet J, Moreau JF, Roux JP, Delmas P, Banchereau J (1994) Interleukin-4 inhibits bone resorption through an effect on osteoclasts and proinflammatory cytokines in an *ex vivo* model of bone resorption in rheumatoid arthritis. *Arthritis Rheum* 37: 1715–1722

64 Bendixen AC, Shevde NK, Dienger KM, Willson TM, Funk CD, Pike JW (2001) IL-4 inhibits osteoclast formation through a direct action on osteoclast precursors *via* peroxisome proliferators-activated receptor gamma 1. *Proc Natl Acad Sci USA* 98: 2443–2448

65 Abu-Amer Y (2001) IL-4 abrogates osteoclastogenesis through STAT6-dependent inhibition of NF-κB. *J Clin Invest* 107: 1375–1385

66 Wei S, Wang MW, Teitelbaum SL, Ross FP (2002) Interleukin-4 reversibly inhibits osteoclastogenesis *via* inhibition of NF-κB and mitogen-activated protein kinase signaling. *J Biol Chem* 277: 6622–6630

67 Moreno JL, Kaczmarek M, Keegan AD, Tondravi M (2003) IL-4 suppresses osteoclast development and mature osteoclast function by a STAT6-dependent mechanism: irreversible inhibition of the differentiation program activated by RANKL. *Blood* 102: 1078–1086

68 Mirosavljevic D, Quinn JM, Elliott J, Horwood NJ, Martin TJ, Gillespie MT (2003) T-cells mediate an inhibitory effect of interleukin-4 on osteoclastogenesis. *J Bone Miner Res* 18: 984–993

69 Horwood NJ, Kartsogiannis V, Quinn JM, Romas E, Martin TJ, Gillespie MT (1999) Activated T lymphocytes support osteoclast formation *in vitro*. *Biochem Biophys Res Commun* 265: 144–150

70 Takayanagi H, Ogasawara K, Hida S, Chiba T, Murata S, Sato K, Takaoka A, Yokochi T, Oda H, Tanaka K et al (2000) T cell-mediated regulation of osteoclastogenesis by signaling cross-talk between RANKL and IFNγ. *Nature* 408: 600–605

71 Huang S, Hendriks W, Althage A, Hemmi S, Bluethmann H, Kamijo R, Vilcek J, Zinkernagel RM, Aguet M (1993) Immune response in mice that lack the interferon-gamma receptor. *Science* 259: 1742–1745

72 Manoury-Schwartz B, Chiocchia G, Bessis N, Abehsira-Amar O, Batteux F, Muller S, Huang S, Boissier MC, Fournier (1997) High susceptibility to collagen-induced arthritis in mice lacking IFNγ receptors. *J Immunol* 158: 5501–5506

73 Vermeire K, Heremans H, Vandeputte M, Huang S, Billiau A, Matthys P (1997) Accelerated collagen-induced arthritis in IFNγ receptor-deficient mice. *J Immunol* 158: 5507–5513

74 Takayanagi H, Kim S, Matsuo K, Suzuki H, Suzuki T, Sato K, Yokochi T, Oda H, Nakamura K, Ida N et al (2002) RANKL maintains bone homeostasis through c-Fos-dependent induction of interferon-beta. *Nature* 416: 744–749

75 Takai H, Kanematsu M, Yano K, Tsuda E, Higashio K, Ikeda K, Watanabe K, Yamada Y (1998) Transforming growth factor beta stimulates the production of osteoprotegerin/osteoclastogenesis inhibitory factor by bone marrow stromal cells. *J Biol Chem* 273: 27091–27096

76 Murakami T, Yamamoto M, Ono K, Nishikawa M, Nagata N, Motoyoshi K, Akatsu T (1998) Transforming growth factor-beta1 increases mRNA levels of osteoclastogenesis inhibitory factor in osteoblastic/stromal cells and inhibits the survival of murine osteoclast-like cells. *Biochem Biophys Res Commun* 252: 747–752

77 Thirunavukkarasu K, Miles RR, Halladay DL, Yang X, Galvin RJS, Chandrasekhar S, Martin TJ, Onyia JE (2001) Stimulation of osteoprotegerin (OPG) gene expression by transforming growth factor-beta (TGFβ). *J Biol Chem* 276: 36241–36250

78 Quinn JM, Itoh K, Udagawa N, Hausler K, Yasuda H, Shima N, Mizuno A, Higashio K, Takahashi N, Suda T et al (2001) Transforming growth factor beta affects osteoclast differentiation *via* direct and indirect actions. *J Bone Miner Res* 16: 1787–1794

79 Fox SW, Fuller K, Bayley KE, Lean JM, Chambers TJ (2000) TGF-β1 and IFNγ direct macrophage activation by TNFα to osteoclastic or cytocidal phenotype. *J Immunol* 165: 4957–4963

80 Yan T, Riggs BL, Boyle WJ, Khosla S (2001) Regulation of osteoclastogenesis and RANK expression by TGF-β1. *J Cell Biochem* 83: 320–325

81 Chabaud M, Durand JM, Buchs N, Fossiez F, Page G, Frappart L, Miossec P (1999) Human interleukin-17. A T cell-derived proinflammatory cytokine produced by the rheumatoid synovium. *Arthritis Rheum* 42:963–970

82 Yao Z, Painter SL, Fanslow WC, Ulrich D, Macduff BM, Spriggs MK, Armitage RJ (1995) Human IL-17: a novel cytokine derived from T cells. *J Immunol* 155: 5483–5486

83 Yao Z, Fanslow WC, Seldin MF, Rousseau AM, Painter SL, Comeau MR, Cohen JI, Spriggs MK (1995) Herpesvirus Saimiri encodes a new cytokine IL-17, which binds to a novel cytokine receptor. *Immunity* 3: 811–821

84 Fossiez F, Djossou O, Chomarat P, FloresRomo L, Ait-Yahia S, Maat C, Pin JJ, Garrone P, Garcia E, Saeland S et al (1996) T cell IL-17 induces stromal cells to produce proinflammatory and hematopoietic cytokines. *J Exp Med* 183: 2593–2603

85 Jovanovic DV, DiBattista JA, Martel-Pelletier J, Jolicoeur FC, He Y, Zhang M, Mineau F, Pelletier JP (1998) IL-17 stimulates the production and expression of proinflammatory cytokines, IL-1β and TNFα, by human macrophages. *J Immunol* 160: 3513–3521

86 Miossec P (1993) Interleukin-17 in rheumatoid arthritis. If T cells were to contribute to inflammation and destruction through synergy. *Arthritis Rheum* 48:594–601

87 Lubberts E, Joosten LAB, Oppers B, Van den Bersselaar L, Coenen-de Roo CJ, Kolls JK,

Schwarzenberger P, Van de Loo FA, van den Berg WB (2001) IL-1-independent role of IL-17 in synovial inflammation and joint destruction during collagen-induced arthritis. *J Immunol* 167: 1004–1013
88 Benderdour M, Tardif G, Pelletier JP, Di Battista JA, Reboul P, Ranger P, Martel-Pelletier J (2002) Interleukin 17 (IL-17) induces collagenase-3 production in human osteoclastic chondrocytes *via* AP-1 dependent activation: differential activation of AP-1 members by IL-17 and IL-1β. *J Rheum* 29: 1262–1272
89 Chabaud M, Lubberts E, Joosten L, van den Berg W, Miossec P (2001) IL-17 derived from juxta-articular bone and synovium contributes to joint degradation in rheumatoid arthritis. *Arthritis Res* 3: 168–177
90 Van Bezooijen R, Farih-Sips HC, Papapoulos SE, Lowik CW (1999) Interleukin-17: a new bone acting cytokine *in vitro*. *J Bone Min Res* 14: 1513–1521
91 Kotake S, Udagawa N, Takahashi N, Matsuzaki K, Itoh K, Ishiyama S, Saito S, Inoue K, Kamatani N, Gillespie MT et al (1999) IL-17 in synovial fluids from patients with rheumatoid arthritis is a potent stimulator of osteoclastogenesis. *J Clin Invest* 103: 1345–1352
92 Bush KA, Farmer KM, Walker JS, Kirkham BW (2002) Reduction of joint inflammation and bone erosion in rat adjuvant arthritis by treatment with interleukin-17 receptor IgG1 Fc fusion protein. *Arthritis Rheum* 46: 802–805
93 Nakae S, Saijo S, Horai R, Sudo K, Mori S, Iwakura Y (2003) IL-17 production from activated T cells is required for the spontaneous development of destructive arthritis in mice deficient in IL-1 receptor antagonist. *Proc Natl Acad Sci USA* 100: 5986–5990
94 Lubberts E, Joosten LAB, Van de Loo FA, Schwarzenberger P, Kolls JK, van den Berg WB (2002) Over-expression of IL-17 in the knee joint of collagen type II immunized mice promotes collagen arthritis and aggravates joint destruction. *Inflamm Res* 51: 102–104
95 Lubberts E, Van den Bersselaar L, Oppers-Walgreen B, Schwarzenberger P, Coenen-de Roo CJJ, Kolls JK, Joosten LAB, van den Berg WB (2003) IL-17 promotes bone erosion in murine collagen-induced arthritis through loss of the RANKL/OPG balance. *J Immunol* 170: 2655–2662
96 Lubberts E (2003) The role of IL-17 and family members in the pathogenesis of arthritis. *Curr Opin Investig Drugs* 4: 572–577
97 Lubberts E, Koenders MI, Oppers-Walgreen B, van den Bersselaar L, Coenen-de Roo CJJ, Joosten LAB, van den Berg WB (2004) Treatment with a neutralizing anti-murine interleukin-17 antibody after the onset of collagen-induced arthritis reduces joint inflammation, cartilage destruction, and bone erosion. *Arthritis Rheum* 50: 650–659

Index

The PIR-Series
Progress in Inflammation Research

Homepage: http://www.birkhauser.ch

Up-to-date information on the latest developments in the pathology, mechanisms and therapy of inflammatory disease are provided in this monograph series. Areas covered include vascular responses, skin inflammation, pain, neuroinflammation, arthritis cartilage and bone, airways inflammation and asthma, allergy, cytokines and inflammatory mediators, cell signalling, and recent advances in drug therapy. Each volume is edited by acknowledged experts providing succinct overviews on specific topics intended to inform and explain. The series is of interest to academic and industrial biomedical researchers, drug development personnel and rheumatologists, allergists, pathologists, dermatologists and other clinicians requiring regular scientific updates.

Available volumes:

T Cells in Arthritis, P. Miossec, W. van den Berg, G. Firestein (Editors), 1998
Chemokines and Skin, E. Kownatzki, J. Norgauer (Editors), 1998
Medicinal Fatty Acids, J. Kremer (Editor), 1998
Inducible Enzymes in the Inflammatory Response,
D.A. Willoughby, A. Tomlinson (Editors), 1999
Cytokines in Severe Sepsis and Septic Shock, H. Redl, G. Schlag (Editors), 1999
Fatty Acids and Inflammatory Skin Diseases, J.-M. Schröder (Editor), 1999
Immunomodulatory Agents from Plants, H. Wagner (Editor), 1999
Cytokines and Pain, L. Watkins, S. Maier (Editors), 1999
In Vivo *Models of Inflammation*, D. Morgan, L. Marshall (Editors), 1999
Pain and Neurogenic Inflammation, S.D. Brain, P. Moore (Editors), 1999
Anti-Inflammatory Drugs in Asthma, A.P. Sampson, M.K. Church (Editors), 1999
Novel Inhibitors of Leukotrienes, G. Folco, B. Samuelsson, R.C. Murphy (Editors), 1999
Vascular Adhesion Molecules and Inflammation, J.D. Pearson (Editor), 1999
Metalloproteinases as Targets for Anti-Inflammatory Drugs,
K.M.K. Bottomley, D. Bradshaw, J.S. Nixon (Editors), 1999
Free Radicals and Inflammation, P.G. Winyard, D.R. Blake, C.H. Evans (Editors), 1999
Gene Therapy in Inflammatory Diseases, C.H. Evans, P. Robbins (Editors), 2000
New Cytokines as Potential Drugs, S. K. Narula, R. Coffmann (Editors), 2000
High Throughput Screening for Novel Anti-inflammatories, M. Kahn (Editor), 2000
Immunology and Drug Therapy of Atopic Skin Diseases,
C.A.F. Bruijnzeel-Komen, E.F. Knol (Editors), 2000
Novel Cytokine Inhibitors, G.A. Higgs, B. Henderson (Editors), 2000
Inflammatory Processes. Molecular Mechanisms and Therapeutic Opportunities,
L.G. Letts, D.W. Morgan (Editors), 2000

Cellular Mechanisms in Airways Inflammation, C. Page, K. Banner, D. Spina (Editors), 2000
Inflammatory and Infectious Basis of Atherosclerosis, J.L. Mehta (Editor), 2001
Muscarinic Receptors in Airways Diseases, J. Zaagsma, H. Meurs, A.F. Roffel (Editors), 2001
TGF-β and Related Cytokines in Inflammation, S.N. Breit, S. Wahl (Editors), 2001
Nitric Oxide and Inflammation, D. Salvemini, T.R. Billiar, Y. Vodovotz (Editors), 2001
Neuroinflammatory Mechanisms in Alzheimer's Disease. Basic and Clinical Research, J. Rogers (Editor), 2001
Disease-modifying Therapy in Vasculitides, C.G.M. Kallenberg, J.W. Cohen Tervaert (Editors), 2001
Inflammation and Stroke, G.Z. Feuerstein (Editor), 2001
NMDA Antagonists as Potential Analgesic Drugs, D.J.S. Sirinathsinghji, R.G. Hill (Editors), 2002
Migraine: A Neuroinflammatory Disease? E.L.H. Spierings, M. Sanchez del Rio (Editors), 2002
Mechanisms and Mediators of Neuropathic pain, A.B. Malmberg, S.R. Chaplan (Editors), 2002
Bone Morphogenetic Proteins. From Laboratory to Clinical Practice, S. Vukicevic, K.T. Sampath (Editors), 2002
The Hereditary Basis of Allergic Diseases, J. Holloway, S. Holgate (Editors), 2002
Inflammation and Cardiac Diseases, G.Z. Feuerstein, P. Libby, D.L. Mann (Editors), 2003
Mind over Matter – Regulation of Peripheral Inflammation by the CNS, M. Schäfer, C. Stein (Editors), 2003
Heat Shock Proteins and Inflammation, W. van Eden (Editor), 2003
Pharmacotherapy of Gastrointestinal Inflammation, A. Guglietta (Editor), 2004
Arachidonate Remodeling and Inflammation, A.N. Fonteh, R.L. Wykle (Editors), 2004
Recent Advances in Pathophysiology of COPD, P.J. Barnes, T.T. Hansel (Editors), 2004

MIX
Papier aus verantwortungsvollen Quellen
Paper from responsible sources
FSC® C105338

If you have any concerns about our products,
you can contact us on
ProductSafety@springernature.com

In case Publisher is established outside the EU,
the EU authorized representative is:
Springer Nature Customer Service Center GmbH
Europaplatz 3, 69115 Heidelberg, Germany

Printed by Libri Plureos GmbH
in Hamburg, Germany